QUANTITATIVE MICROBIAL RISK ASSESSMENT

QUANTITATIVE MICROBIAL RISK ASSESSMENT

CHARLES N. HAAS
JOAN B. ROSE
CHARLES P. GERBA

JOHN WILEY & SONS, INC.

New York / Chichester / Weinheim / Brisbane / Singapore / Toronto

Library of Congress Cataloging in Publication Data:

Haas, Charles N.
    Quantitative microbial risk assessment / Charles N. Haas, Joan B.
Rose, Charles P. Gerba.
      p.   cm.
    Includes index.
    ISBN 0-471-18397-0 (cloth : alk. paper)
    1. Communicable diseases—Epidemiology—Methodology. 2. Health
risk assessment. 3. Infection—Mathematical models.
    4. Environmental health—Mathematical models. I. Rose, Joan B.
II. Gerba, Charles P., 1945–   III. Title.
    RA643.H22   1999
    615.9'02—DC21   98-45800

Printed in the United States of America

10 9 8 7 6 5 4 3 2 1

To our husband and wives, our parents, our children,
our mentors, and our students

CONTENTS

PREFACE

Cryptosporidium, E. coli O157:H7, drug-resistant pathogens—these agents have entered the public consciousness and necessitated a better understanding of the transmission of infectious agents from contaminated food, water, soil, and air. In the field of environmental protection (from chemical hazards), the risk assessment methodology has been a central paradigm for some time. The objective of this book is to set forth risk assessment methodologies suitable for application to infectious microorganisms. Many similarities and differences exist.

We have been working in the areas covered here for almost 20 years. Quantitative microbial risk assessment, while it represents a shift in thinking about hazards of infection, is a process that we believe will continue to advance and develop. This book should be suitable for a reference for the practitioner in the area, and as a textbook for advanced students in environmental sciences, environmental engineering, public health, and microbiology. Some prior statistical background would be useful in approaching the material, but it is not necessary; the key requirement for any risk assessor is the absence of fear from mathematical constructs and concepts.

There is a danger in putting forth this book, which all three of us are aware of: to be the first is to set up a target for challenge and attack. Nevertheless, by doing so we hope to advance the underlying science and practice.

This work could not have been accomplished without the interaction with a number of professional colleagues and students over the many years of our practice. We also thankfully acknowledge funding from many sources, corporate and public, including in particular the American Water Works Association Research Foundation (AWWARF) and the International Life Sciences Institute (ILSI).

Finally, we would note that doing high-quality risk assessments is of necessity a team sport, requiring individuals with different skills and interests. We have learned this directly among ourselves from many years of experience. Practitioners of the art of quantitative microbial risk assessment should be advised to cast a wide net with respect to colleagues and collaborators to perfect their craft.

We encourage comments and feedback from users of this work, and look forward to observing and participating in developments in coming years.

CHARLES N. HAAS, *Philadelphia, PA*

JOAN B. ROSE, *St. Petersburg, FL*

CHARLES P. GERBA, *Tucson, AZ*

January 1999

INTRODUCTION

The prevention of infectious disease transmission from human exposure to contaminated food, water, soil, and air remains a major task of environmental and public health professionals. Indeed, some have argued that the property of virulence of human pathogens is one that is favored by evolutionary interactions between pathogens and host populations and therefore will always be of important concern (21). The objective of this book is to set forth comprehensively the methods for assessment of risk from infectious agents transmitted via these routes in a framework that is compatible with the framework for other risk assessments (e.g., for chemical agents) as set forth in standard protocols (24). The framework for microorganisms has evolved with recent proposals by federal advisory groups. Although there is as yet no federal consensus, other scientific expert panels are attempting to develop consensus approaches (17). In this introduction, information is presented on the occurrence of infectious disease in broad categories along with a historical background on prior methods for assessment of microbial safety of food, water, and air.

PREVALENCE OF INFECTIOUS DISEASE

Outbreaks of infectious waterborne illness continue to occur, although it remains impossible to identify the infectious agent in all cases. For example, in 1991 a waterborne outbreak in Ireland resulting from sewage contamination of water supplies infected about 5000 persons. However, the infectious agent responsible for this outbreak could not be determined (11).

1

In the United States there have typically been three to five outbreaks per year in community drinking water systems involving infectious microorganisms, with perhaps up to 10,000 cases annually. The 1993 Milwaukee *Cryptosporidium* outbreak with over 400,000 cases (7,22) was a highly unusual event among these statistics (Table I-1). Prior to the 1993–1994 reporting period, the largest number of cases were due to unidentified organisms.

There are substantially more outbreaks and cases of foodborne infectious diseases that are reported. Table I-2 summarizes reports of U.S. cases of principal microbial infectious foodborne illnesses. An average of about 7000 cases per year in 120 outbreaks occur, the dominant organism being *Salmonella*. It is interesting in comparison with waterborne outbreaks that in the case of foodborne illnesses, essentially all the agents (in reported outbreaks) are identified.

It is generally recognized that reported outbreaks, either of water- or foodborne infectious disease, represent only a small fraction of the total population disease burden. However, particularly in the United States, voluntary reporting systems and the occurrence of mild cases (for which no medical attention is sought but which are, nevertheless, frank cases of disease) have made it difficult to estimate the total case load.

In the United Kingdom, comparisons between the number of confirmed cases in infectious disease outbreaks and total confirmed laboratory illnesses (occurring in England and Wales) have been made (Table I-3). This suggests that the ratio of reported outbreak cases to total cases that may seek medical attention may be from 10 to 500 : 1, with some dependency on the particular agent.

Bennett et al. (3) estimated the total disease burden from infectious agents in the United States in 1985 (Table I-4). The endemic estimates in Table I-4 have also been supported by estimates of 68.7 million to 275 million cases per year of diarrheal diseases from all causes in the United States (1). It is noteworthy that several agents that are now believed to be highly important in food and/or water were not recognized as completely 10 years ago: for example, *Cryptosporidium* (28) and enterohemorrhagic *Escherichia coli* (18). However, even more striking is the ratio between the disease burden estimated and the outbreak cases from food or water (Tables I-1 and I-2).

For example, there are about 4200 cases of *Salmonella* from foodborne outbreaks in the United States per year (Table I-2). The estimated endemic rate (Table I-4) implies that there is a ratio of 500:1 between the outbreak cases and the endemic case burden. This is substantially greater than the ratio between outbreaks and total laboratory isolations—15.5—observed in the United Kingdom (Table I-3).

PRIOR APPROACHES

Concerns for the microbial quality of food, water, and other environmental media have long existed. In the early twentieth century, the use of indicator

TABLE I-1 Summary of Major Reported Infectious Diseases in Community Drinking Water Supplies

Agent	Reporting Period (Binannual)						Average per Year	
	1989–1990		1991–1992		1993–1994			
	Cases	Outbreaks	Cases	Outbreaks	Cases	Outbreaks	Cases	Outbreaks
AGI[a]	894	4	10,077	3	0	0	1,829	1.17
Giardia	503	4	95	2	385	5	164	1.83
Hepatitis	3	1					0.5	0.17
Escherichia coli O157:H7	243	1					41	0.17
Cryptosporidium			3,000	2	403,237	3	67,706	0.83
Campylobacter					172	1	28.7	0.17
Vibrio					11	1	1.83	0.17

[a] AGI is acute gastroenteritis of unknown etiology.

Source: Data from Refs. 15, 16, 19, 20, and 23.

TABLE I-2 Summary of Major Reported Infectious Diseases in Foodborne Outbreaks

| | Year | | | | | | | | | | Annual Average | |
| | 1988 | | 1989 | | 1990 | | 1991 | | 1992 | | | |
Agent	Cases	Outbreaks	Cases	Outbreaks	Cases	Outbreaks	Cases	Outbreak	Cases	Outbreaks	Cases	Outbreaks
Campylobacter	134	4	61	3	72	3	93	6	138	6	99.6	4.4
Escherichia coli	109	2	3	1	80	2	33	3	19	3	48.8	2.2
Salmonella	2987	94	4920	117	6290	136	4146	122	2834	80	4235.4	109.8
Shigella	3581	6	257	6	834	8	112	4	4	1	957.6	5
Staphyloccus aureus	245	8	524	12	372	12	331	9	206	6	335.6	9.4
Hepatitis	795	12	329	7	452	9	114	7	419	8	421.8	8.6
Listeria monocytogenes			2	1							0.4	0.2
Giardia			21	1	129	3	32	2	2	1	36.8	1.4
Norwalk agent			42	1					250	1	58.4	0.4
Vibrio (all)					49	6	6	2	2	1	11.4	1.8

Source: Ref. 2.

4

TABLE I-3 Comparison of Laboratory Isolations and Outbreak Cases in England and Wales, 1992–1994

Agent	All Laboratory Reports	Confirmed Outbreak Cases	Ratio
Campylobacter	122,250	240	509.4
Rotavirus	47,463	127	373.7
Shigella sonnei	29,080	847	34.3
Salmonella	92,416	5,960	15.5
Cryptosporidium	14,454	1,066	13.6
Escherichia coli O157	1,266	128	9.9

Source: Modified from Ref. 31.

microorganisms was developed for the control and assessment of the hygienic quality of such media and the adequacy of disinfection and sterilization processes. The coliform group of organisms was perhaps first employed for this purpose (13,26,29). Indicator techniques have also found utility in the food industry, such as the total count for milk and other more recent proposals (30). Other indicator groups for food, water, or environmental media have been examined, such as enterococci (5,6,10), acid-fast bacteria (8), bacteriophage (12,14,25), and *Clostridia* spores (4,25,27).

The use of indicator organisms was justified historically because of difficulty in enumerating pathogens. However, with the increasing availability of modern microbial methods (e.g., polymerase chain reaction (PCR), immu-

TABLE I-4 Estimated Total Disease Burden in the United States by Various Infectious Agents for 1985[a]

Agent	Cases	Deaths	Percent Attrtibutable to[b] Food	Water
Campylobacter	2,100,000	2,100	100	15
Escherichia coli	200,000	400	25	75
Salmonella (nontyphoid)	2,000,000	2,000	96	3
Shigella	300,000	600	30	10
Vibrio (noncholera)	10,000	400	90	10
Cryptosporidium	50	25	NI	NI
Giardia	120,000	0	NI	60
Hepatitis A	48,000	144	10	NI
Rotavirus	8,000,000	800	NI	NI
Norwalk and related agents	6,000,000	6	NI	NI

Source: Modified from Ref. 3.

[a] Includes community and noncommunity drinking water and recreational water.
[b] Percentages summing to more than 100% are from original reference. NI, not indicated.

noassay, etc.) for direct pathogen assessment, this justification has become less persuasive. In addition, to develop health-based standards from indicators, extensive epidemiologic surveillance is often necessary. The use of epidemiology has limitations with respect to detection limits (for an adverse effect) and is also quite expensive to conduct. Indicator methods are also limited in that many pathogens are more resistant than indicators to die-off in receiving environments or source waters or have greater resistance than indicators to removal by treatment processes (8,9,14,25). Thus the absence of indicators may not suffice to ensure the absence of pathogens.

The use of quantitative microbial risk assessment will enable direct measurement of pathogens to be used to develop acceptance and rejection guidelines for food, water, and other vehicles that may be the source of microbial exposure to human populations. The objective of this book is to present these methods in a systematic and unified manner.

REFERENCES

1. Archer, D. L., and J. E. Kvenberg. 1985. Incidence and cost of foodborne diarrheal disease in the United States. J. Food Prot. 48(10):887–894.
2. Bean, N. H., J. S. Goulding, C. Lan, and F. J. Angulo. 1996. Surveillance for foodborne-disease outbreaks: United States, 1988–1992. Morbid. Mortal. Wkly. Rep. 45(SS-5):1–66.
3. Bennett, J. V., S. D. Holmberg, M. F. Rogers, and S. L. Solomon. 1987. Infectious and parasitic diseases. Am. J. Prev. Med. 3(5 Suppl.):102–114.
4. Cabelli, V. J. 1977. *Clostridium perfringens* as a water quality indicator. *In* A. Hoadley and B. Dutka, eds., Bacterial indicators/health hazards associated with water. ASTM, Philadelphia.
5. Cabelli, V. J., A. P. Dufour, L. J. McCabe, and M. A. Levin. 1982. Swimming-associated gastroenteritis and water quality. Am. J. Epidemiol. 115:606–616.
6. Dufour, A. P. 1984. Health effects criteria for fresh recreational waters. U.S. Environmental Protection Agency, Washington, D.C.
7. Edwards, D. D. 1993. Troubled Waters in Milwaukee. ASM News 59(7):342–345.
8. Engelbrecht, R. S., C. N. Haas, J. A. Shular, D. L. Dunn, D. Roy, A. Lalchandani; B. F. Severin, and S. Farooq. 1979. Acid-fast bacteria and yeasts as indicators of disinfection efficiency. EPA-600/2-79-091. U.S. Environmental Protection Agency, Washington, D.C.
9. Engelbrecht, R. S., B. F. Severin, M. T. Masarik, S. Farooq, S. H. Lee, C. N. Haas, and A. Lalchandani. 1977. New microbial indicators of disinfection efficiency. EPA-600/2-77-052. U.S. Environmental Protection Agency, Washington, D.C.
10. Fleisher, J. M., F. Jones, and D. Kay. 1993. Water and non-water-related risk factors for gastroenteritis among bathers exposed to sewage-contaminated marine waters. Int. J. Epidemiol. 22(4):698–708.

11. Fogarty, J., L. Thornton, and R. Corcoran. 1995. Illness in a community associated with an episode of water contamination with sewage. Epidemiol. Infect. 114(2): 289–295.

12. Grabow, W. O. K., et al. 1983. Inactivation of hepatitis A virus and indicator organisms in water by free chlorine residuals. Appl. Environ. Microbiol. 46:619.

13. Greenwood, M., and G. U. Yule. 1917. On the statistical interpretation of some bacteriological methods employed in water analysis. J. Hyg. 16:36–56.

14. Helmer, R. D., and G. R. Finch. 1993. Use of MS2 coliphage as a surrogate for enteric viruses in surface waters disinfected with ozone. Ozone Sci. Eng. 15:279–293.

15. Herwaldt, B. L., G. F. Craun, S. L. Stokes, and D. D. Juranek. 1992. Outbreaks of waterborne disease in the United States: 1989–90. J. Am. Water Works Assoc. (April):129–135.

16. Herwaldt, B. L., G. F. Craun, S. L. Stokes, and D. D. Juranek. 1991. Waterborne-disease outbreaks, 1989–1990. Morb. Mortal. Wkly. Rep. 40(SS-3):1–21.

17. ILSI Risk Science Institute Pathogen Risk Assessment Working Group. 1996. A conceptual framework to assess the risks of human disease following exposure to pathogens. Risk Anal. 16(6):841–848.

18. Knight, P. 1993. Hemorrhagic E. coli: the danger increases. ASM News 59(5): 247–250.

19. Kramer, M. H., B. L. Herwaldt, G. F. Craun, R. L. Calderon, and D. D. Juranek. 1996. Surveillance for waterborne-disease outbreaks: United States, 1993–1994. Morbid. Mortal. Wkly. Rep. 45(SS-!):1–33.

20. Kramer, M. H., B. L. Herwaldt, G. F. Craun, R. L. Calderon, and D. D. Juranek. 1996. Waterborne disease: 1993 and 1994. J. Am. Water Works Assoc. 88(3):66–80.

21. Levin, B. R. 1996. The evolution and maintenance of virulence in microparasites. Emerg. Infect. Dis. 2(2):93–102.

22. MacKenzie, W. R., N. J. Hoxie, M. E. Proctor, M. S. Gradus, K. A. Blair, D. E. Peterson, J. J. Kazmierczak, K. R. Fox, D. G. Addias, J. B. Rose, and J. P. Davis. 1994. Massive waterborne outbreak of Cryptosporidium infection associated with a filtered public water supply, Milwaukee, Wisconsin, March and April 1993. N. Engl. J. Med. 331(3):161–167.

23. Moore, A. C., B. L. Herwaldt, G. F. Craun, R. L. Calderon, A. K. Highsmith, and D. D. Juranek. 1993. Surveillance for waterborne disease outbreaks: United States, 1991–1992. Morb. Mortal. Wkly. Rep. 42(SS-5):1–22.

24. National Academy of Sciences. 1983. Risk assessment in the federal government: managing the process. National Academy Press, Washington, DC.

25. Payment, P., and E. Franco. 1993. Clostridum perfringens and somatic coliphages as indicators of the efficiency of drinking water treatment for viruses and protozan cysts. Appl. Environ. Microbiol. 59(8):2418–2424.

26. Phelps, E. 1909. The disinfection of sewage and sewage filter effluents. USGS Water Supply Paper 229. U.S. Geological Survey, Washington, DC.

27. Rice, E. W., K. R. Fox, R. J. Miltner, D. A. Lytle, and C. H. Johnson. 1996. Evaluating plant performance with endospores. J. Am. Water Works Assoc. 88(9): 122–130.

28. Rose, J. B. 1988. Occurrence and significance of *Cryptosporidium* in water. J. Am. Water Works Assoc. 80(2):53–58.

29. Rudolfs, W., and H. W. Gehm. 1935. Multiplication of total bacteria and *B. coli* after sewage chlorination. Sewage Works J. 7:991–996.

30. Subcommittee on Microbiological Criteria. 1985. An evaluation of the role of microbiological criteria for foods and food ingredients. National Academy Press, Washington, DC.

31. Wall, P. G., J. de Louvois, R. J. Gilbert, and B. Rowe. 1996. Food poisoning: notifications, laboratory reports and outbreaks—where do the statistics come from and what do they mean? Communicable Dis. Rep. Rev. 6(7):unnumbered.

CHAPTER 1

SCOPE OF COVERAGE

Quantitative microbiological risk assessment (QMRA) is the application of principles of risk assessment to the estimate of consequences from a planned or actual exposure to infectious microorganisms. In performing a QMRA, the risk assessor aims to bring the best available information to bear in understanding the nature of the potential effects from a microbial exposure. Since the information (such as dose–response relationships, exposure magnitudes, etc.) is almost invariably incomplete, it is also necessary to ascertain the potential error involved in the risk assessment. With such information, necessary steps to mitigate, control, or defend against such exposures may be developed.

At the outset of performing a risk assessment, a scoping task should be undertaken. This task should set forth the objectives of the analysis and the principal issues to be addressed. Items such as consideration of secondary cases, individual versus population risk, agent or agents to be examined, exposure routes, and/or accident scenarios must be stipulated. However, this scoping may be changed during the course of a QMRA, to reflect the input derived from the risk manager(s) and other potentially interested parties.

POTENTIAL OBJECTIVES OF A QMRA

There may be diverse objectives for a QMRA. These objectives relate to the rationale for the performance of the assessment as well as the methods to be employed. Broadly, the different objectives reflect different scales at which a risk assessment may be performed. The step of problem formulation is critical to any risk estimate (13). The problem should be formulated so as to respond directly to the needs of a decision maker, as well as addressing concerns of relevance to interested publics. In general, the problems posed are of several types.

Site-Specific Assessment

The simplest type of QMRA that may be performed involves one site or exposure scenario. The following are typical of the questions that might be asked:

1. If a water treatment plant is designed in a certain way (with given removals of pathogens), what is the risk that would be placed on the population served?
2. A swimming outbreak (in a recreational lake) has just occurred, believed to have resulted from a short-duration contamination event. What pathogen levels would be consistent with the attack rate observed?
3. Microbial sampling of a finished food product has found certain pathogens. What level of risk does this pose to consumers of the product?

Note that there are certain other contrasts in the objectives of the risk assessments to be posed. In questions 1 and 3, a before-the-fact computation is desired, whereas in question 2, an after-the-fact computation is described (the fact being an episode of illness). Also in question 1 and 3, pathogen levels are available (or somehow are estimated), whereas in question 2, an inverse computation is needed given an observed attack rate.

Basically in performing this risk assessment, the relationship between an exposure or technological metric and a risk measurement must be ascertained, and then the particular point of correspondence determined (Figure 1-1). In question 1 and 3, for a known (or assumed) exposure (on the x-axis), the corresponding range of risks on the y-axis is sought. In question 2, for known or assumed risks (on the y-axis), the corresponding range of exposures (or level of technological protection) is to be determined (on the x-axis).

Ensemble of Sites

A somewhat more complex situation occurs if the risk for a set of events or sites must be estimated. Basically, this now includes the necessity to incorporate site-to-site factors in the assessment. Some examples of this are as follows:

1. If we want to keep the risk to a population served by multiple water treatment plants at a given level (or better), what criteria should we use (microbial levels)?

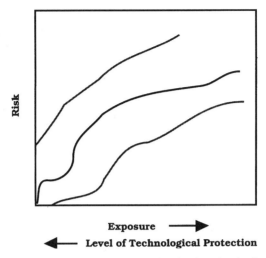

Figure 1-1 Relationship between exposure, level of technological protection, and microbial risk. The middle curve indicates the best estimate. The other two curves indicate the upper and lower confidence regions.

2. For a food product subject to contamination by pathogens, what would be an acceptable treatment specification (e.g., heating time, holding period) to ensure microbial acceptability?

3. A water quality standard for recreational bathing waters is being designed. If a uniform (e.g., national) standard is to be developed, what standard would ensure that average risk was acceptable in the context of keeping the risk of a large "cluster" of illnesses low?

In addition to incorporating a measure of ensemble average risk, in general it is also desired to ensure that no single member of the ensemble be unacceptably extreme. For example, consider the evaluation of three options of disease control among three communities, as indicated in Table 1-1. This table indicates the number of cases and the rate among the three communities. The

TABLE 1-1   Effect of Different Hypothetical Policy Options on Distribution of Risk Among Communities (for a Fixed Total Risk)

| | | Policy Option A | | Policy Option B | | Policy Option C | |
| | Exposed | | Incidence | | Incidence | | Incidence |
Community	Population	Cases	(no./10,000)	Cases	(no./10,000)	Cases	(no./10,000)
A	100,000	20	2	6	0.6	24	2.4
B	50,000	10	2	18	3.6	7	1.4
C	10,000	2	2	8	8	1	1
TOTAL	160,000	32	2	32	2	32	2

three policy options yield the same number of expected cases. However, there are differences in the allocation of risk among communities of different sizes. In option A, all communities have an identical level of estimated risk; in option B, the risk increases as community size decreases; in option C, the risk increases as community size increases. This distribution of risk among affected subsets of the ensemble being considered adds an additional dimension for consideration by a risk manager, which may be termed *risk equity*.

SECONDARY TRANSMISSION

Infectious microbial diseases are different in terms of risk to a population from chemical agents in that a person who may become infected (with or without illness) can then proceed to infect additional people. These secondary (tertiary, etc.) cases may be persons who had no direct contact with the initial vehicle of exposure; nevertheless, when accounting fairly for the public health impact, they should be considered.

Secondary cases may arise by a variety of mechanisms. Particularly among close family members, household secondary cases can arise by direct or indirect (e.g., surface contamination) contact; this is particularly so when the primary case, or one household secondary case is a child (5,6,9). Table 1-2 summarizes secondary case statistics obtained from a variety of outbreaks.

Presumably, secondary cases may also arise from close contact with an asymptomatic person (in the "carrier" state). This is well known for highly acute and (now) uncommon illnesses (such as typhoid). Excretion of Norwalk virus following recovery (and resulting in additional cases) has been documented to occur for as long as 48 hours postrecovery (15).

OUTBREAKS VERSUS ENDEMIC CASES

As noted in the Introduction, there may be a substantial difference between reported outbreak cases and total disease burden in a community. For a disease case to receive recognition by the public health authorities, the following specific and sequential steps must occur (4):

1. An ill person must seek medical care.
2. Appropriate clinical tests (e.g., blood, stool) must be ordered by the attending physician.
3. The patient must comply with obtaining the sample.
4. The laboratory must be capable of detecting the relevant pathogens.
5. The clinical test must be positive.
6. The test result must be reported to the health agency in a timely manner.

TABLE 1-2 Summary of Secondary Case Data in Outbreak Situations[a]

Organism	Secondary Attack Ratio[b]	Secondary Prevalence in Households[c]	Remarks	Reference
Cryptosporidium parvum	0.33	0.33	Outbreak in contaminated apple cider	10
	NA	0.042	Drinking waterborne outbreak (Milwaukee)	9
Shigella	0.28	0.26	Day-care-center outbreaks in children	12
Rotavirus	0.42	0.15	Day-care-center outbreaks in children	12
Giardia lamblia	1.33	0.17	Day-care-center outbreaks in children	12
Viral gastroenteritis	0.22	0.11[d]	Drinking waterborne outbreak	11
	0.56	NA	drinking waterborne outbreak (Denmark)	8
Norwalk virus	0.5–1.0	0.19	Swimming outbreak	1
	1.1	0.29	Swimming outbreak in children	7
	NA	0.44	Foodborne outbreak in children and teachers	6
	0.4	NA	Foodborne outbreak	15
Escherichia coli O157:H7	NA	1.18[d]	Day-care-center outbreak in children	14
Unidentified day-care diarrheal diseases	1.38	0.09[d]		3

[a]NA, not available.
[b]Ratio of secondary cases to primary cases.
[c]Proportion of households with one or more primary cases who have one or more secondary cases.
[d]Proportion of persons in contact with one or more primary cases who have a secondary case.

13

If any of the links in this sequential process are broken, a disease case will not enter the records maintained by health authorities. For example, with increasing controls on medical care, stool samples may not be obtained from mild cases of illness. Some organisms may only be present sporadically, or may be difficult to test in a stool or blood sample. Patients may not seek medical attention for mild cases of illness. Furthermore, in the United States in particular, the surveillance of environmentally induced disease is done on a passive basis, and hence the number of actual illness clusters that are actually compiled into recorded statistics is only a small fraction of such clusters of illness that occur (4).

From a more fundamental point of view, an outbreak of illness is generally defined as occurrence of the illness at a level greater than normal or anticipated. This definition recognizes that there is a level of illness ("endemic") that may exist under usual circumstances. The detection of such outbreaks poses a particular challenge. The problem is illustrated conceptually in Figure 1-2.

Additional complications arise from the different patterns of illness in a community, including definite periodicities as well as temporal trends, and from the presence of reporting lags associated with laboratory analysis and time for patients to seek medical attention. Figure 1-3 illustrates the different patterns of illness in the case of six pathogens for England and Wales (2).

In the case of water- and foodborne illnesses, it is highly likely that the level of such endemic illnesses is substantially greater than those occurring during outbreaks (even accounting for unrecognized outbreaks). As a result,

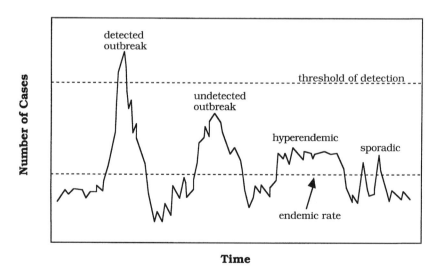

Figure 1-2   Schematic of disease occurrence in a hypothetical community. (Modified from Ref. 4.).

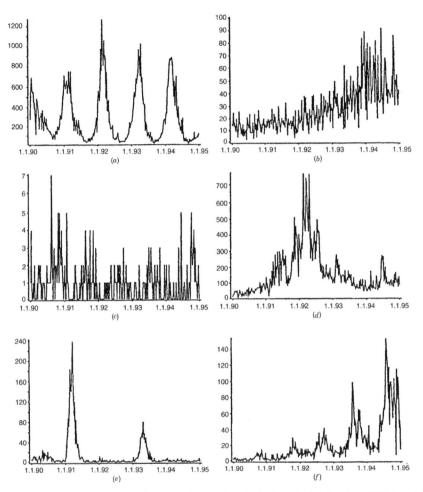

Figure 1-3 Weekly count of reported organism isolations in England and Wales: (*a*) rotavirus; (*b*) *Clostridium difficile;* (*c*) *Salmonella derby;* (*d*) *Shigella sonnei;* (*e*) *influenza B;* (*f*) *Salmonella typhimurium* DT104. (From Ref. 2.) (1996).

there are often many cases of environmentally caused (water, air, food) infectious disease that are unrecognized. One example of this, noted in the Introduction, is *Camplyobacter.* There have been an average of about 200 cases per year of water- and foodborne illness in outbreaks of this organism, yet estimates of the disease burden suggest about 2,100,000 cases per year (i.e., approximately 10,000 cases per case of detectable outbreak illness). Therefore, it will be important to assess the factors that may influence outbreak detection. These issues are discussed in subsequent chapters.

REFERENCES

1. Baron, R. C., F. D. Murphy, H. B. Greenberg, C. E. Davis, D. J. Bregman, G. W. Gary, J. M. Hughes, and L. B. Schonberger. 1982. Norwalk gastrointestinal illness: an outbreak associated with swimming in a recreational lake and secondary person to person transmission. Am. J. Epidemiol. 115(2):163–172.

2. Farrington, C. P., N. J. Andrews, A. D. Beale, and M. A. Catchpole. 1996. A statistical algorithm for the early detection of outbreaks of infectious disease. J. R. Stat. Soc. A 159(3):547–563.

3. Ferguson, J. K., L. R. Jorm, C. D. Allen, P. K. Whitehead, and G. L. Gilbert. 1995. Prospective study of diarrhoeal outbreaks in child long daycare centres in Western Sydney. Med. J. Aust. 163(August 7): 137–140.

4. Frost, F. J., G. F. Craun, and R. L. Calderon. 1996. Waterborne disease surveillance. J. Am. Water Works Assoc. 88(9):66–75.

5. Griffin, P. M., and R. V. Tauxe. 1991. The epidemiology of infections caused by *Escherichia coli* O157:H7, other enterohemorrhagic *E. coli* and the associated hemolytic uremic syndrome. Epidemiol. Rev. 13:60–98.

6. Heun, E. M., R. L. Vogt, P. J. Hudson, S. Parren, and G. W. Gary. 1987. Risk factors for secondary transmission in households after a common source outbreak of Norwalk gastroenteritis. Am. J. Epidemiol. 126(6):1181–1186.

7. Kappus, K. D., J. S. Marks, R. C. Holman, J. K. Bryant, C. Baker, G. W. Gary, and H. B. Greenberg. 1982. An outbreak of Norwalk gastroenteritis associated with swimming in a pool and secondary person to person transmission. Am. J. Epidemiol. 116(5):834–839.

8. Laursen, E., O. Mygind, B. Rasmussen, and T. Ronne. 1994. Gastroenteritis: a waterborne outbreak affecting 1600 people in a small Danish town. J. Epidemiol. Community Health 48:453–8.

9. MacKenzie, W. R., W. L. Schell, B. A. Blair, D. G. Addiss, D. E. Peterson, N. J. Hozie, J. J. Kazmierczak, and J. P. Davis. 1995. Massive outbreak of waterborne *Cryptosporidiam* infection in Milwaukee, Wisconsin: recurrence of illness and risk of secondary transmission. Clin. Infect. Dis. 21:57–62.

10. Millard, P., K. Gensheimer, D. G. Addiss, D. M. Sosin, G. A. Beckett, A. Houck-Jankoski, and A. Hudson. 1994. An outbreak of cryptosporidiosis from fresh-pressed apple cider. J. Am. Med. Assoc. 272(20):1592–1596.

11. Morens, D. M., R. M. Zweighaft, T. M. Vernon, G. W. Gary, J. J. Eslien, B. T. Wood, R. C. Holman, and R. Dolin. 1979. A waterborne outbreak of gastroenteritis with secondary person to person spread. Lancet (May 5):964–966.

12. Pickering, L. K., D. G. Evans, H. L. DuPont, J. J. Vollet, and D. J. Evans, Jr. 1981. Diarrhea caused by *Shigella,* rotavirus and *Giardia* in day care centers: prospective study. J. Pediatr. 99(1):51–56.

13. Presidential/Congressional Commission on Risk Assessment and Management. 1997. Framework for environmental health risk management. U.S. Government Printing Office, Washington, DC.

14. Spika, J. S., J. E. Parsons, and D. Nordenberg. 1986. Hemolytic uremic syndrome and diarrhea associated with *Escherichia coli* O157:H7 in a day care center. J. Pediatr. 109:287–291.

15. White, K. E., M. T. Osterbolm, J. A. Mariotti, J. A. Korlath, D. H. Lawrence, T. L. Ristinen, and H. B. Greenberg. 1986. A foodborne outbreak of Norwalk virus gastroenteritis. Am. J. Epidemiol. 124(1):120–126.

CHAPTER 2

MICROBIAL AGENTS AND THEIR TRANSMISSION

MICROBIAL TAXONOMY

No classification system of microorganisms is accepted by all biologists. Protista is a kingdom commonly used to describe microorganisms. It is divided into eukaryotic organisms, which possess a cell nucleus, and prokaryotic organisms, which possess no organized nucleus. Eukaryotic organisms include helminths (worms), protozoans, fungi, and algae (Figure 2-1). Prokaryotic organisms include the bacteria, blue-green algae (cyanobacteria), and rickettsia (Figure 2-2). Viruses that are obligate intercellular parasites are not members of either group. They are usually composed only of a nucleic acid surrounded by a protein coat. Eukaryotes are much more complex than procaryotic cells in structure and nutrition requirements. Procaryotes divide by simple binary fission whereas eucaryotes divide by a more complex process called *mitosis*.

Eucaryotes

Fungi include both multicellular organisms (mushrooms) and unicellular organisms (*Penicillium*). Fungi obtain nutrients solely by absorption of organic matter from dead organisms. Even when they invade living tissues, they typically kill cells and then adsorb the nutrients from them. They are capable of forming long filaments called *hyphae,* which form a mass called *mycelium.* In most fungi, the hyphae have separations that divide the filament into separate cells containing one nucleus each. Some fungi are capable of forming environmentally resistant spores. Many are capable of living under very ad-

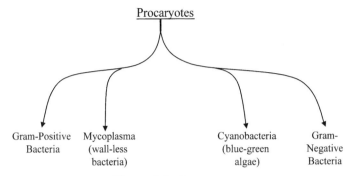

Figure 2-1 Procaryotes.

verse conditions such as cold temperatures and freezing conditions. Most fungi are aerobic, but some are facultatively anaerobic. A limited number are pathogenic to plants and animals. Fungal diseases in humans are referred to as *mycoses*. Airborne fungal spores are responsible for allergies in humans. Yeasts are classified with the fungi and some are pathogenic to humans *(Candida albicans)*. Some fungi may produce toxins (i.e., aflatoxins) when growing in food that are mutagenic or carcinogenic in animals. Water and food do not appear to be significant routes of transmission of pathogenic fungi for humans. However, inhalation is an important method of initiating an infection for some fungal diseases, such as *Aspergillosis, Blastomycosis, Coccidioidomycosis,* and *Histoplasmosis.* These fungi may be found growing in animal feces, soil, or decaying organic matter (e.g., compost, manure) and may be transmitted through the air when this material is disturbed. Some fungi may be transmitted by direct contact with the skin.

Protozoans are unicellular organisms that are surrounded by a cytoplasmic membrane that is covered by a protective structure called a *pellicle.* Most protozoans are free-living. Some of these can cause diseases, whereas others are obligate parasites. They are capable of forming structures called *cysts, oocysts,* or *spores* (depending on the life cycle of the organism) that are very

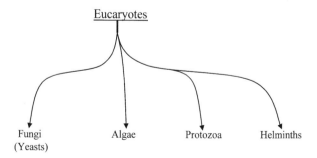

Figure 2-2 Eucaryotes.

resistant to adverse environmental conditions such as desiccation, starvation, high temperatures, and disinfectants. Some pathogenic protozoa are found naturally in the soil (*Naegleria*) or water (*Acanthamoeba*). Others, such as *Giardia* and *Cryptosporidium,* whose natural habitat is the intestines of warm-blooded animals, are capable of causing disease in both humans and animals. Others, such as *Entamoeba histolytica,* are only capable of causing disease in humans.

Five groups of protozoans containing members that are pathogenic to humans and having the potential for transmission through food and/or water are:

1. amoebas (*Rhizopoda*)
2. flagellates (*Mastlgohora*)
3. ciliates (*Cilratea*)
4. *Sporozoa*
5. *Microsporidia*

Pathogenic enteric protozoans are usually excreted in the feces or urine and can be transmitted by fecally contaminated food and water. Some protozoans may be transmitted through the air (*Acanthamoeba*) or through *fomites* (inanimate objects). Some important environmentally transmitted protozoans are shown in Table 2-1.

The clinically important intestinal protozoa are divided taxonomically into the amoebas (i.e., *Entamoeba* sp.), flagellates (*Giardia*), and coccidia, meaning "globose in shape" (*Cryptosporidium* and *Cyclospora*). These are single-celled microscopic animals, which reproduce in the intestinal tract of the host and as a result of their life cycles produce environmental stages that are excreted in the feces.

They are obligate parasites (which means that they can only reproduce in the host) and are transmitted via the infectious cyst and oocyst stages by the fecal–oral route. The life cycles of the protozoa are similar. The initial stage occurs during ingestion of the cyst or oocyst. These are egglike structures that are produced as a result of the life cycle and are excreted in the feces of infected persons (thus the term *fecal–oral*). Body temperature and passage through the stomach cause the cyst or oocyst to open up in a process known as excystation. Excystation can also be accomplished in the laboratory in a test tube (with trypsin, bile salts, and temperature at 37°C). This is one measure of viability, whether the cyst or oocyst is alive.

The *Giardia* cyst contains a chitinlike substance, which gives it its resistance. The cyst houses two trophozoites, which is the stage that attaches and grows in the intestinal tract through asexual reproduction. As the trophozoites begin to cover the wall of the intestinal tract, diarrhea can result (Figure 2-3). As the trophozoites are released, some encyst (form cysts) as they pass through the bowels.

TABLE 2-1 Some Medically Important Protozoans Transmitted Through the Environment

Protozoans	Environmentally Resistant Stage (Size, μm)	Illness or Disease	Transmission Route
Amoebas			
Entamoeba histolytica	Cyst (10–20)	Diarrhea, abscesses in intestinal tissues (e.g., liver)	Water, food
Naegleria fowlei	Cyst	Primary meningoencephalitis (PAM)	Water, swiming, no person-to-person route
Acanthamoeba	Cyst	Meningoencephalitis, eye lesions, respiratory and skin lesions	Water, aerosols, no person-to-person route
Flagellate			
Giardia lamblia	Cyst (8–15)	Diarrhea	Water, food
Cilate			
Balantidium coli	Cyst (45–65)	Diarrhea	Water
Sporozoans			
Isopora spp.	Oocyst (10–33)	Diarrhea	Water?
Cryptosporidium parvum	Oocyst (3–8)	Diarrhea	Food, water, fomites
Toxoplasma gondii	Oocyst (10–12)	Congenital infection with onset in utero, febrile illness, encephalitic infection	Food, water, fomites
Cyclospora	Oocyst (8–10)	Diarrhea	Water, food
Microsporidia			
Enterocytozoan bieneus	Spore (1–3)	Diarrhea	Water?
Encephalitzoan cuniculi	Spore	Disseminated disease of lungs and liver	Water?
E. intestinalis			

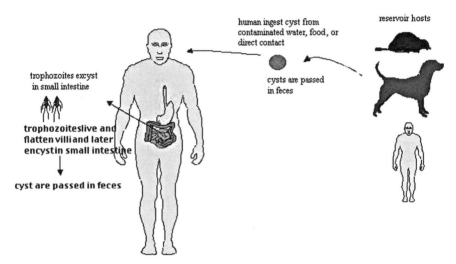

Figure 2-3 Life cycle of *Giardia lamblia*.

Cryptosporidium and *Cyclospora* both produce oocysts. The life cycle of *Cyclospora* is not as well described but is thought to be like other coccidia. The oocyst contains complex carbohydrates and lipids and is therefore acid-fast (a type of stain used to differentiate microorganisms). This is used as a stain to detect the oocysts in feces in the clinical laboratory. *Cryptosporidium* oocysts (Figure 2-4) are about 5 μm in diameter and house four sporozoites. The sporozoite stage enters into the intestinal cell and begins the infection process. There are many stages of the parasite that form in the cell, a process

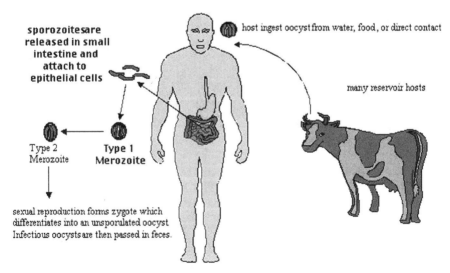

Figure 2-4 Life cycle of *Cryptosporidium parvum*.

that can take up to 18 hours. One stage (the merozoites) starts an autoreinfective process, causing other cells to become infected. As more cells become infected, the illness begins and oocysts are excreted in high numbers (about 100,000/per gram). The oocysts are produced as a result of sexual reproduction. The oocyst of *Cyclospora* is 10 μm in diameter and houses two sporocysts, two oblong structures in the oocyst, which in turn house the sporozoites. The sporocysts are not formed until the oocyst undergoes maturation in the environment. After the maturation, the oocyst is referred to as being sporulated, and only then is it infectious. After ingestion and excystation, the life cycle appears to proceed as in *Cryptosporidium*.

Worms or helminths are small multicellular animals that are parasitic in humans and animals. They include flukes, tapeworms, and roundworms (*Ascaris lumbricoides*). The ova (eggs) are excreted in the feces of an infected animal or human and spread by wastewater, soil, or food. Some major environmentally transmitted helminths are listed in Table 2-2.

Algae are capable of obtaining their energy needs through photosynthesis. All algae contain chlorophyll. Most are unicellular organisms, but some others are colonial. They are the primary producers in most aquatic environments and are referred to as *phytoplankton.* In drinking water they are often responsible for taste and odor problems. Some forms, such as dinoflagellates, are capable of producing nerve toxins which result in fish kills and human deaths associated with fish and shellfish consumption. They do not cause disease by invasion of animal tissues.

Prokaryotes

Bacteria occur in three basic shapes: spherical (*cocci*), rod (*bacilli*), and *spirilla* (corkscrews) and generally range in size from 0.3 to 2 μm. The Gram stain was among the first criteria used to classify bacteria. Gram-positive

TABLE 2-2   Major Parasitic Helminths

Organism	Disease (Main Site Affected[a])
Nematodes (roundworms)	Ascariasis: intestinal obstruction in children (small
Ascaris lumbricoides	intestine)
Trichuris trichiura	Whipworm: trichuriasis (intestine)
Hookworms	
Necator americanus	Hookworm disease (GI tract)
Ancylostoma duodenale	Hookworm disease (GI tract)
Cestodes (tapeworms)	
Taenia saginata	Beef tapeworm: abdominal discomfort, hunger pains, chronic indigestion (GI tract)
T. solium	Pork tapework (GI tract)
Trematodes (flukes)	Schistosomiasis [complications in liver (cirrhosis),
Schistosoma mansoni	bladder, and large intestine]

[a]GI, gastrointestinal tract.

bacteria stain purple, and Gram-negative bacteria pink using a differential staining technique. Gram-negative bacteria have a lipid-polysaccharide in their cell wall, which is responsible for fever production in animals. Many pathogenic bacteria are capable of producing a wide range of toxins that may be responsible for the symptoms associated with the disease they cause. Bacteria may cause illness by direct killing of cells or through the production of toxins. Some forms of bacteria produce spores (*Bacillus, Clostridium*) that are very heat resistant and capable of surviving for prolonged periods of time in the environment (i.e., years). Some of the most important environmentally transmitted bacteria are shown in Table 2-3.

Many human pathogenic bacteria infect both animals and humans and are capable of growth in the environment outside the host organism. Examples are common among the enteric bacteria (bacteria that normally infect the intestinal tract) such as *Salmonella*, which causes diarrhea in chickens, cattle, and humans. It is also capable of growing to large numbers in contaminated food (e.g., meat and eggs). In contrast, some human enteric pathogens only infect humans (e.g., *Shigella*). In some cases only certain strains of bacteria produce illness. *Escherichia coli* is a common inhabitant of all warm-blooded animals, but only certain strains are capable of producing illness in humans (O157:H7, for example).

Most human enteric bacterial pathogens do not survive long in aquatic environments or the soil. There are exceptions such as *Vibro cholerae*, whose natural habitat is the marine environment. It is introduced into the human population through seafood, and under conditions of poor sanitation is readily spread by the water route.

Rickettsias and chlamydias are bacteria that are obligate intracellular parasites. Many are human pathogens. They often infect both animals and arthropods (ticks and lice). They can be transmitted to humans by ticks and arthropods or in food products such as milk from an infected cow. They are associated with illnesses such as Q fever and Rocky Mountain spotted fever.

Mycoplasmas are small bacteria (0.3 to 0.8 μm) that are pleomorphic in shape and lack a cell wall like most bacteria. Mycoplasmas can be grown on agar, where they form colonies with a "fried egg" appearance. Some are parasitic to plants and animals, while others live on decaying organic matter. *Mycoplasma pneumonia* is the leading cause of pneumonia among college students and is common among military recruits. The disease is common and difficult to diagnose—hence the popular name "walking pneumonia." It does not spread quickly among populations, and contact with droplets generated by sneezing and coughing may be the primary route of transmission (92).

Blue-green algae are classified with the bacteria and referred to as *cyanobacteria* since, unlike the green algae, they do not have a nucleus. They contain chlorophyll and carry out photosynthesis. Cyanobacteria are capable of survival under extreme conditions and can grow in hot springs. They are responsible for algal blooms in lakes and other aquatic environments. Some types (microcysts) are capable of producing neuotoxins, potential carcinogens,

TABLE 2-3 Selected Bacteria of Medical Importance Transmitted Through the Environment

Family, Genus, Species	Clinical Infection	Route of Transmission (Source)
Spirochetes Gram-negative helical bacteria		
Leptospiraceae		
Leptospira interrogans	Leptospirosis	Water (urine)
Aerobic/microaerophilic, helical gram-negative bacteria		
Campylobacter jejuni	Gastroenteritis	Water, food
Gram-negative aerobic rods and cocci		
Pseudomonodaceae		
Pseudomonas aeruginosa	Wound, burn, urinary tract infections	Water, food, air
Legionellaceae		
Legionella pneumophilia	Pneumonia (Legionnaire's disease)	Air
Other genera		
Brucella abortus	Brucellosis	Food, air
B. melitensis	Brucellosis	Food, air
B. suis	Brucellosis	Air
Bordetella pertussis	Pertussis (whooping cough)	Air
Francisella tularensis	Tularemia	Food, water, direct contact, insects
Facultative anaerobic Gram-negative rods		
Enterobacteriaceae		
Escherichia coli	Opportunistic infections; some strains cause diarrhea	Water, food
Enterobacter aerogenes	Opportunistic infections	
Shigella dysenteriae	Dysentry	Water, food
Salmonella typhi	Typhoid fever	
Klebsiella pneumoniae	Pneumonia	
Proteus mirabilis	Urinary tract infections	Unknown
Serratia marcescens	Opportunistic infections	
Yersina enterocolitica	Gastroenteritis	Water, food
Vibronaceae		
Vibrio cholerae	Cholera	Water, food
V. parahaemolyticus	Gastroenteritis from seafood	
Pasteurellaceae		
Haemophilus influenzae	Meningitis, other pediatric diseases	Unknown

TABLE 2-3 (*Continued*)

Family, Genus, Species	Clinical Infection	Route of Transmission (Source)
Rickettsias and chlamydias, Gram-negative		
Coxiella burnetti	Q fever	Air, food
Chlamydia psittaci	Psittacosis	Air
Mycoplasmas (cell wall–less bacteria)		
Mycoplasma pneumoniae	Primary, atypical pneumonia	Air
Gram-positive cocci		
Micrococcaceae		
Staphylococcus aureus	Food intoxification	Food
	Skin infections	Direct contact
Other genera		
Streptococcus pyogenes	Pharyngitis, skin infections	Direct contact
S. faecalis	Opportunistic infection	Unknown
Endospore-forming Gram-positive rods and cocci		
Bacillus anthracis	Anthrax	Air, water, food, soil
Clostridium botulinum	Botulism	Soil
	Food intoxifications	Food
Cl. perfringens	Gas gangrene, food intoxifications	Food, soil
Cl. difficile	Gastroenteritis, colitis	Unknown
Regular, nonsporing, Gram-positive rods		
Listeria montocytogenes	Meningitis	Food, air
Irregular, nonsporing, Gram-negative rods		
Coryhebacterium diphtheria	Diphtheria	Air?
Actinomycetes israelli	Actinomycosis	Air soil?
Mycobacteria (acid-fast rods)		
Mycobacterium tuberculosis	Tuberculosis	Food, air water
M. avium	Pulmonary disease, disseminated disease in immunocompromised	

and taste and odor problems in drinking water. They do not cause infections in animals.

Opportunistic pathogens are organisms that ordinarily do not cause disease in their normal habitat in normal healthy persons. For example, microbes that gain access to the broken skin or mucous membranes can cause opportunistic infections. Opportunistic infections are, however, of greatest concern in persons already weakened (e.g., severe burns on the skin, diabetes, cystic fibrosis, existing infection by a pathogen) or those who have an impaired immune system. Opportunistic infections are of greatest concern in hospitals (nosocomial), where there is a risk of serious illness which may be life threatening because of the weakened condition of the patient due to preexisting illness. Many types of organisms in our environment have the ability to cause opportunistic infections, particularly in seriously debilitated patients; some, however, such as *Escherichia coli, Pseudomonas* spp., and *Staphylococcus aureus,* are common causes of such infections in hospitals (134). Sources of opportunistic pathogens include patients and their contacts: medical devices, water, and food. While hospital staff have been well documented as the source of some opportunistic organisms (e.g., *S. aureus*) and medical devices, the role of water and food is not as well documented. Many opportunistic pathogens occur in drinking water, and this has often been suspected; however, little evidence is available to demonstrate this link (234). Opportunistic infections that have been linked to drinking water are *Mycobacterium avium* and *Legionella pneumophila* (234).

Viruses

Viruses are usually 200 nm or smaller and cannot be seen with a light microscope. They are not living cells and do not need food, nor do they carry out independent metabolic processes. Viruses are obligate intracellular parasites; they replicate or multiply only inside a living host. Viruses differ from living cells in several important ways. Whereas cells contain both RNA and DNA and both grow and divide, viruses contain only RNA or DNA, do not grow, and never divide. Viruses multiply by directing synthesis and assembly of viral components inside cells to form new viruses. When not in a host cell, a virus is an inert macromolecule (i.e., nucleic acid and proteins). The major components of viruses are a central core nucleic acid and a protein coat or capsid. Certain viruses contain enzymes, and some have a lipid coat. Viruses die or become inactivated (i.e., no longer capable of replicating in a host cell) by damage to the structural integrity of the protein, nuclei acid, or lipid envelope (Figures 2-5 and 2-6). A complete virus particle, including its lipid envelope if it has one, is called a *virion*.

The most widely used taxonomic criteria for animal viruses are based on four characteristics:

1. The nature of the nucleic acid: DNA or RNA, single or double stranded

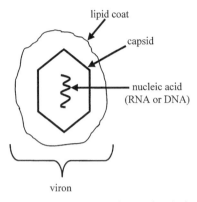

Figure 2-5 Structure of an animal virus.

2. Structure: spherical (polyhedral), cylindrical (rod-shaped), or complex
3. Presence or absence of a viral envelope
4. Size of the viral particle

Beyond these physical characteristics, other criteria, such as immunological typing and effects on host cells, are used. Based on these criteria, animal

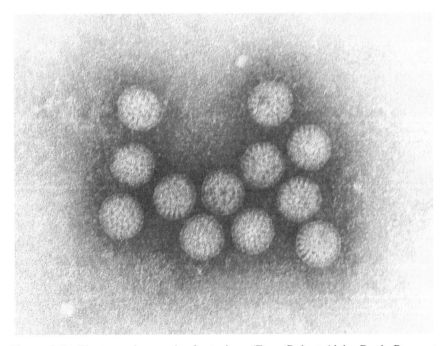

Figure 2-6 Electron micrograph of rotavirus. (From Robert Alain, Perrie Payment, Institut Armand-Frappier, Center for Research in Virology, Montreal, Quebec, Canada.)

viruses are divided into a number of families whose names end in "-viridae": for example, Picornaviridae. Each family contains numerous genera and species whose names end in "-virus": for example, rotavirus. The species name is often the disease that the virus causes, in contrast to bacterial nomenclature, in which an organism is referred to by its genus and species name. The characteristics of the major types of animal viruses are shown in Table 2-4.

Viruses are fairly host specific: for example, many humans and primates are the only hosts for many viruses (e.g., Norwalk virus). Other viruses, such as foot-and-mouth disease, infect both cattle and humans. Other viruses, such as some togaviruses, are capable of infecting arthropods, which in turn bite humans, thus transmitting several kinds of encephalitis and tropical fevers such as dengue and yellow fever. The virus replicates in the insect but does not cause illness. Genera of the same virus may infect both animals and humans but seldom, if ever, cross species barriers. For example, rotaviruses are a common cause of diarrhea in almost all infant animals, including humans. However, the rotavirus types that infect animals usually do not infect humans in the natural environment. This may occur rarely and probably explains how viruses evolve to adapt from one species to another. Other viruses, such as reoviruses, may cause a serious illness in one species (e.g., chickens) and infect humans without definable or reproducible ill effects. The virus may infect the host, but no apparent illness is observed.

The viruses that infect plants, fungi, algae, and bacteria are classified in a manner similar to the animal viruses. Viruses that infect these groups or organisms do not infect animals. Viruses that infect bacteria are known as *bacteriophages* or simply *phages*. Almost all groups of bacteria have phages. Phages that infect coliform bacteria are called *coliphage*. Coliphages have been used extensively as models for studies of virus transport and behavior in the environment because they are inexpensive to grow and assay compared to animal viruses. Also, they can be grown in much higher numbers and are not infectious to humans.

The most environmentally significant viruses are the enteric viruses, transmitted by the fecal–oral route, and the respiratory viruses, transmitted by aerosols or droplet contamination of fomites. There are more than 120 known enteric viruses, which cause a wide range of illnesses (Table 2-5). Some may be transmitted by both fecal matter or aerosols (e.g., Coxsackievirus B3). A large number of respiratory viruses are also known to exist (Table 2-6).

Prions

A group of exceedingly small infectious agents known as *prions* are responsible for neurological diseases in humans and animals (222). They appear to be infectious proteins without nucleic acid, either DNA or RNA. The agent is resistant to enzymes that digest nucleic acids, cooking, and normal autoclaving. At least nine animal diseases now fall into this category, including mad cow disease which has been plaguing Great Britain since 1987. All are neurological diseases called *spongiform encephalopathies* because of the for-

TABLE 2-4 Classification of Animal Viruses

Class	Nature of Nucleic Acid	Envelope and Shape	Typical Size (nm)	Example	Diseases
			RNA Viruses		
1a	Positive, single-stranded RNA	Naked, polyhedral	30	Picornaviruses	Paralysis, common cold, myocarditis
1b	Positive, single-stranded RNA	Enveloped, polyhedral	40–70	Togarviruses	Encephalitis, yellow fever
II	Negative, single-stranded RNA	Enveloped, helical	12–15 × 150–300	Paramyxoviruses, rhabdoviruses	Measles, mumps, rabies
III	Negative, segmented, single-stranded RNA	Enveloped, helical	70–85 × 130–380 90–120 50–300	Orthomyxoviruses, arenaviruses	Influenza, hemorrhagic fevers
IV	Segmented, double-stranded RNA	Naked, polyhedral	60–80	Rotaviruses	Respiratory and gastrointestinal infections
V	Positive, single-stranded RNA, two strands	Enveloped, helical	80–130	Retroviruses	Leukemia, tumors, AIDs
			DNA Viruses		
1a	Double-stranded, linear DNA	Naked, polyhedral	70–90	Adenoviruses	Respiratory infections, gastroenteritis
1b	Double-stranded, linear DNA	Enveloped, polyhedral	150–200	Herpesviruses	Oral and genital herpes, chickenpox, shingles, mononucleosis
1c	Double-stranded linear DNA	Enveloped, complex shape	160–260 × 250–450	Poxviruses	Smallpox, cowpox
II	Double-stranded circu lar DNA	Naked, polyhedral	45–55	Papovaviruses	Warts
III	Single-stranded linear DNA	Naked, polyhedral	18–26	Parvoviruses	Roseola in children, aggravates sickle cell anemia

TABLE 2-5   Some Human Enteric Viruses

Virus Group	Serotypes	Some Diseases Caused
Enteroviruses		
Poliovirus	3	Paralysis, aseptic meningitis
Coxsackievirus		
A	23	Herpangia, aseptic meningitis, respiratory illness, paralysis, fever
B	6	Pleurodynia, aseptic meningitis, pericarditis, myocarditis, congenital heart disease, anomalies, nephritis, fever
Echovirus	34	Respiratory infection, aseptic meningitis, diarrhea, pericarditis, myocarditis, fever, rash
Enteroviruses (68–71)	4	Meningitis, respiratory illness
Hepatitis A virus (HAV) and E (HEV)		infectious hepatitis
Reoviruses	3	Respiratory disease
Rotaviruses	4	Gastroenteritis
Adenoviruses	49	Respiratory disease, acute conjunctivitis, gastroenteritis
Norwalk agent (calicivirus)	1	Gastroenteritis
Astroviruses	5	Gastroenteritis
Calcivirus	?	Gastroenteritis
Cornavirus	?	Gastroenteritis

mation of large vacuoles that develop in the brain. The human diseases are kuru, Creutzfeldt–Jakob disease (CJD), Gerstmann–Straüssler–Scheinker syndrome, and fatal insomnia. They appear to be transmitted by ingestion of infected food. Kuru was discovered to be transmitted among cannibals in New Guinea. An association with CJD and consumption of certain foods (e.g., sheep, oxen, hog brains) has been observed. The epidemic of bovine spongiform encephelopath (BSE) or mad cow disease was caused by the feeding

TABLE 2-6   Some Respiratory Viruses

Virus Group	Serotypes	Some Diseases Caused
Rhinoviruses	>100	Common cold
Parainfluenza	4	Flulike illness
Influenza	4	Flu
Respiratory syncytial virus	1	Severe respiratory illness
Coronaviruses	?	Illness in young children
Adenoviruses	41	Nose, throat, and eye infection

of scrapie (a prion infection)-infected sheep meat to cattle. Recent studies in Europe have strongly implicated a new variant of CJD in humans to consumption of BSE-infected meat. A recent study assessed the risk of BSE transmission by water (98). These diseases are characterized by a progressive dementia, leading to death. The course of the illness may take many years or decades before clinical signs develop.

CLINICAL CHARACTERIZATION

Although humans are continually exposed to a vast array of microorganisms in the environment, only a small proportion of those microbes are capable of interacting with the host in such a manner that infection and disease will result. *Infection* is the process in which a microorganism multiplies or grows in or on a host. Infection does not necessarily result in disease. In the case of most enteric infections, usually only half of those infected develop clinical signs of illness. In the case of measles, almost 100% of those infected develop the disease. A person who is infected but does not develop clinical signs of illness is referred to as *asymptomatic* (Figure 2-7). Although not ill, these persons can still excrete large numbers of pathogens back into the environment. The ability of an organism to produce *symptomatic* disease (a *clinical illness*) is related to a number of factors, generally related to the type and strain of microorganism and the age of the host. For example, hepatitis A

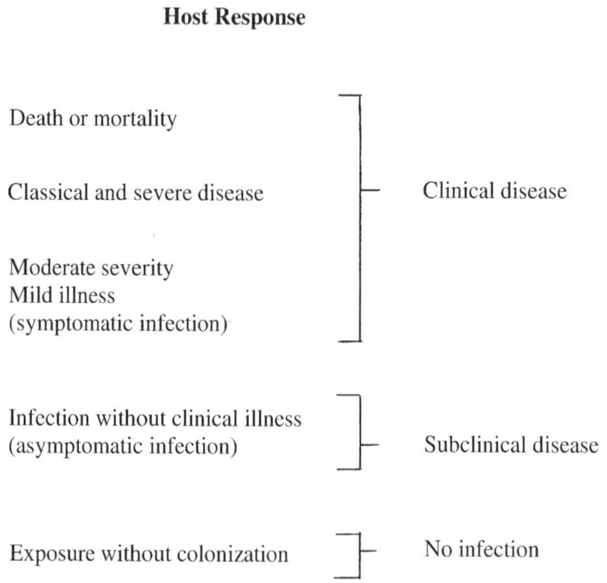

Host Response

Death or mortality

Classical and severe disease ⊢ Clinical disease

Moderate severity
Mild illness
(symptomatic infection)

Infection without clinical illness
(asymptomatic infection) ⊢ Subclinical disease

Exposure without colonization ⊢ No infection

Figure 2-7 Outcomes of exposure to a microbial infection.

infections are generally asymptomatic in small children (<10% are symptomatic), while 75% or more of adults are symptomatic.

Although infection usually precedes the causation of disease, some microbes can produce toxins in the environment, and when exposure occurs through ingestion or other routes, illness can result. Examples are *Clostridium perfringens* or *Staphylococcus aureus,* which produce toxins when growing in food. Ingestion of the food results in gastroenteritis due to the toxin in the food, not the growth of the organism in the host.

The capacity of a microorganism to cause disease is determined by the production of a variety of virulence factors possessed by the infecting microorganism. *Virulence* is the degree of intensity of the disease produced by a pathogen. Factors that determine virulence include the ability of the organisms to adhere to certain cells within the body, to produce toxins, to penetrate living cells, to produce substances that prevent attack by the host immune system, and to produce enzymes that help the organism spread through tissues. Since virulence factors are controlled by genetic traits, they vary between the types and strains of microorganisms. Thus the virulence of influenza may vary from one season to the next, depending on the genetic characteristics that govern virulence. A number of host factors may also influence the outcome of infection and the severity of disease. These are listed in Table 2-7. Generally, infections are more severe in the elderly, the very young, the immunocompromised, and those with poor nutrition.

A *frank pathogen* is a microorganism capable of producing disease in both healthy and compromised persons. *An opportunistic pathogen* is usually capable of causing infection only in compromised persons (e.g., burn patients, patients on antibiotics, those with an impaired immune system). Opportunistic pathogens are often common in the environment and may be present in the human gut or skin but usually do not cause illness.

To cause illness the pathogen must first grow within the host. The time between infection and the appearance of clinical signs and symptoms (diarrhea, fever, rash, etc.) is the *incubation time* (Table 2-8). This time may be as short as 6 to 12 hours in the case of Norwalk virus gastroenteritis to 30 to 60 days for hepatitis A virus. At any time during the infection the pathogen may be excreted into the environment by the host in the feces, urine, respi-

TABLE 2-7   Host Factors That Can Influence Susceptibility and Severity of Disease

Age (often most severe in the very young and the elderly)
Alcoholism
Chronic diseases (e.g., diabetes)
Double infection (i.e., infection by more than one organism)
Immune state
Nutritional status (poor nutrition increases severity)
Presence of receptors on cells for attachment or entry of organisms

TABLE 2.8 Incubation Time for Common Enteric Pathogens

Agent	Incubation Period	Modes of Transmission	Duration of Illness
Adenovirus	8–10 days	Unknown	8 days
Aeromonas species	Unknown	Food ingestion	1–7 days
Campylobacter jejuni	3–5 days	Food ingestion, direct contact	2–10 days
Cryptosporidium	2–14 days	Food/water ingestion, direct and indirect contact	Weeks–months (immunocompromised)
Entamoeba histolytica	7–14 days	Food/water ingestion, direct and indirect contact	Variable (weeks–months)
Escherichia coli			
ETEC	16–72 hours	Food/water ingestion	3–5 days
EPEC	16–48 hours	Food/water ingestion, direct and indirect contact	5–15 days
EIEC	16–48 hours	Food/water ingestion?	2–7 days
EHEC	72–120 hours	Food ingestion, direct or indirect contact?	2–12 days
Giardia lamblia	7–14 days	Food/water ingestion, direct and indirect contact	Weeks–months
Hepatitis A	30–60 days	Hepatitis	2–4 weeks
Listeria monocytogenes	3–70 days	Food digestion, direct or indirect contact?	Variable
Norwalk agent(s)	24–48 hours	Food/water ingestion, direct and indirect contact, aerosol?	1–2 days
Rotavirus	24–72 hours	Direct and indirect contact, aerosol?	4–6 days
Salmonellae	16–72 hours	Food ingestion, direct and indirect contact	2–7 days
Shigellae	16–72 hours	Food/water ingestion, direct and indirect contact	2–7 days
Yersinia enterocolitica	3–7 days	Food ingestion, direct contact	1–3 weeks

ratory excretions, and so on. Although maximum release may occur during the height of the clinical illness, it may also precede the first signs of clinical illness. In the case of hepatitis A virus, maximum excretion in the feces occurs before the onset of the first signs of clinical illness. The concentration of organisms excreted into the environment varies with the type of organism and the route of transmission. The concentration of enteric viruses during gastroenteritis may be as high as 10^{10} to 10^{12} per gram of feces (Table 2-9).

Diseases are often described by their signs or symptoms. A *sign* is a characteristic of a disease that can be observed by examining a patient. A *symptom* is a characteristic of a disease that can be observed or felt only by the patient. Signs of disease include swelling, redness, coughing, runny nose, diarrhea, and fever. Symptoms include such things as pain, nausea, and malaise. Even after apparent recovery of some diseases, aftereffects called *sequelae* can develop. For example, some enteric bacterial infections can result in reactive arthritis. An *acute disease* develops rapidly and runs its course quickly. Gastroenteritis and colds are examples of acute diseases. A *chronic disease* develops more slowly and persists for a longer period of time. Tuberculosis is an example of a chronic disease.

MICROORGANISMS OF INTEREST

Viruses

Picronavirus. The picronaviruses include several groups of important viruses transmitted by air, food, and water. These include the enteroviruses, hepatitis A, and foot-and-mouth disease of cattle. The enteroviruses are perhaps the most diverse in terms of the range of clinical diseases they cause. Diseases caused by enteroviruses range from severe paralysis (paralytic poliomyelitis) to aseptic meningitis, hepatitis, myocarditis, skin rashes, and the common cold. The initial site of enterovirus replication is usually the intestinal tract, although it may quickly spread to other organs, particularly nerve tissue. This is why of all the viruses transmitted by water, this group probably causes the most serious infections in terms of the body organs affected and the seriousness of outcomes. Fortunately, serious outcomes represent only a small proportion of all infections. The best known members of this group are the polioviruses, which before the development of a vaccine in the 1950s were considered a major public health problem because poliomyelitis is a crippling disease. Widespread use of the vaccine has resulted in the eradication of this disease from most of the world.

Enteroviruses (polio, echo, coxsackie) cause a wide variety of illnesses in humans (194). Poliomyelitis, a once common crippling disease in the United States, has largely been eliminated around the world due to the development of the polio vaccine in the 1950s. Because enteroviruses were the first human

TABLE 2-9 Concentration of Enteric Pathogens in Feces

Organism	Concentration Per Gram of Feces
Protozoan parasites	10^6–10^7
Enteric viruses	
Enteroviruses	10^3–10^{7a}
Rotavirus	10^{10b}
Adenovirus	10^{12b}
Enteric bacteria	
Salmonella spp.	10^4–10^{11}
Shigella	10^5–10^9
Ascaris	10^4–10^5
Indicator bacteria	
Coliform	10^7–10^9
Fecal coliform	10^6–10^9

Source: Ref. 85.
[a]Cell culture assays.
[b]Electron microscope observations of viral particles.

viruses easily grown in cell culture, a great deal has been learned about their epidemiology. Infections are most common in childhood. Isolation of echo- and Coxsackievirus from stools of children may be as high as 8 to 10% during the summer months (89). Fecal–oral is believed to be the main route of transmission, although respiratory transmission may also be significant for some types (205). It is believed that almost all enteroviruses (except, possibly, enterovirus type 70, which causes eye infections) can be transmitted by the fecal–oral route; however, it is not known how many can be transmitted by the respiratory route (205). Secondary transmission in the household is very high and may exceed 90%, although it is typically lower (18 to 75%), often depending on type and strain of virus and sanitary conditions (205). Incubation periods vary greatly with the type of virus and may be as short as 12 hours for Coxsackievirus type A24 eye infections, to 35 days for poliovirus. The presentation of symptomatic illness is also highly type and strain dependent. Some infections may pass through a community with no illness observable (89). The only evidence of infection may be the isolation of the virus in the stool or an antibody response. However, typically half of those infected will be symptomatic (89). Echoviruses usually cause milder infections than those caused by Coxsackieviruses.The case:fatality ratio in recognized cases of enterovirus illness has been reported to range from 0.01 to 0.94% (9).

Coxsackie viruses, named after Coxsackie, New York, where they where first isolated are the enteric viruses that have most commonly been isolated from sewage-contaminated water. They have been associated with recreational waterborne outbreaks and numerous common source outbreaks (Table 2-10).

TABLE 2-10 Common Source Outbreaks of Coxsackievirus Aseptic Meningitis

Type	Location	Source	Number III	Attack Rate (%)	Hospitalization Rate (%)	Reference
B2	Alabama	Sharing water bottles	81	25 (entire school) 53 (football team)	No date	3
B4	Missouri	Sharing water cups	21	23 (football tam)	24	203
B4, B5	Vermont	Lake swimming	21	64 (campers/counselors)	1	126
B5	North Carolina	Close human contact; sharing water bottles	49	2 (entire school) 16 (football team)	10 (entire school) 33 (football team)	203

Most infections due to Coxsackievirus are symptomatic; according to Cherry (45), 50% of infections of Coxsackievirus type A and 80% of type B result in the development of symptoms in the host; however, asymptomatic epidemics, where there is little or no acute illness, can occur. Table 2-11 lists the various diseases caused by enteroviruses. The most common diseases associated with the Coxsackieviruses reported during the period were cardiovascular disease, respiratory illness, gastroenteritis, and central nervous system disorder. Coxsackieviruses cause the greatest variety of illness types of any waterborne agent (Figure 2-8).

Coxsackievirus type B is now recognized as the most common viral etiological agent associated with heart disease (26,40,87,136) and causes greater than 50% of all cases (240) (Table 2-12). Lasting or chronic heart damage has been reported from Coxsackievirus infections of the heart. In children less than 3 months of age the onset of myocarditis from Coxsackievirus infection often leads to death.

Approximately 70% of all aseptic meningitis cases are due to enteroviruses, particularly Coxsackievirus types A7, A9, and B2–5 (199). Documented outbreaks of aseptic meningitis have been associated with recreational water activity and school groups who were sharing water cups (Table 2-10). In one outbreak among a football team, the attack rate was 16%, with a hospitalization rate of 33% (203).

Enterovirus infections are common among neonates (99,151,153,154). Neonatal infection may occur transplacentally or through contact after birth. The range of response is from inapparent to severe and even fatal. Among mothers of 41 documented fatal cases only half were symptomatic, mostly mild, before birth. Some studies have also suggested a possible role of enteroviruses in developmental defects of the fetus (205).

The role of Coxsackieviruses in insulin-dependent diabetes mellitus has been the subject of many epidemiological and animal studies (13,39,62,97,195,273). Diabetes can be produced in mice during Coxsackievirus infection, and several serological studies have demonstrated an associ-

TABLE 2-11   Enterovirus-associated Disease by Serotype, WHO, 1975–1983 (%)

Disease	Total	Coxsackie A	Coxsackie B	Echo
Respiratory illness	14.0	9.0	34.9	56.0
Central nervous system disorder	38.4	7.2	19.2	73.5
Cardiovascular disorder	1.6	6.2	64.8	28.9
Gastroenteritis	16.4	8.5	25.0	66.4
Skin disorder	4.4	48.5	13.9	37.5
Other	21.8	8.6	29.8	61.5

Source: Ref. 110.

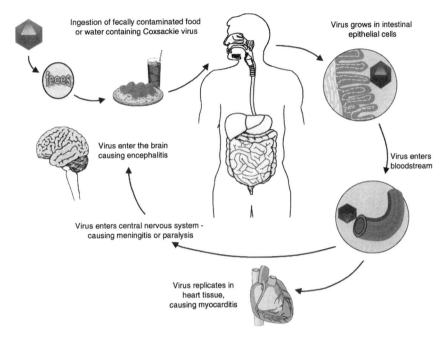

Figure 2-8   Phases of infection with Coxsackievirus.

TABLE 2-12   Diseases/Outcomes Caused by Coxsackievirus

Type	Disease/Outcome
A	Common cold, fever, herpangina, paralysis, aseptic meningitis
A10	Hand-foot-and-mouth disease (HFMD) (147)
A21	Respiratory illness
B	Pericarditis, pleurodynia, meningitis, myalgia, orchitis, fever
B	Systemic infection of infants, fever, paralysis, aseptic meningitis, pleurodynia (183)
B	Myocarditis
B	Neonatal myocarditis
B	Encephalitis, myocarditis, fulminant infection in newborns
B	Aseptic meningitis, myocarditis, diabetes, myalgia, respiratory illness
B	Spontaneous abortion, stillbirth (96)
B	Myopericarditis
B	Miscarriage
B1–5	Myocarditis
B1–5	Encephalitis, paralysis, sepsis, meningitis, carditis symptoms
B	Diabetes

(204, 231)

ation of diabetes and previous Coxsackievirus infection. Recent work suggests that Coxsackievirus infection may be related to the age of infection with the virus, with a 1 to 2-year period after infection before the development of diabetes (256).

Enteroviruses also play a minor role in respiratory disease (22) and fever, and at least several types cause eye infections or conjunctivitis (120). Gastroenteritis appears only rarely to result from enteroviruses infections. Although one waterborne outbreak of gastroenteritis in a swimming pool was believed to be caused by echovirus type 30 (159). Pleurodynia (sharp-intensity chest pains) is a common Coxsackievirus infection that has been associated with a common-source outbreak involving sharing water bottles and ice (145).

Hepatitis A virus (HAV) or infectious hepatitis (to differentiate it from serum hepatitis or hepatitis B) was first identified in the early 1970s, although its existence was known based on epidemiological evidence (169). Hepatitis A is a common infection throughout the world, with incidence related to socioeconomic conditions. In developing countries, prevalence of antibodies exceeds 90% in adults, while in developed countries it varies from 10 to 50% (157). The agent is transmitted only by the fecal–oral route, and humans and primates are the only known hosts with an incubation period averaging 30 days. Transmission by water and food (particularly shellfish) is well documented (59,127). Secondary transmission among susceptible household contacts is 10 to 20% from a family member with acute illness (169,263). Transmission of HAV among contacts of persons with hepatitis A in classroom settings, offices, factories, or hospitals is rare (37). In the United States the highest incidence is in children and young adults. Development of clinical disease is very age dependent. Only 5% of the cases are symptomatic in children under 5, while in adults it may be as high as 75% (157). Hepatitis A accounts for 1.5 to 27% of the cases of fulminant hepatitis in developed countries (187). Hepatitis surveillance has shown that the fatality rate among reported cases is age dependent. The highest rates are among children under age 5 (0.015%) and in persons over 40(0.27%). Overall mortality in the United States is 0.3% (42). There is no evidence that HAV causes chronic infection. However, relapses with production of infectious virus have been reported in 15 to 20% of the illnesses (107). The infection in adults is characterized by malaise, nausea, loss of appetite, and jaundice. Costs of hepatitis A outbreaks can be significant, ranging from $1882 for mild cases to $86,894 (average $7899) for hospitalized cases (69).

Rhinoviruses produce a typical common cold characterized by nasal obstruction, sneezing, pharyngeal discomfort, and cough. The length of infection in young adults is about 7 days, with peak symptomatology occurring on the second and third days. Symptoms last up to two weeks and in some cases as long as a month. Rhinovirus infections are less severe than those of influenza and do not result in mortality. More than 100 serotypes of rhinoviruses are known. Infections are most common in children: children under 1 year of age, 1.2 infections per year; children 2 to 9 years, 0.54 infection; and adults, 0.2 infection per year (111). In temperate climates rhinoviruses reach their peak incidence in September and early fall. In the tropics, the respiratory disease season coincides with the rainy season. The majority of rhinovirus infections are associated with symptomatic respiratory illness. The percentage of rhinovirus infection associated with clinical disease ranges from 63% in families to 74 to 88% in military trainees and medical students (148,160). Secondary attack rates are related to the titer of preexisting antibodies and have been reported to vary from 21 to 70% (111). Maximum virus excretion occurs 2 to 3 days after infection. Evidence suggests that transmission of the virus is due to direct contact with nasal secretions and not by inhalation (129). In nasopharyngeal washes, concentrations peaked at 832 $TCID_{50}$ (50% tissue culture infectious dose) on the third day of infection (in one study).

Caliciviruses. Within the *caliciviruses* group of viruses are the Norwalk virus and an array of related viruses causing gastroenteritis. Hepatitis E is also now considered a calcivirus. These viruses are commonly associated with water- and foodborne outbreaks in developed countries. Norwalk virus is the virus most often associated with water- and foodborne disease outbreaks in the United States. In the United States it has recently been estimated that perhaps 33 to 65% of nonbacterial gastroenteritis outbreaks may be associated with Norwalk or related viruses (109,152). In England caliciviruses are the third leading identified cause of gastroenteritis, usually being associated with the consumption of shellfish (73). The Snow Mountain agent has been identified with several food- and waterborne outbreaks in the United States (181). Like hepatitis A virus (HAV), hepatitis E virus (HEV) is transmitted by the fecal–oral route and has caused large waterborne outbreaks in developing countries. However, no waterborne outbreaks have been observed in developed countries, where evidence of infection by serological studies indicates that it is an uncommon infection. Epidemiological studies have also suggested foodborne outbreaks.

The *Norwalk virus* was first identified in an outbreak of gastroenteritis in 1972 in Norwalk, Ohio. Norwalk virus produces a mild to moderate diarrhea lasting 18 to 36 hours. The diarrhea is often accompanied by vomiting. The virus occurs in the vomit and epidemiological evidence suggests that it may be transmitted via aerosols generated by vomiting. The virus has not been detected, however, in nasopharyngeal washings of infected volunteers. In addition, many patients infected with Norwalk virus will also have headaches and malaise. The symptoms do not usually persist beyond 72 hours. The occurrence of chronic infection or recurrent gastroenteritis has not been reported. Studies in human volunteers suggest that reinfection with Norwalk virus occurs, indicating no long-term protection against symptomatic reinfection. Norwalk virus has been observed in AIDS patients but does not appear to result in life-threatening illness. Deaths from Norwalk agent appear uncommon. However, two deaths were reported in a nursing home outbreak, suggesting that the elderly may be at increased risk of severe infection and mortality (152).

Hepatitis E virus infections are characterized by a high case fatality rate among pregnant women (Table 2-13). Mortality has ranged from 17 to 33% in pregnant women, while case-fatality rates among nonpregnant women do not differ for men. The highest rate of fulminate hepatitis and death occur during weeks 20 to 32 of gestation, as well as during labor. Hemorrhage is the most common cause of death with fulminant hepatitis E during pregnancy. Mortality in the nonpregnant population is also significant, ranging from 2 to 3%. Person-to-person transmission has been demonstrated in case control studies, and secondary attack rates range from 0.7 to 8.0% in households (209,262). The asymptomatic rate is currently not known. Because adults are affected primarily during outbreaks, it has been suggested that the infection is largely asymptomatic in children. The high incidence among adults may be due to the loss of immunity acquired during childhood in developing countries where sanitation is poor. HEV has also been shown to infect pigs, which may be a reservoir of this agent (50).

Adenoviruses. *Adenoviruses* cause a wide variety of illnesses in humans, from eye infections to diarrhea (Table 2-14). There are currently 49 known serological types that infect humans.

Adenovirus is second only to rotavirus in terms of its significance as a pathogen of childhood gastroenteritis (2,28,68,125,130,143). Approximately one-third of the 49 serotypes are associated with illness (139), with serotypes 31, 40, and 41 associated with enteric illness.

Several seroepidemiology studies have been conducted to determine prevalences of various serotypes of adenovirus. It has been estimated from studies conducted in the United States that 40 to 60% of children have antibodies to serotypes 1, 2, and 5 (29,142,149). In a cross-sectional community study of Norwegian adults, antibodies to adenovirus [serotype(s) not given] were the second most frequent virus antibodies detected (212). The latter study also demonstrated that risk factors such as age, smoking, and occupational exposure to gas or dust were useful predictors for detecting levels of respiratory virus antibodies in this adult population.

Approximately 2 to 7% of all lower respiratory tract illness in children is due to adenoviruses (29). Symptoms such as fever, chills, headache, malaise, and myalgia are commonly observed during adenovirus respiratory infections. It has been estimated that 10% of childhood pneumonia cases are attributable to an adenovirus infection (184). Adenovirus type 7 has been implicated in severe respiratory infections that resulted in mortality (70,140,208). In one such study, 10 of 29 child cases (age <5 years) of adenovirus type 7 died from subsequent pneumonia and necrotizing bronchiolitis (208).

Adenovirus infections can also result in gastroenteritis, and this has been shown to be particularly significant in children (30,200,259,272). Although it

TABLE 2-13 Waterborne Outbreaks of Hepatitis E and Mortality in Pregnant Women[a]

Site	Dates	Number of Cases	Highest Age Incidence (years)	Mortality in Pregnant Women (%)
New Delhi, India	12/55–1/56	29,000	15–39	10.5
Ahmedabad City, India	12/75–1/76	2,572	16–30	High
Mandalay, Burma	6/76–8/77	20,000	20–29	18
Rangoon, Burma	1982–1983	399	20–40	12
Algeria	10/80–1/81	788	Young adults	100
Eastern Sudan	8/85–9/85	2,012	Young adults	NA
Tog Wajale, Somalia	1985–1986	NA	Young adults	17
Huitzililla, Mexico	6/86–10/86	94	15–45	0
Kirghiz Republic, USSR	1955–1956	10,812	NA	18
Islamabad, Pakistan	1993	3,827	11–30	10.9

Source: Data from Refs. 27 and 223.

[a]NA, not available.

TABLE 2-14   Diseases Associated with Adenovirus Serotypes

Disease	Serotypes
Acute respiratory disease	1–7, 14, 21
Pharyngoconjunctival fever	3, 7, 14
Acute febrile pharyngitis	1–3, 5–7
Pneumonia	1–4, 7
Keratoconjunctivitis	8, 11, 19, 37
Acute hemorrhagic cystitis	11, 21
Urinary tract complications	34, 35
Gastroenteritis	31, 40, 41

Sources: Ref. 140, 150.

is not as prevalent as rotavirus, outbreaks of gastroenteritis among day-care centers and orphanages for children are well documented (47,173). Adenoviruses can be shed for extended periods of time in the feces (as well as respiratory secretions) (259), with as many as 10^{11} particles per gram in feces reported in children (264). During an evaluation often separate outbreaks involving day-care centers in Houston, 38% (94 of 249) of the children were infected with enteric adenovirus types 40 and 41 (260). However, 43 of these 94 cases (46%) were asymptomatic.

Adenoviruses have been responsible for numerous outbreaks among military personnel (51,72,193), within hospitals (31,51,191), and within facilities for children (day-care centers and schools) (191,218,260). In most of the outbreak situations, the clinical outcomes were acute respiratory disease, keratoconjunctivitis, and conjunctivitis. The economic costs of these cases have been estimated at $2134 per case in 1995 (141).

Attack rates during adenovirus gastroenteritis outbreaks among young children have been reported as high as 70% (232). There have been several documented waterborne outbreaks (35,66,91,186,190,213,215), all involving swimming in contaminated recreational waters (Table 2-15). Attack rates have been documented as high as 67%, with secondary attack rates of 19% for adults and 63% for children (90,91).

The severity of the disease resulting from an adenovirus infection depends on the host's immune system and adenovirus infections in immunocompromised hosts (including AIDS patients and transplant recipients) have been well documented (5,21,79,132,241,267,275). Infections with adenovirus serotypes 1, 5, and 7 and the enteric adenoviruses have been documented among this population. Adenovirus serotypes 43 to 47, and more recently serotypes 48 and 49, were first identified in patients with AIDS (132,133,236). Devastating outcomes can result from adenovirus infections in immunosuppressed persons, among whom mortality ratios have been reported as high as 50 to 60% (241,275).

TABLE 2-15 Waterborne Outbreaks of Adenovirus

Type	Source	Population	Disease	Attack Rate (%)	Reference
3	Swimming pool	Swim team (ages 8–10 years)	Pharyngoconjunctival fever	67	91
7	Swimming pool	Swim team (ages 10–18 years)	Conjunctivitis	65 33.3	35
4	Swimming pool	Family open swim (ages not given) Swim team (ages not given)	Pharyngoconjunctival fever	52.6	66
3	Swimming pool	Community (ages 1–47 years)	Conjunctivitis	32	186
3	Pond	Campers (ages 7–16 years) staff (ages 17–22 years)	Pharyngoconjunctival fever	52 (overall)	190
Not reported	Drinking water	Municipality, contaminated groundwater	Gastroenteritis	25–50 (multiple agents)	166

Paramyxoviridae. The human *parainfluenza viruses* are responsible for a major component of acute respiratory illness in childhood. They are exceeded only by respiratory syncytial virus (RSV) as important causes of lower respiratory disease in young children, and they commonly reinfect older children and adults to produce upper respiratory diseases. There is considerable diversity in both epidemiological and clinical manifestations of infections by parainfluenza viruses (Table 2-16). Parainfluenza type 1 is the principal cause of croup in children and parainfluenza 3 is second only to RSV as a cause of pneumonia and bronchiolitis in children. Limited mortality data are available and consist of sporadic case reports (77) and reports of fatal pneumonia in severely immunocompromised patients (93,269). The failure to identify more fatal cases may be a result of the liability of these viruses in postmortem specimens (105). Most deaths caused by parainfluenza virus infections are probably related to parainfluenza type 3 in young infants, but even in this group, the etiology of fatal cases is not documented with sufficient frequency to allow a reasonable estimate of the mortality rate in the general population.

Transmission of parainfluenza viruses is by direct person-to-person contact or large-droplet spread. These viruses are known not to persist long in the environment, but parainfluenza type 1 virus has been recovered from air samples collected in the vicinity of infected patients (189), and from 1 to 10% of parainfluenza type 3 virus particles in aerosol may be viable after 1 hour (196). Adult volunteers who have had prior natural infection have been reinfected experimentally by inoculation of the upper airway with nasal drops. The viruses are probably transmitted by both inhalation and contact with contaminated surfaces and self-inoculation. The incubation period ranges from 3 to 6 days. Adult volunteers with preexisting antibodies were reinfected with as little as 1500 $TCID_{50}$ (257). More than half of the volunteers developed signs and symptoms of infection. The duration of virus shedding ranges from 8 to 11 days, which is similar for influenza and RSV (Figure 2-1). Fewer days of shedding (1 to 3 days) may occur with reinfection (128).

Mumps and *measles virus* are paramyxoviruses that infect only humans. Aerosols and direct contact with saliva are the routes of transmission for these viruses. Although these viruses once caused serious morbidity and mortality in the United States, they have largely been eliminated as a serious public health problem through the development of effective vaccines.

TABLE 2-16   **Percentage of Lower Respiratory Illness in Children Due to Parainfluenza (%)**

Virus Type	Group	Pneumonia	Bronchiolitis
1	18–21	3	1–3
2	4–8	1	1
3	9–13	3–13	5–14
Total	31–42	7–17	7–18

Source: Ref. 128.

Orthomyxovirdae. The influenza viruses are important viruses with regard to the epidemiology of worldwide respiratory disease. These viruses have been causing recurrent epidemics of febrile respiratory disease every 1 to 3 years for at least 400 years. Pandemics, or epidemics of worldwide influenza-related illness, have occurred every 10 to 20 years and have accounted for up to 21 million deaths. In addition to mortality, influenza viruses account for major morbidity in children and adults and can place the elderly at high risk for serious complications. The cost that can be attributed to these illnesses is in the billions of dollars annually.

The spectrum of illness associated with influenza virus ranges from inapparent infection to fatal pneumonia. The severity of the illness may depend on previous infection with antigenically related variants. When a large proportion of the population has at least partial immunity, about 20% of infections will be asymptomatic, and about 30% will show symptoms and signs of upper respiratory tract involvement without fever (104). Febrile upper respiratory illness of flulike illness (sudden onset of fever, chills, sore throat, myalgia, malaise, headache, and hacking, nonproductive cough) will occur in about 50% of the infected patients. About 5% will progress to frank involvement of the lower respiratory tract with probable development of pneumonia. Illness typically lasts 2 to 7 days. Influenza illness is associated with a number of potential complications, including secondary bacterial infection, pneumonia, myocarditis, pericarditis, meningitis, encephalitis, and Guillain–Barré syndrome. During epidemic periods, 11 to 14% of persons afflicted may seek medical attention (104). During the 1980–1981 outbreak in the United States, 16.9 per 10,000 cases were hospitalized (11). Certain patient populations are at increased risk for mortality due to influenza. Persons over 65 and persons with underlying illness (Table 2-17) are at increased risk of death. Typically, 80 to 90% of the deaths occur in persons over 65. Outbreaks of influenza B infections are also associated with excess mortality. Influenza C illness tends to be less severely symptomatic than influenza A and B infection. Children tend to be more susceptible to influenza C.

Transmission of influenza viruses from person to person is thought to occur through several mechanisms. Small-particle aerosols appear to be partially responsible for transmission of infection virus from one person to another (4,114,175). These aerosols are produced while talking, coughing, or sneezing. Influenza virus survives best when humidities and temperatures are low, making winter an ideal time for transmission of the virus. In addition to small-particle aerosols, dried secretions (with infectious virus) on fomites may also be responsible for transmission of influenza viruses. It has been demonstrated that influenza A and B viruses remain viable and infectious for up to 24 hours on hard, smooth surfaces, and for up to 4 to 6 hours on more porous surfaces (14).

In experimental influenza in volunteers, inoculation with small-particle aerosols produces an illness that more closely mimics natural disease than does inoculation with large drops into the nose (4,175). Aerosols are probably also the main route of transmission as it has been demonstrated that the $TCID_{50}$

TABLE 2-17 Categories of Patients at High Risk for Death Due to Pneumonia or Influenza During 1968–1969 and 1972–1973 Influenza A Outbreaks

High-Risk (HR) Category	Deaths per 100,000 Population
Cardiovascular	104
Pulmonary	240
Cardiovascular and diabetes	1040
Cardiovascular and pulmonary	920
Age ≥45	
Without Hr	4
With 1 HR	157
With ≥2 Hr	615
Age ≥65	
Without HR	9
With 1 HR	217
With ≥2 HR	797

Source: Adapted from Ref. 11.

rate for nasal drops is 137 to 300 times that required for the aerosol route (0.6 to 3.0 $TCID_{50}$) (4,55,56).

The incubation period ranges from 2 to 3 days, with virus usually being detected within 24 hours before the onset of illness. Virus rises rapidly to a peak of 3.0 to 7.0 $\log_{10} TCID_{50}$ per milliliter in nasal washing and remains elevated for 24 to 48 hours (55). Usually, influenza virus is no longer detectable after 5 to 10 days of virus shedding (Figure 2-9). In young children, shedding high titers of virus is prolonged (119). The severity of the illness correlates temporally with the quantities of virus shed in experimental influenza in volunteers (55). It is not known whether such a correlation holds for natural influenza, since some severely ill persons shed only small amounts of virus.

Respiratory Syncytial Virus. *Respiratory syncytial virus* (RSV) is the major cause of lower respiratory tract illness in children. (241,243) So effectively does RSV spread that essentially all persons have experienced an infection within the first few years of life. Immunity, however, is not complete, and reinfection in older children and adults is common. In one study in which children were followed from birth, the infection rate was 68.8 per 100 children in the first year of life, and during the second year at least half were reinfected (109). Although life-threatening infections generally occur only during the first few years of life, there is growing concern that contracting bronchiolitis and lower respiratory tract illness during infancy may contribute to the development of chronic lung disease in later life (116,221).

The average incubation period ranges from 2 to 8 days. The eye and nose appear to be equally sensitive portals of entry, and the mouth, a very sensitive

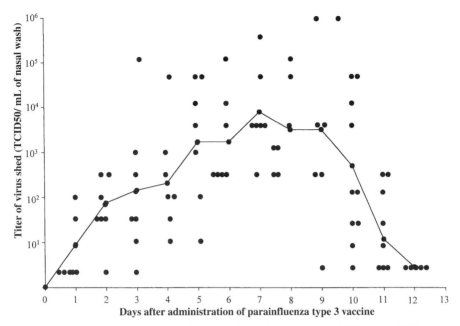

Figure 2-9   Pattern of virus shedding with parainfluenza type 3 $TCID_{50}$ = (50% tissue culture infective dose). (From Ref. 15, with permission.)

means of inoculation (115). RSV infection is usually confined to the respiratory tract. Among children, the clinical syndromes caused by RSV include febrile upper respiratory disease, bronchiolitis, pneumonia, and bronchitis. RSV is the major cause of bronchiolitis and pneumonia in early childhood (Table 2-18). Otitis media (ear infection) may be caused primarily by RSV or by secondary bacterial infection.

RSV is clearly spread by infected respiratory secretions. The major mode of transmission appears to be by large droplets and/or fomite contamination rather than by inhalation of aerosols (115,118,119). Spread requires either close contact with infected infants or contamination of the hands with respiratory secretions and subsequent contact between fingers and nasal or conjunctival mucosa.

TABLE 2-18   Proportion of Respiratory Illnesses Caused by RSV in Children

Syndrome	Percentage Caused by RSV
Bronchiolitis	43–90
Pneumonia	5–40
Tracheobronchitis	10–30
Croup	3–10

Source: Modified from Ref. 113.

Among seven adults caring for infected infants for 2 to 4 hours, five became infected (115). Among 10 adults who touched surfaces contaminated with disease from infected infants and who then autoinoculated their conjunctival or nasal membranes, four became infected. In another study, none of 14 adults sitting at a distance of greater than 6 ft from RSV-infected infants became infected (118). These studies suggest that fomites are responsible for transmission of RSV to adults. RSV can be recovered from countertops for up to 6 hours and from other surfaces such as cloth gowns and paper tissue for up to 45 minutes and from skin for up to 20 minutes. The virus replicates in the nasopharyhx, commonly reaching titers of 10^4 to 10^6 $TCID_{50}$ per milliliter of nasal secretion in young infants (116,117) and somewhat lower in adult volunteers (197). Some infants may shed virus for as long as 3 weeks after infection. In one study there was some correlation between severity of disease and duration of virus shedding.

Death from RSV is usually associated with underlying illness. Children who are immunosuppressed for treatment of malignancy or immunodeficiency syndromes and congenital heart disease are more susceptible to severe RSV illness (202). In adults, the elderly, the immunocompromised, and military recruits are the most severely affected. Mortality rates in the United States and Canada among hospitalized cases range from 0.3 to 1.0% (Table 2-19).

Bacteria

Enterobacteriaceae. More than 2000 *Salmonella* serotypes are known to exist, with the number of nontyphoid salmonellosis cases in the United States per year estimated to be between 2 million and 5 million (17). It is the second most common cause of diagnostic cases of gastrointestinal illness in the United States (88). Morbidity and severity of the disease is perhaps more

TABLE 2-19   Risk of Hospitalization with Respiratory Syncytial Virus Infection Among Children Less Than 12 Months Old

Rate per 1000 Infants/Year	Country (Location)
5.8–11.1	Sweden
9–38[a]	England (Newcastle)
2.8–9.2	United States (Houston, Texas)
	United States (Huntington, West Virginia)
12[b]	United States (Washington, DC)
10[c]	

Source: Ref. 15.

[a] Urban greater than rural.
[b] No difference between urban and rural.
[c] Bronchiolitis only.

varied for this bacterium than any other group of microorganism. A study of *Salmonella* outbreaks reported that the numbers of infected persons with resulting illness may range from a low of 6% to a high of 80% (43). The type of illness may be mild diarrhea lasting for a few days to severe gastrointestinal illness. Diarrhea usually lasts 3 to 7 days, but may be prolonged. In outbreaks the typical case-fatality rate is 0.1% of affected persons (217). It has now been demonstrated that *Salmonella* infections are associated with reactive arthritis in 2.3% of the cases and Reiter's syndrome in 0.23% (243). Strearh and Roberts (250) reported that 15,408 cases of *Salmonella* infections result in hospitalization annually. Perhaps 90% of all cases are foodborne; however, waterborne outbreaks have been documented. Two of the most notable waterborne outbreaks were in Riverside, California (24) and the 1976 outbreak in Suffolk County, New York (57). Immunity to *Salmonella* infections is partial at best. In volunteer experiments, patients rechallenged with the homologous strain became ill a second time, although with somewhat milder symptoms than on initial exposure (188).

Shigellosis is an acute infectious dysentery manifested by bloody mucoid stools and abdominal cramps. It is much more severe than *Salmonella* or *Campylobacter* infections. The case/fatality ratio is about 0.2% (17) and hospitalization 5.9%. *Shigella* spp. infect humans only and are easily transmitted by person-to-person contact, food, and water. Usually, food handlers are implicated in foodborne outbreaks. In recent years *Shigella* spp. have been a common cause of documented swimming recreational waterborne outbreaks in the United States in lakes and rivers (38). *Shigella* is more infective than most enteric bacteria, and secondary attack rates are high among children (Table 2-20). The spread of shigellosis has been related to the occurrence of household flies, which are a vector of transmission (172).

Escherichia coli is a normal inhabitant of the gastrointestinal tract of humans and other warm-blooded animals. There are, however, at least four types of *E. coli* pathogenic to humans (Table 2-21): (1) enterotoxigenic (ETEC), (2) enterohemorrhagic (EHEC)–O157:H7, enteropathogenic (EPEC), and enteroinvasive (EIEC). Enterotoxigenic *E. coli* causes gastroenteritis with profuse water diarrhea accompanied by nausea, abdominal cramps, and vomiting. Water and food are important in the transmission of this organism, and it is a common cause of traveler's diarrhea in Mexico. Approximately 2 to 8% of

TABLE 2-20 Secondary Attack Rates Among Household Contacts of *Shigellosis* Patients

Age of Exposed Contact	Secondary Attack Rate (%)	
	S. flexneri	*S. sonnei*
Child	42	42
Adult	14	20

Source: Data from Ref. 121 and 255.

TABLE 2-21 Features of *E. coli* Infections

	EPEC	ETEC	EIEC	EHEC
Prevalence	10–40% of hospitalized diarrhea	Common cause of traveler's diarrhea:40–60%	Rare	Developed countries
Outbreaks	Nurseries	Travelers, infants	Foodborne	Food- and waterborne
Age	<2 years	Adults, infants	Unknown	Adults, children
Illness	Diarrhea respiratory symptoms	Mild, choleralike	*Shigella*-like	Bloody diarrhea
Length of illness	7 days	5 days	7 days	8 days
Fatalities	High in young children: average 5–6%, neonates 16%	≤0.1%	≤0.1%	0.2% average; 10% results in hemolytic uremic syndrome if untreated

Source: Data from Ref. 78.

the *E. coli* present in water have been found to be enteropathogenic *E. coli* (171). In ETEC, the disease resembles a mild cholera lasting usually 5 days with rare fatalities. EPEC strains are similar to ETEC isolates but contain toxins similar to those found in shigellae. Although EPEC strains appear to produce illness in adults, the disease is primarily of newborns and infants. The age distribution suggests that protective immunity develops. Overt illness develops in about 25% of the children under 1 year of age (244). Illness usually lasts about 1 week, yet protracted illness as well as fatalities occur (25). ETEC strains produce an illness indistinguishable from shigellosis. As in shigellosis, the disease is usually self-limiting, and fatalities are unusual. EHEC starts as a water diarrhea and abdominal cramps and progresses to grossly bloody diarrhea with little or no fever. EHEC infections may lead to hemolytic uremic syndrome, which is a common sequel to this organism (Figure 2-10). This sequelae is severe in young children and the elderly, with a significant (10%) mortality. Cattle appear as a reservoir of this organism and red meat products are associated with outbreaks. It has been detected in water and drinking water (100), and recreational waterborne outbreaks have also been documented.

Campylobacter is now the most common identifiable cause of gastroenteritis in the United States (88) and Europe (192), with a mortality rate of 0.1% (88). Illness usually lasts up to 10 days, but up to 20% of the patients studied in a hospital-based study had prolonged and severe illness, with persistently high fever (20). In a case study in Norway it was observed that 13% of the

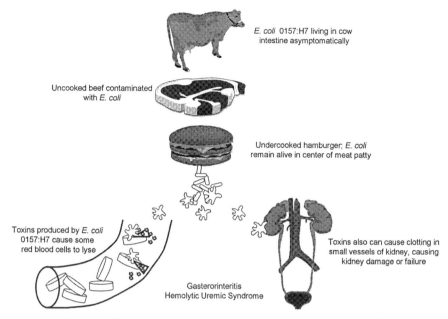

E. coli 0157:H7 living in cow
intestine asymptomatically

Uncooked beef contaminated
with *E. coli*

Undercooked hamburger; *E. coli*
remain alive in center of meat patty

Toxins produced by *E. coli*
0157:H7 cause some
red blood cells to lyse

Toxins also can cause clotting in
small vessels of kidney, causing
kidney damage or failure

Gasterorinteritis
Hemolytic Uremic Syndrome

Figure 2-10   Sequence of events leading to infection by EHEC.

ill were hospitalized (155). In this study prenatal transmission was observed. *Campylobacter* infections are also associated with reactive arthritis and Reiter's syndrome (243). *Campylobacter* has also been associated with Guillain-Barré syndrome, which causes acute neuromuscular paralysis; 20 to 40% of patients with Guillain-Barré syndrome had *Campylobacter* infections. In one study 0.17% of the patients developed Guillain-Barré syndrome (198). The infection/case ratio is not known for certain, but in one large outbreak, 75% of those who became infected developed symptoms (221). Epidemiological studies have shown a strong association between acquiring a *Campylobacter* infection and consumption of poultry and raw milk (122,270). Widespread contamination of poultry with *Campylobacter* in the United States is well documented (56). Ingestion of contaminated drinking water (especially untreated surface water) and recreational exposure through swimming are also known routes of transmission (138). Person-to-person transmission appears to be very rare.

Cholera. *Vibrio cholerae* is the agent of cholera, which has plagued humans for centuries. Cholera is an acute epidemic dehydrating diarrhea, usually sudden in onset, and if not properly treated, fatal in as many as half those affected. Cholera may be present in an asymptomatic state, as a mild diarrhea, or as the typical "full-blown" syndrome. In its extreme manifestation, cholera is one of the most rapidly fatal illnesses known. A healthy person may die within 3 to 4 hours of the onset of symptoms due to rapid fluid loss if no treatment is provided. More commonly, the disease progresses from the first liquid stool to shock in 4 to 12 hours, with death following 18 hours to several days later. All the signs and symptoms of cholera are derived from the rapid loss of liquid and salt. The rapid loss of fluid is due to an enterotoxin produced by the organism. The illness is more severe among pregnant women and as many as 50% of fetuses are stillborn for those infected during the second trimester. Lower incidence in adults in endemic areas suggests that immunity to cholera is important (162). However, second attacks have been documented 11 to 60 months after the initial illness (271). With proper replacement of fluids lost during infection and use of antibiotics, mortality is low.

Cholera, which appears to be a native marine organism, is spread rapidly by contaminated food and water under unsanitary conditions. Cholera is currently endemic in the Indian subcontinent, from where it has spread around the world in several pandemics, with the first documented one in 1817. The seventh pandemic, which is the current one, originated in 1958 in Indonesia. It spread to Europe and Africa by 1965, and reached South America in the 1990s. *Vibrio cholerae* is believed to be a native marine organism, and cases occur in the United States on a regular basis, usually traced to seafood. From this route it is probably introduced within the human population on a regular basis, but becomes a significant public health problem only when poor sanitary conditions exist. Secondary attack rates within families in the developing world have been observed to vary from 3.5 to 44.4% depending on age, children having the higher rate (16,162,253).

Listeria monocytogenes. *Listeria monocytogenes* infections resulting in disease occur most often in an immunocompromised host. Disease occurs in the very young and very old as well as in patients with neoplastic disease and in organ transplant patients. Alcoholism and diabetes mellitus are also frequent associated factors. The types of infection vary. Severe, usually lethal, disseminated, multiorgan disease is seen in neonates following intrauterine infection. In adults, there is associated meningitis subacute cerebritis with headache, fever, and varying degrees of paralysis. Focal infections have been associated with skin, eye, heart, spine, and arthritis (8). Infections are more likely to occur in persons over 40 years of age. The incidence rate in the United States and Europe is estimated at 2 to 11 cases per million. Prenatal rates in the United States have been found to vary from 6.8 to 51.4, depending on demographics and racial/ethnic backgrounds (8).

Mortality rates vary considerably with the type of host infected. In utero infection mortality rates of from 33 to 100% have been observed (23,108). In persons with meningitis, mortality has ranged from 12.5 to 43% (8). Overall mortality rates have been estimated at 19.1%, with the rate increasing with increasing age (48). Foodborne outbreaks involving milk, cheese, and cole slaw have been well documented, and the infection is considered a zoonosis and is commonly isolated in the environment, including sewage and soil. Epidemiological data suggest that person-to-person spread is not significant.

Legionella. The genus *Legionella* contains more than a dozen species that are pathogenic to humans. Water appears to be the natural habitat for these organisms, although an outbreak has been linked to potting mixtures used in greenhouses. They appear capable of causing disease only after growing some period of time at 20°C or above. This adaptation to growth at higher temperatures enhances their virulence to humans. Thus outbreaks have usually been associated with hot water systems, such as cooling towers, ornamental fountains, misters, and hot water taps. No person-to-person transmission has been documented and environmental exposures to aerosols is believed to be the route of transmission.

This group of organisms was first recognized in 1976 when an outbreak of pneumonia occurred in a hotel at the site of an American Legion convention in Philadelphia. A total of 220 persons contracted pneumonia, and 34 died. The cause of the outbreak was unknown at the time and the disease was dubbed Legionnaire's disease. The outbreak was traced to the growth of the organism named *Legionella pneumophilia* in the cooling tower and dissemination by the hotel's air-conditioning system. Legionnaire's disease is characterized by pneumonia, high fever, diarrhea, cough, headache, chest pain, and mild renel disease or renal failure, necessitating temporary dialysis in 3% of the patients (95). A milder form of *L. pneumophilia* known as Pontiac fever (Table 2-22) is an acute, self-limiting, flulike illness without pneumonia (158). It is characterized by high fever, headache, and myalgia. The illness is debilitating for 2 to 7 days, but all persons recover completely.

The majority of confirmed cases of Legionnaire's disease have occurred in middle-aged or elderly adults. The risk of disease appears to increase with age and may be related to underlying medical conditions (immunosuppression, smoking, cancer, diabetics). Numerous outbreaks have been documented

TABLE 2-22 Clinical and Epidemiological Characteristics of Legionnaire's Disease and Pontiac Fever

Characteristic	Legionnaire's Disease	Pontiac Fever
Attack rate	0.1–5%	46–100%
Incubation period	2–10 days	1–2 days
Symptoms	Fever, cough, myalgia, chills, headache, chest pain, sputum, diarrhea	Fever, chills, myalgia, headache
Lung	Pneumonia	No pneumonia
Other organ systems	Kidney, liver, gastrointestinal tract, central nervous system	None
Case/fatality ratio	15–20$	0%

among hospitalized patients receiving immunosuppressive therapy for cancer, organ transplants, and others. Travel also appears to be a significant risk factor, apparently related to exposure at hotels (251). A recent study indicated that 40% of the sporidic cases may result from exposure to water taps in the home (252). Disease and seropositivity have also been shown to be more common among power plant workers who are exposed to cooling towers (33,206). The incidence of the infection in a community is probably related to the degree of aerosol exposure and may be expected to vary greatly from community to community. In prospective community-acquired studies the incidence of Legionnaire's disease has ranged from 1 to 15% of cases, while for nosocomial pneumonia among hospital patients the incidence has ranged from 1 to 40% (274). Without treatment, mortality averages 15 to 20%, and with antibiotic treatments, 5 to 10%. Weakness and shortness of breath may persist for 5 months in some patients.

Protozoa

Giardia lamblia is the cause of giardiasis, which is characterized by an acute onset of diarrhea, foul-smelling stools, abdominal cramps, bloating, and flatulence. The patient usually expresses feelings of malaise, nausea, and sulfuric belching. One of the most important distinguishing features is the prolonged duration of diarrhea with giardiasis, usually continuing for 2 to 4 weeks. Weight loss is common among infected persons. Although infection is self-limiting in the majority of healthy individuals, as many as 30 to 50% will go on to have persistent diarrhea lasting for months. Periods of diarrhea may be interrupted by periods of constipation or normal bowel movements. Recurrent infection with *Giardia,* which often occurs in the developing world, can retard the growth and development of children because of nutritional deficiency. It has an incubation period of 1 to 2 weeks. Of 100 people ingesting *Giardia* cysts, an estimated 5 to 15% will become asymptomatic cyst passers, 25 to 50% will become symptomatic with an acute diarrheal syndrome, and the remaining 35 to 70% will have no trace of infection (135). Asymptomatic cyst passage has been documented for children in day-care centers to last as long as 6 months (220).

The life cycle of *Giardia lamblia* is composed of two stages: an actively multiplying trophozite and a resistant cyst stage. Cysts are excreted in the feces of infected persons, and when ingested, the acid environment of the stomach triggers excystation. The released trophozoites attach to the intestinal lining of the intestine via a ventral disk and replicate to large numbers by binary division. Cyst formation takes place as the trophozoites move through the colon.

Giardiasis is the most common protozoal infection of the human intestine worldwide. It occurs throughout temperate and tropical locations, prevalence varying between 2 and 5% in the industrialized world and up to 20 to 30% in the developing world (95,156). Outbreaks of giardiasis are well recognized in day-care centers, residential institutions, and schools, where prevalence may be as high as 35% (239). In recent years *Giardia* has been the agent most commonly identified with waterborne disease outbreaks in the United States when an agent can be identified (58,163). Outbreaks have also been associated with swimming pools and food (102,219). Rodents such as beavers and muskrats are believed to be potential sources of *Giardia* that infect humans.

It has only been in the last 15 years that *Cryptosporidium* has gained recognition as a significant pathogen of humans. Although it was first described in laboratory mice at the turn of the century, it was not until it was found to cause serious illness in AIDS patients that its significance was recognized. The incubation period is 5 to 28 days, with a mean of 7.2 days. Diarrhea, which characteristically may be choleralike, is the most common symptom. Other symptoms include abdominal pain, vomiting, nausea, fever, and fatigue. In immunocompetent persons, the infection is self-limiting and is usually resolved within 10 to 14 days. However, the severity and duration of the disease is prolonged in immunocompromised persons, in whom 2 to 3 L of diarrhetic stool is common, with significant dehydration and malnutrition and accompanying profound weight loss. With AIDS patients the disease becomes chronic in nature and can lead to death. In waterborne outbreaks, approximately half of infected AIDS patients die (163,178).

Infection begins after ingestion of the environmentally resistant thick-walled oocyst, which excysts within the small intestine and releases four sporozites which parasitize the cells lining the gastrointestinal tract. The sporozites penetrate these cells and develop into trophozoites, which divide to form merozites, which can reproduce sexually or asexually. During sexual reproduction, oocysts form which are released in the feces. Many species of *Cryptosporidium* are known that are capable of infecting a wide variety of animals (reptiles, birds, rodents, mammals), but *Cryptosporidium parvum* is the species responsible for infections in humans. It is also especially prevalent among ruminants, such as cattle. Contact with calves, from animals and petting zoos, are sources of *Cryptosporidium* infection (170). Food- and waterborne transmission are well documented (242,254). The source in most of these outbreaks has been contamination by cattle feces. The prevalence rate for cryptosporidium is estimated to be 1 to 3% in North America and Europe, but is considerably higher in underdeveloped countries, ranging from 5% in Asia to 10% in Africa (60).

Amebic dysentery (amebiasis) caused by *Entamoeba histolytica* is the third-leading cause of parasitic death worldwide (265). It has a simple life cycle consisting of a cyst, which is excreted into the environment, and a trophozoite form, which establishes itself in the intestine. The cysts of *E. histolytica* are not as environmentally stable or resistant to disinfectants as *Giardia* cysts or *Cryptosporidium* oocysts. The incubation period may vary from a few days to months, ranging from 1 to 4 months. *E. histolytica* is unique among the intestinal amebae because it is able to invade tissues. The presentation of infection may range from an asymptomatic infection to a disseminated fatal disease. The major clinical syndromes fall into four groups: asymptomatic cyst passers, acute colitis, fulminant colitis, and amboina (229). It is estimated that 90% of the persons infected are asymptomatic carriers (265). Acute amebic colitis usually begins with the gradual onset of abdominal pain and frequent, loose, watery stools containing blood and mucus. Fulminant colitis is an infrequent complication with mortality rates approaching 60% (7), with a predisposition for pregnant women, recipients of corticosteroids, the malnourished, and the very young (225). Amebic liver abscess is the most common complication of invasive amebiasis. It can occur with colitis, but most frequently there is no evidence or recent history of infection by *E. histolytica*. An enlarged painful liver is the most common sign of infection, along with fever. Complications occur in 12 to 15% of those diagnosed with liver abscess (229). Death from uncomplicated amebic abscess is less than 1%, but mortality ranges from 6 to 30% if rupture occurs.

It is estimated that more than 10% of the world's population is infected with *E. histolytica* and that more than 100,000 deaths occur annually (265). In some underdeveloped areas of the world the prevalence of infection may be as high as 50%. In the United States, the overall prevalence is approximately 4% (229). Humans are the only known source of *E. histolytica*. Food- and waterborne outbreaks are common modes of transmission, but outbreaks are now rare in the United States.

Like *Cryptosporidium, Cyclospora cayetanensis* is a coccidian pathogen of humans, causing gastroenteritis. Cyclosporiasis is an illness characterized by mild to severe nausea, abdominal cramping, and watery diarrhea, with symptoms persisting for an average of 7 weeks (214). Most adults report a weight loss of 5 to 10%. In AIDS patients, infections tend to last up to 4 months (124), but the infection is treatable with drugs. Diagnostic methods are not well established and the prevalence of cyclosporiasis is not known, although it appears to be largely an infection of the developing world. Currently, humans are the only identified host for this organism. Several water- and food-borne outbreaks have been associated with *Cyclospora,* particularly from imported fruits and vegetables (41,137,214).

The term *microsporidia* is a nontaxonomic term used to describe organisms in the order *Microsporidia* of the phylum Microspora. This phylum contains over 100 genera and 1000 species that are ubiquitous in nature and infect a wide range of vertebrate and invertebrate hosts. Awareness of the significance of these organisms as human pathogens has been lacking until recently. Since the first documented case in 1985, five genera have been implicated in human disease: *Encephalitozoon, Enterocytozoon, Septata, Pleistophora,* and *Vittaforma. Septata intestinalis* has recently been placed in the genus *Encephalitozoon* (123). Host specificity is considered only moderate selective, since human infections with non-mammalian genera have been documented. *Microsporidium* is a term used to accommodate microsporidia that have not yet been identified. Table 2-23 lists some microsporidia identified as causes of human diseases.

The life cycle of microsporidia contains three stages: the environmentally resistant spore, merogony, and sporogony. The spore may be ingested by the host, or possibly inhaled (exact routes of transmission have not been elucidated) in some cases. Once in the body, it infects cells and goes through merogony followed by sporogony, which results in resistant infective spores. The spores are shed via bodily fluids such as urine and excreta (268). Microsporidia infections have been most commonly associated with AIDS patients, although recent studies suggest that antibodies to *Encephalitozoon* in blood donors in France and the Netherlands (261) indicate that infections may be common, but unrecognized. The organism has been isolated from 18 to 70% of the stool specimens and intestinal biopsy specimens from otherwise unexplained chronic diarrhea in AIDS patients (185). It has also been reported as a cause of chronic traveler's diarrhea (67).

Chronic diarrhea, dehydration, and weight loss greater than 10% of body weight are the most common symptoms. Prolonged diarrhea in AIDS patients up to 48 months has been reported, and although it is not always the sole cause of death, mortality rates for infected patients are often greater than 56% (32,224). Infection usually does not spread to other organs outside the gastrointestinal tract, but it has been found in the lungs and sinuses, causing bronchitis, pneumonia, and sinusitis (237). It may be excreted in both the feces and urine. Microsporidia have been identified in sewage and in surface and ground waters, but no water- or foodborne outbreaks have been identified (76,245).

TRANSMISSION ROUTES

The major routes of microbial transmission are shown in Table 2-24. Many organisms have more than one route of transmission (e.g., air and water). The sequence of spread usually involves the exit of the organisms from the infected host and transport through the environment until they come in contact with another susceptible host. Organisms transmitted through the environment must be able to survive long enough to be transmitted from one host to another. Many microorganisms can only survive short periods of time and are unlikely to be transmitted through the environment. Examples are venereal diseases such as syphilis and gonorrhea, whose survival is so short outside the host that direct person-to-person contact is the only natural mode of transmission. Other organisms, such as hepatitis A virus, may survive for many months in water under the right conditions. Many microorganisms may have more than one route of transmission, such as some enteroviruses, which may be transmitted by both the respiratory route or through ingestion. Others, such as *Legionella,* are largely transmitted through inhalation. Rhinoviruses may

TABLE 2-23   Some Microsporidia Species Causing Human Infections

Species	Disease	Possible Environmental Transmission Routes
Enterocytozoon bieneusi	Chronic diarrhea, traveler's diarrhea	Feces
Encephalifozoon cuniculi	Eye, kidney, and urinary infections; hepatitis	Feces, urine
Encephalitozoon hellem	Eye, kidney, prostate, and lung infections	Urine
Encephalitozoon intestinalis (Septata)	Chronic diarrhea, renal failure, nose and sinus infections	Feces and urine
Vittaforma corneae	Eye infection	Fomites

be transmitted by contact of virus-contacted fingers with the eye or nose. The infectivity of an organism may be different for one route over another (53). Often, the respiratory route requires fewer organisms for infection than does ingestion.

Routes of transmission may be simple, such as the direct inhalation of *Legionella*-contaminated aerosols or complex such as the transfer of *Salmonella* from contaminated food products to a surface in a kitchen, to the fingers of the food handler, to a salad, and finally to the mouth of the person consuming the salad (Figure 2-11).

Transmission routes may occur over prolonged distances in time and space. Foot-and-mouth disease was transmitted over 100 km between Denmark and Sweden (86). Enteric viruses have been observed to travel at least 317 km downstream from a discharge source (61). In 1908 in Lancashire, England, suspicion was voiced that a smallpox outbreak could be traced to imported raw cotton contaminated with virus in crusts or scabs (80). In more recent years, importation of food from Central America has been implicated in outbreaks of cyclosporidia diarrhea (214).

Transmission routes may be complex and involve more than one medium or route. For example, enteroviruses released into marine waters via sewage outfalls may adsorb to sediments and later be taken up by shellfish during feeding. Blue crabs may then become contaminated by consumption of the shellfish (Figure 2-12).

Inhalation

Airborne transmission of organisms can occur via release from the host in droplets (e.g., coughing) or through natural (surf at a beach) or human-made activities (cooling towers, showers). Their success in reaching a susceptible

TABLE 2-24 Routes of Transmission

Route of Exit	Route of Transmission[a]	Examples	Factors	Route of Entry
Respiratory	Nasal discharges, fomites	Rhinovirus	Household contact	Respiratory, eye
	RS → air, fomites	Influenza	Direct contact	Respiratory
	RS → air	Plague (pneumonia)	Pneumonia case	Respiratory
	RS → droplets	Streptococcal	Close contact, carrier	Respiratory
	RS → droplet nucleic	Tuberculosis	Household contact, carrier	Respiratory
Skin squames	Respiratory, direct contact, fomites	Nosocomial bacterial infections	Hospitalization, surgery	Nose, respiratory, skin
	Direct contact, fomites	Impetigo due to staph and/or strep infection	Low socioeconomic level, tropics	Skin
	Fomites, direct contact	Papillomavirus, warts	Bathroom floors	Skin
Fecal–oral	Stool → water, food	Cholera	Water, food	Mouth
	Stool → food, water	Salmonellosis	Food, animal contact	Mouth
	Stool → water, fomites, food, vomit	Rotavirus	Day-care centers, water, food	Mouth
Urine	Water (swimming)	Leptospirosis	Infected animals	Skin
Inaninate sources	Soil, air, water, food	Legionnaire's disease	Warmth, humidity, water coolers, air conditioners, potable water supplies	Respiratory
		Acantamabea	Water, swimming	Eye
		Nagelaria	Swimming	Nose

[a]RS, respiratory system.

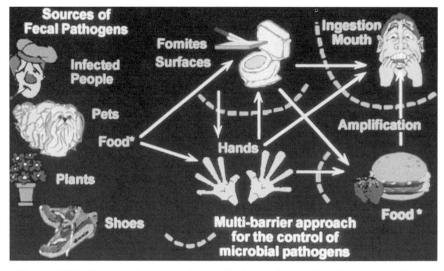

Figure 2-11 Routes of transmission in the home for fecal–oral microorganisms.

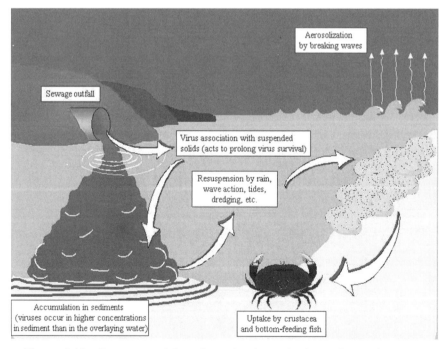

Figure 2-12 Transport and fate of enteric viruses in the marine environment.

host depends on the droplet size, the force with which they are propelled into the air, their resistance to drying, the temperature, the humidity, air currents, ultraviolet (UV) light, and distance to the host. Some infections may be carried great distances e.g., Legionnaire's and foot-and-mouth disease. Virus transmission by the airborne route may be both direct and indirect. Infection of a host may be by direct inhalation of infectious droplets or through contact with inanimate objects (*fomites*) on which the airborne droplets have settled. Hand or mouth contact with the organism on the surface results in transfer of the organism to the *portal of entry* (i.e., nose, mouth, or eye). Although respiratory infection or the presence of the organism in the respiratory tract is often taken as evidence that the agents are transmitted primarily by this route, actual experiment evidence is often lacking.

The most important determinant of the probability that a disease will be transmitted by aerosols is the ability of the microorganism to survive in aerosolized droplets or particulates. The most significant variables controlling survival are the nature of the suspending solution (mucus, water), air temperature, relative humidity, and sunlight (ultraviolet light is detrimental to the survival of many organisms). For example, some viruses survive better at low relative humidity than at high relative humidity (235). This has been used to explain the seasonality of some virus infections. Influenza virus spreads rapidly in the winter months when indoor relative humidity is low because of heating. It survives best in aerosols and fomites when the relative humidity is low (246). Enteroviruses survive best at the higher relative humidities characteristic of late summer in the United States, which corresponds to their peak incidence (89).

Any activity that produces droplets has the potential to produce microbial aerosols. Some examples that have been shown to produce potentially infectious aerosols are shown in Table 2-25. Many activities, both natural and human-made, are capable of producing microbial aerosols. The significance of the activity is the amount of material aerosolized, the concentration of pathogens in the original material, survival of the pathogens in the aerolized

TABLE 2-25 Sources of Microbial Aerosols

Wastewater treatment plants	Bathing
Compost operations	Talking
Domestic waste transfer stations	Faucets
Sewage sludge application	Toilet flushing
Spray irrigation with sewage	Air humidifiers
Cooling towers	Water fountains
Earthmoving during construction	Showers
Sneezing	Waves breaking on a beach
Coughing	Spills of infectious materials
Release from clothing and skin	Medical devices (suction, syringes)
during walking	Dental instruments

state, and dispersion (dilution in the air). In outdoor environments aerosols can be dispersed very rapidly, however, death of the pathogens can also be rapid because of the presence of ultraviolet light. Because microorganisms tend to accumulate at the air–water interface, the concentration of microbes above the surface of a liquid can be greater than in the solution (13,19).

The size of the aerosolized particle is important because that will determine how long the particle will be suspended in the air and if it can be inhaled. Sneezing and coughing produce aerosol droplets varying in size from about 1 μm to more than 20 μm. The dispersion of an aerosol depends on air movement. In still air, a spherical particle 100 μm in diameter requires 10 seconds to fall the height of the average room (3 m); 40-μm particles, 1 minute; 20-μm particles, 4 minutes; and 10-μm particles, 17 minutes (157). This means that particles under 10 μm have a relatively long circulation time in an ordinary room. Particles 6 μm or more in diameter are usually trapped in the nose, whereas those 0.6 to 6.0 μm in diameter are deposited on sites along the upper and lower respiratory tract.

Generation of infectious aerosols by large-scale handling of wastes during wastewater treatment (activated sludge process), land application of wastewater and sludge by spraying, compost operation (generation of fungal aerosols) (49), and handling and recycling of domestic wastes have received a significant amount of study because of the potential exposure of workers and nearby communities (18,207). Biological aerosol monitoring is costly and labor intensive, and large volumes of air must often be sampled. To assess concentrations to which people may be exposed, dispersion models that integrate biological and meteorological data have been developed to predict the downwind concentrations of aerosolized microorganisms from known sources (e.g., activated sludge units, spray irrigation sites) (174,201). Certain model components are site specific and need to be determined at the site under study. These include meteorological data (wind direction and speed), microbial concentration in the liquid, flow rate (spraying operations), and background aerosol concentration. Non-site-specific model components include aerolization efficiency, microbial decay rate, and survival factor, which represents the initial shock during the aerosolization process.

Many pathogens transmitted by the aerosol route originate from human activities. For example, fungal pathogens such as *Histoplasma capsulatum* spores may be spread by aerosols when construction activities disturb sites where bird manure has accumulated (176). Composting activities may also result in the large-scale release of spores of *Aspergillus fumigatus,* a pathogen that can produce allergies when inhaled by workers (49). Water-based pathogens such as *Legionella pneumophila* and *Mycobacterium avium* are capable of growth in drinking water distribution systems and are believed spread by aerosols generated during showering and tap water use (18).

How important inhalation is for the transmission of some respiratory infections versus contact with aerosol-contaminated surfaces is often subject to debate. Evidence for the common cold suggests that surfaces contaminated by droplets or direct contact may be more important than inhalation (129). In one study, airborne transmission of rhinovirus did not occur across a wire mesh barrier from infected to susceptible volunteers in a closed barracks (74). In another, infected volunteers who engaged in singing and other activities designed to create infectious aerosols failed to spread rhinovirus to subjects confined in the same closed room (63). Contamination of the hands and environment is common in persons with rhinovirus infections. Rhinovirus has been recovered from the hands of 40 to 90% of adults with colds and from 6 to 15% of selected objects in the room with persons with colds (64,112,228). The number of viral particles in sneezes and coughs have varied from study to study depending on the methodology employed. In one study, 1,940,000 particles were present in sneezes and 90,765 in coughing, a ratio of 2.14:1 (103).

Aerosolization of organisms may occur from suction devices and from catheters in intensive care units and from blood products in dialysis units. These not only include respiratory and intestinal agents, but also bloodborne agents such as hepatitis B. Finally, *Hantavirus* and other area viruses may be spread by aerosols created from soil containing the rodent urine in which those viruses are excreted.

Dermal Exposure

The skin is another point of entry and exit for viral and bacterial infections. Transmission may occur from release of microorganisms from the unbroken skin, but more likely from skin vesicles, lesions, boils, pustules, and so on. Small cuts or abraded skin may serve routes of release and entry into the bloodstream for bloodborne pathogens. Transmission of *Staphylococcus aureus* skin infections has been demonstrated by fomites, including clothing (131,247). Plantar warts, for which a papovavirus is responsible, generally are contracted by walking barefoot in swimming areas, gymnasia, barracks, or other public places (83). Hepatitis B virus has been shown to be transmitted through activities that may cause small abrasions to the skin. One outbreak was traced to computer cards, which inflicted small cuts when handled (216).

Oral Ingestion

Microorganisms transmitted by the fecal–oral route are usually referred to as enteric pathogens because they infect the gastrointestinal tract and are often excreted in large numbers in the feces. They are characteristically stable in water and food and in the case of bacteria may be capable of growth. *Waterborne pathogens* are those excreted or secreted by man or other animals. *Water-based pathogens* are those whose natural habit is a water environment or which are capable of growth in a water environment. Examples of water-based pathogens are *Legionella* and *Pseudomonas aerugonisa. P. aeruginosa* may be grown to large numbers in hot tubs, if not properly disinfected, resulting in skin and eye infections. Certain illnesses produced by *Clostridium perfrigens* and *Staphylococcus aureus* are not considered infections since they produce illness by production of a toxin during growth in the food. Ingestion of live organisms is not necessary to produce illness.

The origin of enteric pathogens in the environment may result form many sources, including direct release of feces into the environment, sewage discharges, septic tanks, land application of sewage sludge or wastewater, solid waste disposal, and so on. In most of the world, sewage does not receive treatment prior to discharge, or if treated, disinfection is not practiced. In the United States, sewage is disinfected, which largely reduces the risk of waterborne bacteria pathogens. However, significant concentrations of enteric viruses and protozoa remain. Upon reaching the water environment, many routes of transmission to humans are possible (Figure 2-13).

Drinking Water. Contaminated drinking water has the greatest impact on human health worldwide. Drinking nontreated or improperly treated water is a major cause of illness in developing countries. In 1980, 25,000 persons per day died worldwide as a result of consumption of contaminated water. A quarter of all hospital beds were occupied by persons who had become ill after consuming water (83). In the United States most waterborne outbreaks are due to untreated or inadequately treated (lack of filtration or disinfection) water, or to subsequent contamination of the distribution network (38,57). However, recent large-scale outbreaks of *Cryptosporidium* in the United States and United Kingdom have shown that conventional drinking water treatment (flocculation, filtration, and disinfection) may not always be adequate to prevent waterborne disease transmission (233). The unusual resistance of *Cryptosporidium* oocysts to conventional disinfectants (e.g., chlorine) allows for penetration of viable oocysts into the treated water supply. Removal by filtration is then the main barrier in the elimination of this organism from

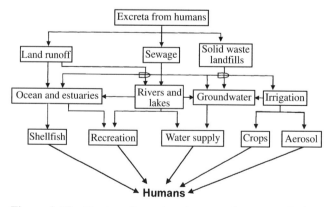

Figure 2-13   Routes of enteric microorganism transmission.

the water supply. The widespread occurrence of protozoan parasites *Giardia* and *Cryptosporidium* in surface waters in the United States (233) demonstrates that any utility which draws its water from a surface source is at risk: human enteric viruses originate only from humans, so they are less frequently encountered in surface waters. However, because of their small size they are more likely to contaminate groundwater supplies, which are often not disinfected. A recent survey detected evidence of human enteric viruses in almost 40% of the drinking water supply wells in the United States (168).

For surface water supplies in the United States, minimum treatment requirements have been established. Treatment processes are required to filter and disinfect the water to reduce the concentration of *Giardia* cysts by 99.9% and enteric viruses by 99.999% from those in the source water. The objective of the reduction requirements are to reduce the risk of infection to 1:10,000 per year for both organisms from the consumption of the tap water, based on average concentrations of these organisms estimated for polluted surface waters in the United States (83,84,179,180,230). Newer rules under development are designed to evaluate these standards, including *Cryptosporidium,* after collection of data or the occurrence of these organisms in supplies used for drinking water sources become available. It is expected that additional treatment may be required by those facilities that extract their water from highly polluted sources (84).

Recreational Activities. Recreational activities associated with swimming, water skiing, scuba diving, surfing, and so on, have frequently been associated with the acquisition of infectious diseases (34,52). The symptoms experienced by ill swimmers include respiratory, gastrointestinal, eye, skin, and allergic. Recreational outbreaks have been associated with many of the enteric waterborne pathogens, including *Cryptosporidium, Giardia,* echovirus, hepatitis A, *Shigella,* and *E. coli* O157:H7 (18,101,233). Eye and throat infections have been associated with the adenoviruses (81,257). Exposure to pathogens may occur due to direct ingestion, inhalation, or contact of contaminated hands with the mouth or eyes. It has been estimated that for each swimming event, for those people who submerge their heads in the water, the exposure is on average 100 milliliters (210). One study showed the risks to be three times greater for children under the age of 2 who immerse their heads in water than for adults (46), probably because of the greater ingestion of water. Sources of pathogens may not only be animal and human wastes but also the bathers themselves. Bathing activity has been found to release large numbers of bacteria, and outbreaks of *Cryptosporidium* have been traced to the accidental release of feces by infants or young children (6,233).

Food. Food represents another major source of human illness. Foods may become contaminated with enteric pathogens during production due to the use of contaminated water for irrigation, or growth in contaminated waters, such as shellfish (e.g., oysters, clams). Contamination may also occur by infected persons or cross contamination during harvesting, processing, and preparation. The problem is most acute with foods that are not traditionally cooked or that involve thermal processing, such as raw vegetables or shellfish. Many outbreaks originate from food handlers who had an intestinal infection and contaminate the food during preparation before serving. Although viral and protozoan pathogens cannot grow in foods, enteric bacterial pathogens are often capable of growth. Thus even a small amount of exposure of food to *Salmonella* can result in a rapid increase in numbers in a few hours under the proper conditions.

Other foodborne pathogens may originate from incorporation of eggs or cysts such as those of helminths and *Toxoplasma gondii* in animal tissues during growth. These pathogens are destroyed in meat products cooked properly.

Contamination of foods can be complex and can occur during the many events that occur between the production of the food and its consumption by the consumer (71). Consider the preparation of meals in the home kitchen. Much of the poultry (30 to 100%) brought into the home today is contaminated with *Salmonella* or *Campylobacter* (56), although when the poultry is properly cooked, all the pathogens are destroyed. However, cross-contamination events can occur at many steps before the food is placed in the oven (Figure 2-11). The cutting board on which the poultry was prepared may become contaminated. If the same cutting board is used to prepare a salad to be eaten raw, exposure to the pathogen through ingestion will result. Contamination of food preparers' hands can also result in widespread contamination of the kitchen area, creating multiple points of exposure. Food handlers can also become infected by bringing their hands to their mouth, resulting in self-inoculation.

Foodborne intoxication is different than foodborne infection in that the organism responsible for the illness grows in the food, producing a toxin that causes the illness. The organism does not grow in the host and does not actually still have to be viable in the food. *Staphylococcus aureus, Clostridium perfringens,* and *C. botulinum* are examples of bacteria that produce illness by toxin production.

Soil and Fomites. Ingestion of soil (pica) and mouth contact directly with inanimate objects (fomites) or indirectly by contamination of fingers which are then brought to the mouth are additional methods of transmission of microorganisms by ingestion. Exposure through soil ingestion could occur from irrigation of fields used for recreation (e.g., football, golf) with reclaimed wastewater. A common route of exposure of enteric pathogens in preschool children is from placing objects into the mouth, or indirectly from placing the fingers in the mouth. A small child may place his or her fingers in his or her mouth as often as once every 3 minutes (248). Enteric pathogens have been detected on fomites (e.g., toys) in day-care centers (161), food service establishments (36), and households (89,238). It has been demonstrated that rotaviruses placed on the person's hands can be transmitted by placement of the fingers (licking) to the mouth (266).

The effectiveness of transmission by fomites may be complex, involving the transfer of a pathogen from a contaminated food to a fomite, such as a kitchen sink, to the fingers, then to the mouth. To model this, an understanding of the number of organisms transferred during each step is necessary, as well as the survival of the organisms on the various surfaces. Many enteric viruses and bacteria may survive for many hours to days on fomites (182), but this depends on many factors, such as:

- Humidity
- Suspending media
- Type of organism
- Temperature
- Type of surface
- Moisture at the surface

These same factors may also influence the degree of transfer (i.e., number of organisms transferred from the fomite to the hand) in addition to the amount of pressure applied and the rubbing action (235).

REFERENCES

1. Abad, F. X., R. M. Pinto, J. M. Diez, and A. Bosch. 1994. Disinfection of human enteric viruses in water by copper and silver in combination with low levels of chlorine. Appl. Environ. Microbiol. 60:2377–2383.

2. Albert, M. J. 1986. Enteric adenoviruses: brief review. Arch. Virol. 88:1–17.

3. Alexander, J. P., L. E. Chapman, M. A. Pallansch, W. T. Stephenson, T. J. Torok, and L. J. Anderson. 1993. Coxsackievirus B2 infection and aseptic meningitis: a local outbreak among members of a high school football team. J. Infect. Dis. 167:1201–1205.

4. Alford, R. H., J. A. Kasel, P. J. Gerone, and V. Knight. 1966. Human influenza resulting from aerosol inhalation. Proc. Soc. Exp. Biol. Med. 122:800–804.

5. Ambinder, R. F., W. Burns, M. Forman, P. Charache, R. Arthur, W. Beschorner, G. Santos and R. Saral. 1986. Hemorrhagic cystitis associated with adenovirus infection in bone marrow transplantation. Arch. Intern. Med. 146:1400–1401.

6. Anderson, M. A., M. Stewart, M. Yates, and C. P. Gerba. 1998. Modeling impact of body contact recreation on pathogen concentrations in a proposed drinking water reservoir. Water Res. 32:3293–3306.

7. Aristizabal, H., J. Acevedo, and M. Botero. 1991. Fulminant amebic colitis. World J. Surg. 15:216–221.

8. Armstrong, D. 1990. *Listeria monocytogenes, In:* Principles and practice of infectious diseases, 3rd ed., G. L. Mandell, R. G. Douglas, Jr., and J. E. Bennett, eds. pp. 1587–1592. Churchill Livingstone, New York.

9. Assaad, F., and I. Borecka. 1977. Nine-year study of WHO virus reports on fatal virus infections. Bull. WHO 55:445–453.

10. Axelsson, C., K. Bondestam, G. Frisk, S. Bergstron, and H. Diderholm. 1993. Coxsackie B virus infections in women with miscarriage. J. Med. Virol. 39:282–285.

11. Barker, W. H., and J. P. Mullooly. 1982. Pneumonia and influenza deaths during epidemics. Arch. Intern. Med. 142:85–89.

12. Barrett-Connor, E. 1985. Is insulin-dependent diabetes mellitus caused by Coxsackievirus B infection? A review of the epidemiologic evidence. Rev. Infect. Dis. 7:207–215.

13. Baylor, E. R., V. Peters, and M. B. Baylor. 1977. Water-to-air transfer of virus. Science 197:763–764.

14. Bean, B., B. M. Moore, B. Sterner, L. R. Peterson, D. N. Gerding, and H. H. Balfour. 1982. Survival of influenza viruses on environmental surfaces. J. Infect. Dis. 146:47–51.

15. Belshe, R. B., and M. A. Mufson. 1991. Respiratory syncytial virus, pp. 388–407. *In* R. B. Belshe, ed., Textbook of human virology, 2nd ed. Mosby, St. Louis, MO.

16. Beneson, A. S., W. H. Mosley, W. H. Fahimuddin, and R. O. Oseasohn. 1968. Cholera vaccine field trials in East Pakistan. 2. Effectiveness in the field. Bull. WHO 38:359–372.

17. Bennett, J. V., S. D. Holmberg, M. F. Rogers, and S. L. Solomon. 1987. Infectious and parasitic diseases. Am. J. Prev. Med. 55:102–114.

18. Bitton, G. 1994. Wastewater microbiology. Wiley-Liss, New York.

19. Blanchard, D. C., and L. Syzdek. 1970. Mechanism for the water-to-air transfer and concentration of bacteria. Science 170:626–628.

20. Blazer, M. J., L. B. Keller, N. W. Levchtefeld, and W. L. L. Wang. 1982. *Campylobacter* enteritis in Denver. West. J. Med. 136:287–290.

21. Blohme, I., G. Nyberg, S. Jeansson, and C. Svalander. 1992. Adenovirus infection in a renal transplant patient. Transplant. Proc. 24:295.

22. Bloom, H. H., K. M. Johnson, M. A. Mufson, and R. M. Chanock. 1962. Acute respiratory disease associated with Coxsackie A21 virus infection. J. Am. Med. Assoc. 179:120–125.

23. Bojsen-Moller, J. 1972. Human listeriosis: diagnostic, epidemiological and clinical studies. Acta Pathol. Microbiol. Scand. (Suppl.) 229:1–157.

24. Boring, J. R., W. T. Martin, and L. M. Elliott. 1971. Isolation of *Salmonella typhimurium* from municipal water, Riverside, CA, 1965. Am. J. Epidemiol. 93:49–54.

25. Bower, J. R., B. L. Congeni, et al. 1989. *Escherichia coli* O114: nonmotile as a pathogen in an outbreak of severe diarrhea associated with a day-care center. J. Infect. Dis. 160:243–247.

26. Bowles, N. E., P. J. Richardson, E. G. J. Olsen, and L. C. Archard. 1986. Detection of Coxsackie-B-virus-specific RNA sequences in myocardial biopsy sam-

ples from patients with myocarditis and dilated cardiomyopathy. Lancet 1:1120–1122.

27. Bradley, D. W., K. Krawczynski, and M. A. Kane. 1991. Hepatitis E, pp. 781–790. *In* R. B. Belshe, ed., Textbook of human virology, 2nd ed. Mosby, St. Louis, MO.

28. Brandt, C. D., H. W. Kim, W. J. Rodriguez, J. O. Arrobio, B. C. Jeffries, E. P. Stallings, C. Lewis, A. J. Miles, M. K. Gardner, and R. H. Parrott. 1985. Adenovirus and pediatric gastroenteritis. J. Infect. Dis. 151:437–443.

29. Brandt, C. D., H. W. Kim, A. J. Vargosko, B. C. Jeffries, J. O. Arrobio, B. Ridge, R. H. Parrott, and R. M. Chanock. 1969. Infections in 18,000 infants and children in a controlled study of respiratory tract disease. I. Adenovirus pathogenicity in relation to serologic type and illness syndrome. Am. J. Epidemiol. 90:484–500.

30. Brandt, C. D., W. J. Rodriguez, H. W. Kim, J. O. Arrobio, B. C. Jeffries, and R. H. Parrott. 1984. Rapid presumptive recognition of diarrhea-associated adenoviruses. J. Clin. Microbiol. 20:1008–1009.

31. Brummet, C. F., J. M. Cherrington, D. A. Katzenstein, B. A. Juni, N. Van Drunen, C. Edelman, F. S. Rhame, and M. C. Jordan. 1988. Nosocomial adenovirus infections: molecular epidemiology of an outbreak due to adenovirus 3a. J. Infect. Dis. 158:423–432.

32. Bryan, R. T. 1995. *Microsporidia,* pp. 2513–2524. *In* C. Mandell, J. E. Bennett, and R. Dolin, eds., Principles and practice of infectious diseases, 4th ed., Churchill Livingstone, New York.

33. Buchler, J. W., R. K. Sikes, J. N. Kuritsky, J. N. Gorman, A. W. Hightower, and C. V. Broome. 1985. Prevalence of antibodies to *Legionella pneumophila* among workers exposed to a contaminated cooling tower. Arch. Environ. Health. 40:207–210.

34. Cabelli, V. J. 1989. Swimming-associated illness and recreational water quality criteria. Water Sci. Technol. 21:13–21.

35. Caldwell, G. G., N. J. Lindsey, H. Wulff, D. D. Donnelly, and F. N. Bohl. 1974. Epidemic of adenovirus type 7 acute conjunctivitis in swimmers. Am. J. Epidemiol. 99:230–234.

36. Candeias, J. A. N., D. DeAlmeida Christovao, and Z. L. G. Cotillio. 1969. Isolation of virus from drinking glasses in coffee shops and bars in the city of São Paulo. Abstr. Hyg. 44:930.

37. Centers for Disease Control. 1990. Protection against viral hepatitis: recommendation of the immunization practices advisory committee. Morb. Mortal. Wkly. Rep. 39 (RR-2):3.

38. Centers for Disease Control. 1990. Waterborne disease outbreaks. U.S. Department of Health and Human Services, Atlanta, GA. Morb. Mortal. Wkly. Rep. 39(SS-1):1–57.

39. Centers for Disease Control. 1993. Diabetes surveillance. U.S. Department of Health and Human Services, Public Health Service, Atlanta, GA.

40. Centers for Disease Control. 1994. Mortality from congestive heart failure: U.S., 1980–1990. Morb. Mortal. Wkly. Rep. 43:77–81.

41. Centers for Disease Control. 1996. Update: outbreaks *of Cyclospora cayetanensis* infection: United States and Canada. Morb. Mortal. Wkly. Rep. 45:611–612.

42. Centers for Disease Control. 1997. Prevention of hepatitis A through active or passive immunization: recommendations of the Advisory Committee on Immunization Practices. Morb. Mortal. Wkly. 45(RR15); 1–30.

43. Chalker, R. B., and M. J. Blaser. 1988. A review of human *salmonellosis*. III. Magnitude of *Salmonella* infection in the United States. Rev. Infect. Dis. 10: 111–124.

44. Cherry, J. D. 1981. pp. 1316–1365. *In* Textbook of pediatric infectious diseases. R. D. Feigin and J. D. Cherry, eds., W. B. Saunders, Philadelphia.

45. Cherry, J. D. 1992. Enteroviruses: poliovirus (poliomyelitis), Coxsackieviruses, echoviruses and enteroviruses, pp. 1705–1753. *In* R. D. Feigin, ed., Textbook of pediatric infectious diseases, 3rd ed., Vols. 1 and 2. W. B. Saunders, Philadelphia.

46. Cheung, W. H. S., K. C. K. Chang, and R. P. S. Hung. 1990. Health effects of beach water pollution in Hong Kong. Epidemiol. Infect. 105:139–162.

47. Chiba, S., I. Nakamura, S. Urasawa, S. Nakata, K. Taniguchi, K. Fujinaga, and T. Nakao. 1983. Outbreak of infantile gastroenteritis due to type 40 adenovirus. Lancet (October 22):954–957.

48. Ciesielski, C. A., A. W. Hightower, S. K. Parsons, and C. V. Broome. 1988. Listeriosis in the United States, 1980–1982. Arch. Intern. Med. 148:1416-1419.

49. Clark, C. S., H. S. Bjornson, J. Schwartz-Fulton, J. W. Holand, and P. S. Gartside. 1989. Biological health risks associated with the composting of wastewater treatment plant sludge. J. Water Pollut. Control Fed. 56:1269–1276.

50. Clayson, E. T., B. L. Innis, K. S. A. Myint, S. Narupiti, D. W. Vaughn, S. Giri, P. Ranabhat, and M. P. Shrestha. 1995. Detection of hepatitis E virus infections among domestic sewage in the Kathmandu Valley of Nepal. Am. J. Trop. Med. Hyg. 53:228–232.

51. Colon, L. E. 1991. Keratoconjunctivitis due to adenovirus type 8: report on a large outbreak. Ann. Ophthalmol. 23:63–65.

52. Corbett, S. J., G. L. Rubin, G. K. Curry, D. G. Kleinbaum, and Sydney Beach Users Study, Advisory Group. 1993. The health effects of swimming at Sydney beaches. Am. J. Public Health 83:1701–1706.

53. Couch, R. B., T. R. Cate, R. G. Douglas, Jr., P. J. Gerone, and V. Knight. 1966. Effect of route of inoculation on experimental respiratory viral disease in volunteers and evidence for airborne transmission. Bacteriol. Rev. 30:517–529.

54. Couch, R. B., R. G. Douglas, D. S. Fedson, and J. A. Kasel. 1971. Correlated studies of a recombinant influenza virus vaccine. III. Protection against experimental influenza in man. J. Infect. Dis. 124:423–480.

55. Couch, R. B., J. A. Kasel, J. L. Gerin, J. L. Schulman, and E. D. Kibourne. 1974. Induction of partial immunity to influenza by a neuraminidase-specific vaccine. J. Infect. Dis. 129:411–420.

56. Council for Agricultural Science and Technology. 1994. Foodborne pathogens: risks and consequences. CAST, Ames, IA.

57. Craun, G. F. 1986. Waterborne diseases in the United States. CRC Press, Boca Raton, FL.

58. Craun, G. F. 1991. Cause of waterborne outbreaks in the United States. Water Sci. Technol. 24:17–20.

59. Cromeans, T., O. V. Nainan, H. A. Fields, M. O. Favorov, and H. S. Margolis. 1994. Hepatitis A and E viruses, pp. 1–56. *In* Y. H. Hui, J. R. Gorham, K. D. Murrell, and D. O. Oliver, eds. Foodborne disease handbook, Vol. 2. Marcel Dekker, New York.

60. Current, W. L. 1994. *Cryptosporidium parvumi* household transmission. Ann. Intern. Med. 120:518–519.

61. Dahling, D. R., and R. S. Safferman. 1979. Survival of viruses under natural conditions in a subarctic river. Appl. Environ. Microbiol. 38:1103–1110.

62. Dahlquist, G. 1991. Epidemiological studies of childhood insulin-dependent diabetes: Review article. Acta Paediatr. Scand. 80:583–589.

63. D'Alessio, D., C. R. Dick, and E. C. Dick. 1972. Transmission of rhinovirus type 55 in human volunteers, pp. 115–116. *In* J. L. Melnick, ed., International virology 2. S Karger, Basel.

64. D'Alessio, D. J., J. A. Peterson, C. R. Dick, and E. C. Dick. 1976. Transmission of experimental rhinovirus colds in volunteer married couples. J. Infect. Dis. 133: 28–36.

65. Dalldorf, G., and J. L. Melnick. 1965. Coxsackie viruses, pp. 474–512. *In* F. L. Horsfall and I. Tamm, eds., *Viral and rickettsial infections of man 4th ed.,* J. B. Lippincott, Philadelphia.

66. D'Angelo, L. J., J. C. Hierholzer, R. A. Keenlyside, L. J. Anderson, and W. J. Martone. 1979. Pharyngoconjunctival fever caused by adenovirus type 4: report of a swimming pool–related outbreak with recovery of virus from pool water. J. Infect. Dis. 140:42–47.

67. Delbac, R. L., F. Debac, V. Broussolle, M. Rabodonirina, V. Girault, M. Wallon, G. Cozon, C. P. Vivres, and F. Peyron. 1998. Identification of *Encephalitozoon intestinalis* in travelers with chronic diarrhea by specific PCR amplification. J. Clin. Microbiol. 36:37–40.

68. deJong, J. C., R. Wigand, A. H. Kidd, G. Wadell, J. G. Kapsenberg, C. J. Muzerie, A. G. Wermenbol, and R.-G. Firtzlaff. 1983. Candidate adenoviruses 40 and 41: fastidious adenoviruses from human infant stool. J. Med. Virol. 11:215–231.

69. Demicheli, V., D. Rivetti, and T. O. Jefferson. 1996. Economic aspects of a small epidemic of hepatitis A in a religious community in northern Italy. J. Infect. 33: 87–90.

70. deSilva, L. M., P. Colditz, and G. Wadell. 1989. Adenovirus type 7 infections in children in New South Wales, Australia. J. Med. Virol. 29:28–32.

71. DeWit, J. C., G. Brockuizen, and E. H. Kampelmacher. 1979. Cross-contamination during the preparation of frozen chickens in the kitchen. J. Hyg. (Camb.) 83:27–32.

72. Dingle, J., and A. D. Langmuir. 1968. Epidemiology of acute respiratory disease in military recruits. Am. Rev. Respir. Dis. 97:1–65.

73. Djuretic, T., P. G. Wall, M. J. Ryan, H. S. Evans, G. K. Ajack, and J. M. Cowden. 1996. General outbreaks of infectious intestinal disease in England and Wales, 1992 to 1994. Communicable Dis. Rep. Rev. 6:R57–R63.

74. Douglas, R. G. 1970. Pathogenesis of rhinovirus common colds in human volunteers. Ann. Otol. Rhinol. Larynogol. 79:563–671.

75. Douglas, R. G., T. R. Cate, P. J. Gerone, and R. B. Couch. 1966. Quantitative rhinovirus shedding patterns in volunteers. Am. Rev. Respir. Dis. 94:159–167.

76. Dowd, S. E., C. P. Gerba, and I. L. Pepper. Confirmation of the human pathogenic microspiridia, *Enterocytozoon bieneusi, Encephalitozoon intestinalis,* and *Vittaforma corneae* in water. Appl. Environ. Microbiol. 64:3332–3335.

77. Downham, M. A., P. S. McQuillen, and P. S. Gardner. 1974. Diagnosis and clinical significance of parainfluenza virus infections in children. Arch. Dis. Child. 49:8–15.

78. DuPont, H. L., and J. J. Mathewson. 1991. *Escherichia coli* diarrhea, pp. 239–254. *In* A. S. Evans and P. S. Brachman, eds., Bacterial infections of humans. Plenum Press, New York.

79. Durepaire, N., S. Ranger-Rogez, J. A. Gandji, P. Wembreck, J. P. Rogez, and F. Denis. 1995. Enteric prevalence of adenovirus in human immunodeficiency virus seropositive patients. J. Med. Virol. 45:56–60.

80. England, B. L. 1982. Detection of viruses on fomites. *In* C. P. Gerba and S. M. Goyal, eds., Methods in environmental virology, Marcel Dekker, New York.

81. Enriquez, C. E., C. J. Hurst, and C. P. Gerba. 1995. Survival of the enteric adenoviruses 40 and 41 in tap, sea and wastewater. Water Res. 29:2548–2553.

82. Environmental Protection Agency. 1989. Health effects of drinking water treatment technologies. Lewis Publishers, Chelsea, MI.

83. Environmental Protection Agency. 1989. National primary drinking water regulations: filtration and disinfection turbidity: *Giardia lamblia;* viruses, *Legionella,* and heterotrophic bacteria. Fed. Reg. 54:27486–27541.

84. Environmental Protection Agency. 1994. National primary drinking water regulations: Enhanced surface water treatment requirements; proposed rule. Fed. Reg. 59:38832–38858.

85. Feachem, R. G., D. J. Bradley, H. Garelick, and D. D. Mara. 1983. Sanitation and Disease. Wiley, NY.

86. Fenner, F., P. A. Bachmann, E. P. J. Gibbs, F. A. Murphy, M. J. Studdert, and D. O. White. 1987. Veterinary Virology. Academic Press, San Diego, CA.

87. Fletcher, E., and C. F. Brennan. 1957. Cardiac complications of Coxsackievirus infections. Lancet 1:913–915.

88. Food Safety and Inspection Service. 1997. FSIS/CDC/FDA site study: the establishment and implementation of an active surveillance system for bacterial foodborne diseases in the United States. Report to Congress. U.S. Department of Agricultural, Washington, DC.

89. Fox, J. P., and C. E. Hall. 1980. Viruses in families. PSG Publishing, Littleton, MA.

90. Foy, H. M. 1989. Adenoviruses, pp. 77–94. *In* A. S. Evans, ed., Viral Infections of humans: epidemiology and control, 3rd ed. Plenum Press, New York.

91. Foy, H. M., M. K. Cooney, and J. B. Hatlen. 1968. Adenovirus type 3 epidemic associated with intermittent chlorination of a swimming pool. Arch. Environ. Health. 17:795–802.

92. Foy, H. M., M. K. Cooney, R. McMahan, and J. T. Grayston. 1973. Viral and mycoplasmal pneumonia in a prepaid medical-care group during an eight-year period. Am. J. Epidemiol. 97:93–102.

93. Frank, J. A., R. W. Warren, J.A. Tucker, J. Zeller, and C. M. Wilfert. 1983. Disseminated parainfluenza infection in a child with severe combined immuno-deficiency. Am. J. Dis. Child. 137:1172–1174.

94. Fraser, D. W. 1991. Legionellosis. In: Bacteria Infections of Humans, 2nd ed. A. S. Evans and P. S. Brachman. pp. 333–347. Plenum Press, New York.

95. Fraser, D. 1994. Epidemiology of *Giardia lamblia* and *Cryptosporidium* infections in childhood. Isr. J. Med. Sci. 30:356–361.

96. Frisk, G., and H. Diderholm. 1992. Increased frequency of Coxsackie B virus IgM in women with spontaneous abortion. J. Infect. 24:141–145.

97. Frisk, G., E. Nilsson, T. Tuvemo, G. Friman, and H. Diderholm. 1992. The possible role of Coxsackie A and echo viruses in the pathogenesis of type I diabetes mellitus studied by IgM analysis. J. Infect. 24:13–22.

98. Gale, P., C. Young, and D. Oakes. 1998. A review: development of a risk assessment for BSE in the aquatic environment. J. Appl. Microbiol. 84:467–470.

99. Gear, J. H. S., and V. Measroch. 1973. Coxsackie virus infections of the newborn. Prog. Med. Virol. 15:42–62.

100. Geldreich, E. E., K. R. Fox, J. A. Goodrich, E. W. Rice, R. M. Clark, and D. L. Swerdlow. 1992. Searching for a water supply connection in the Cabool, Missouri disease outbreak of *Escherichia coli* O157:H7. Water Res. 26:1127–1137.

101. Gerba, C. P., C. E. Enriquez, and P. Gerba. 1997. Virus-associated outbreaks in swimming pools. *In* R. Denkewicz, C. P. Gerba, and J. Q. Hales, eds., Water chemistry and disinfection: swimming pools and spas. National Spa and Pool Institute, Alexandria, VA.

102. Gerba, C. P., and P. Gerba. 1995. Outbreaks caused by *Giardia* and *Cryptosporidium* associated with swimming pools. J. Swim. Pool Spa Ind. 1:9–18.

103. Gerone, P. J., R. B. Couch, G. V. Keefer, R. G. Douglas, E. B. Derrenbacher, and V. Knight. 1966. Assessment of experimental and natural viral aerosols. Bacteriol. Rev. 30:576–584.

104. Glezen, W. P., and R. B. Couch. 1997. Influenza viruses, pp. 473–505. *In* A. S. Evans and R. A. Kaslow, eds., Viral infections of humans, 4th ed. Plenum Press, New York.

105. Glezen, W. P., and F. W. Denny. 1997. Parainfluenza viruses. pp. 551–567. *In* A. S. Evans and R. A. Kaslow, eds., Viral infections of humans, 4th ed. Plenum Press, New York.

106. Glezen, W. P., L. H. Taber, L. H. Frank, and J. A. Kasel. 1986. Risk of primary infection and reinfection with respiratory syncytial virus. Am. J. Dis. Child. 140: 543–546.

107. Glikson, M., E. Galun, R. Oren, R. Tur-kaspa, and D. Shouval. 1992. Relapsing hepatitis. Medicine 7:14–23.

108. Gray, M. L., and A. H. Killinger. 1966. *Listeria monocytogenes and Listeria* infections. Bacteriol. Rev. 30:309–382.

109. Greenberg, H. B., R. G. Wyatt, R. Kalica, R. H. Yolken, R. Black, A. Z. Kapikian, and R. M. Chanock. 1981. New insights in viral gastroenteritis. Perspect. Virol. 11:163–187.

110. Grist, N. R., and D. Reid. 1988. General pathogenicity and epidemiology, pp. 221–239. *In* M. Bendinelli and H. Friedman, eds., Coxsackieviruses: A general update. Plenum Press, New York.

111. Gwaltney, J. M. 1997. Rhinovirus, pp. 815–838. *In* A. S. Evans and R. A. Kaslow, eds., Viral infections of humans, 4th ed. Plenum Press, New York.

112. Gwaltney, J. M., P. B. Moskalski, and J. O. Hendley. 1978. Hand-to-hand transmission of rhinovirus colds. Am. J. Intern. Med. 88:463–467.

113. Hall, C. B. 1990. Respiratory syncytial virus, pp. 1265–1279. *In* G. L. Mandell, R. G. Douglas, Jr., and J. E. Bennett, eds., Principles and practice of infectious diseases, 3rd ed. Churchill Livingstone, New York.

114. Hall, C. B., and R. G. Douglas. 1975. Nosocomial influenza infection as a cause of intercurrent fevers in infants. Pediatrics 55:673–677.

115. Hall, C. B., and R. G. Douglas. 1981. Modes of transmission of respiratory syncytial virus. J. Pediatr. 99:100–103.

116. Hall, C. B., R. G. Douglas, and J. M. Geiman. 1975. Quantitative shedding patterns of respiratory syncytial virus in infants. J. Infect. Dis. 132:151–156.

117. Hall, C. B., R. G. Douglas, and J. M. Geiman. 1976. Respiratory syncytial virus infections in infants: quantitation and duration of shedding. J. Pediatr. 89:11–15.

118. Hall, C. B., R. G. Douglas, and J. M. Geiman. 1980. Possible transmission by fomites of respiratory syncytial virus. J. Infect. Dis. 141:98–102.

119. Hall, C. B., R. G. Douglas, K. C. Schnnbel, and J. M. Geiman. 1981. Infectivity of respiratory syncytial virus by various routes of inoculation. Infect. Immun. 33: 779–783.

120. Hara, J., S. Okamoto, Y. Minekawa, K. Yamazaki, and T. Kase. 1990. Survival and disinfection of adenovirus type 19 and enterovirus 70 in ophthalmic practice. Jpn. J. Ophthalmol. 34:421–427.

121. Hardy, A. V., and J. Watt. 1948. Studies of the acute diarrheal diseases. XVIII. Epidemiology. Public Health Rep. 63:363–378.

122. Harns, N. V., N. S. Weiss, and C. M. Nolan. 1986. The role of poultry and meats in the etiology of *Campylobacter jejuni/coli* enteritis. Am. J. Public Health. 76: 407–411.

123. Harskeeri, R. A., T. Van Gool, A. R. Schuitema, E. S. Didier, and W. J. Terpstra. 1995. Reclassification of the microsporidian *Septata intestinalis* to *Encephalitozoon intestinalis* on the basis of genetic and immunological characterization. Parasitology 110:277–285.

124. Hart, R. H., M. T. Soundarajan, C. S. Peters, A. L. Swiatlo, and F. E. Kocka. 1990. Novel organism with chronic diarrhea in AIDS. Lancet 335:169–170.

125. Hashimoto, S., N. Sakaibara, H. Kumai, M. Nakai, S. Sakuma, S. Chiba, and K. Fujinaga. 1991. Fastidious human adenovirus type 40 can propagate efficiently and produce plaques on a human cell line, A549, derived from lung carcinoma. J. Virol. 65:2429–2435.

126. Hawley, H. B., D. P. Morin, M. E. Geraghty, J. Tomkow, and A. Phillips. 1973. Coxsackievirus B epidemic at a boy's summer camp. J. Am. Med. Assoc. 226: 33–36.

127. Hejkal, T. W., B. Kewsick, R. L. LaBelle, C. P. Gerba, Y. Sanchez, G. Dreesman, B. Hafkin, and J. L. Melnick. 1982. Viruses in a community water supply associated with an outbreak of gastroenteritis and infectious hepatitis. J. Am. Water Works Assoc. 74:318–321.

128. Hendley, J. O. 1990. Para influenza viruses, pp. 1255–1260. *In* G. L. Mandell, R. G. Douglas, Jr., and J. E. Bennett, eds., Principles and practice of infectious diseases, 3rd ed. Churchill Livingston, New York.

129. Hendley, J. O., R. P. Wenzel, and J. M. Gwaltney. 1973. Transmission of rhinovirus colds by self-inoculation. N. Engl. J. Med. 288:1361–1364.

130. Herrmann, J. E., N. R. Blacklow, D. M. Perron-Henry, E. Clements, D. N. Taylor, and P. Echeverria. 1988. Incidence of enteric adenoviruses among children in Thailand and the significance of these viruses in gastroenteritis. J. Clin. Microbiol. 26:1783–1786.

131. Hieber, J. P., A. P. Nelson, and G. H. McCracken. 1977. Acute disseminated staphylococcal disease in childhood. Am. J. Dis. Child. 131:181–185.

132. Hierholzer, J. C. 1992. Adenoviruses in the immunocompromised host. Clin. Microbiol. Rev. 5:262–274.

133. Hierholzer, J. C., R. Wigand, L. J. Anderson, T. Adrian, and J. W. M. Gold. 1988. Adenoviruses from patients with AIDS: a plethora of serotypes and a description of five new serotypes of subgenus D (types 43–47). J. Infect. Dis. 158:804–813.

134. Hierholzer, W. J., Jr., and M. J. Zervos. 1991. Nosocomial bacterial infections. pp. 467–497. *In* A. S. Evans and P. S. Brachman, eds., Bacterial infections of man, 2nd. ed. Plenum Press, New York.

135. Hill, D. R. 1990. *Giardia lamblia,* pp. 2110–2115. *In* G. L. Mandell, R. G. Douglas, Jr., and J. E. Bennett, eds., Principles and practice of infectious diseases. G. L. Mandell, R. G. Douglas and J. E. Bennett, eds. pp. 21 10-2115. Churchill Livingstone, New York.

136. Hirschman, S. Z., and G. S. Hammer. 1974. Coxsackie virus myopericarditis: a microbiological and clinical review. Am. J. Cardiol. 34:224–232.

137. Hoge, C. W., D. R. Shlim, R. Rajah, J. Triplett, M. Shear, J. G. Rabold, and P. Echeverria. 1993. Epidemiology of diarrhoeal illness associated with coccidian-like organism among travelers and foreign residents in Nepal. Lancet 341: 1175–1179.

138. Hopkins, R. S., R. N. Olmsted, and G. R. Istre. 1984. Endemic *Campylobacter jejuni* infection in Colorado: identified risk factors. Am. J. Public Health. 74: 249–250.

139. Horwitz, M. S. 1990. Adenoviruses, pp. 1723–1740. *In* B. N. Fields, ed. Fields virology, 2nd ed. Raven Press, New York.

140. Horwitz, M. S. 1996. Adenoviruses, pp. 2149–2171. *In* B. N. Fields, D. M. Knipe, P. M. Howley, et al., eds., Fields virology, 3rd ed. Lippincott-Raven, Philadelphia.

141. Howell, M. R., R. N. Nanag, C. A. Gaydos, and J. C. Gaydos. 1998. Prevention of adenoviral acute respiratory disease in army recruits: cost-effectiveness of a military vaccination policy. Am. J. Prev. Med. 14:168–175.

142. Huebner, R. J., W. P. Rowe, T. G. Ward, R. H. Parrott, and J. A. Bell. 1954. Adenoidalpharyngoconjunctival agents. N. Engl. J. Med. 251:1077–1086.

143. Hurst, C. J., K. A. McClellan, and W. H. Benton. 1988. Comparison of cytopathogenicity, immunofluorescence and in situ DNA hybridization as methods for the detection of adenoviruses. Water Res. 22:1547–1552.

144. Hyypia, T. 1993. Etiological diagnosis of viral heart disease. Scand. J. Infect. Dis. 88:25–31.

145. Ikeda, R. M., S. F. Kondracki, P. D. Drabkin, G. S. Birkhead, and D. L. Morse. 1993. Pleurodynia among football players at a high school: an outbreak associated with Coxsackie Dl. J. Am. Med. Assoc. 270:2205–2206.

146. Irving, L. G., and F. A. Smith. 1981. One-year survey of enteroviruses, adenoviruses, and reoviruses isolated from effluent at an activated-sludge purification plant. Appl. Environ. Microbiol. 41:51–59.

147. Itagaki, A., J. Ishihara, K. Mochida, Y. Ito, K. Saito, Y. Nishino, S. Koike, and T. Kurimura. 1983. A clustering outbreak of hand, foot, and mouth disease caused by Coxsackie virus A10. Microbiol. Immunol. 27:929–935.

148. Johnson, K. M., H. H. Bloom, B. R. Forsyth, and K. M. Chanook. 1965. Relationship of rhinovirus infection to mild upper respiratory disease. II. Epidemiologic observations in male military trainees. Am. J. Epidemiol. 81:131–139.

149. Jordan, W. S., Jr., G. F. Badger, C. Curtiss, J. H. Dingle, H. S. Ginsberg, and E. Gold. 1956. A study of illness in a group of Cleveland families. X. The occurrence of adenovirus infections. Am. J. Hyg. 64:336–348.

150. Kajon, A. E., C. Larranaga, M. Suarez, G. Wadell, and L. F. Avendano. 1994. Genome type analysis of Chilean adenovirus strains isolated in a children's hospital between 1988 and 1990. J. Med. Virol. 42:16–21.

151. Kapikian, A. Z., and R. G. Wyatt. 1992. Viral gastrointestinal infections, pp. 665–676. In R. D. Feigin and J. D. Cherry, eds., Textbook of pediatric infectious diseases, 3rd ed., Vol. 1. W. B. Saunders, Philadelphia.

152. Kaplan, J. E., G. Gary, R. C. Baron, N. Singh, L. B. Shonberger, L. B. Feldman, and H. B. Greenberg. 1982. Epidemiology of Norwalk gastroenteritis and the role of Norwalk virus in outbreaks of acute nonbacterial gastroenteritis. Ann. Intern. Med. 96:756–761.

153. Kaplan, M. H., 1988. Coxsackievirus infection under three months of age, pp. 241–251. In M. Bendinelli and H. Friedman, eds., Coxsackieviruses: a general update. Plenum Press, New York.

154. Kaplan, M. H., S. W. Klein, J. McPhee, and R. G. Harper. 1983. Group B Coxsackievirus infections in infants younger than three months of age: a serious childhood illness. Rev. Infect. Dis. 5:1019–1032.

155. Kapperud, G., J. Lassen, J. M. Ostroff, and S. Aaqsen. 1992. Clinical features of sporadic *Campylobacter* infections in Norway. Scand. J. Infect. Dis. 24:741–749.

156. Kappus, K. D., R. G. Lundgren, D. Juranek, J. M. Roberts, and H. C. Spener. 1994. Intestinal parasitism in the United States: update on a continuing problem. Am. J. Trop. Med. Hyg. 50:705–713.

157. Kaslow, R. A., and A. S. Evans. 1997. Epidemiologic concepts and methods, pp. 3–88. In A. S. Evans and R. A. Kaslow, eds., Viral infections of humans, 4th ed. Plenum Press, New York.

158. Kaufmann, A., J. McDade, J. E. Patton, C. M. Benett, P. Skality, J. C. Feeley, D. C. Anderson, M. E. Potter, V. F. Newhouse, M. B. Gregg, and P. S. Brachman. 1981. Pontiac fever: isolation of the etiologic agent (*Legionella pneumophila*) and demonstration of its mode of transmission. Am. J. Epidemiol. 114:337–347.

159. Kee, F., G. McElroy, D. Stewart, P. Coyle, and J. Watson. 1994. A community outbreak of echovirus infection associated with an outdoor swimming pool. J. Public Health Med. 16:145–148.

160. Kefler, A., C. E. Hall, J. P. Fox, L. Elreback, and M. K. Cooney. 1969. The virus watch program: a continuing surveillance of viral infections in metropolitan New York families. VII. Rhinovirus infections: observations of virus excretion, intrafamilial spread and clinical response. Am. J. Epidemiol. 90:244–254.

161. Kewsick, B. H., L. K. Pickering, H. L. DuPont, and W. E. Woodward. 1983. Survival and detection of rotaviruses on environmental surfaces in day-care centers. Appl. Environ. Microbiol. 46:813–816.

162. Khan, A. Q. 1967. Role of carriers in the intrafamilial spread of cholera. Lancet 1:245–246.

163. Kramer, M. H., B. L. Herwaldt, G. F. Craun, R. Calderon, and D. D. Juranek. 1996. Surveillance for waterborne disease outbreaks: United States, 1993–1994. Morb. Mortal. Wkly. Rep. 45(SS-1):1–15.

164. Krikelis, V., P. Markoulatos, N. Spyrou, and C. Serie. 1985. Detection of indigenous enteric viruses in raw sewage effluents of the city of Athens, Greece, during a two-year survey. Water Sci. Technol. 17:159–164.

165. Krikelis, V., N. Spyrou, P. Markoulatos, and C. Serie. 1984. Seasonal distribution of enteroviruses in domestic sewage. Can. J. Microbiol. 31:24–25.

166. Kukkula, M., P. Astila, M. L. Klossner, L. Maunula, C. H. Bonsdorff, and P. Vaatinen. 1997. Waterborne outbreak of viral gastroenteritis. Scand. J. Infect. Dis. 29:415–418.

167. Lansdown, A. B. G. 1978. Viral infections and diseases of the heart. Prog. Med. Virol. 24:70–113.

168. LeChevallier, M. W. 1997. What do studies of public water system groundwater sources tell us? pp. 56–73. *In* Under the microscope: examining microbes in groundwater. American Water Works Association, Denver, CO.

169. Lemon, S. M. 1985. Type A viral hepatitis: new developments in an old disease. N. Engl. J. Med. 313:1059–1067.

170. Lengerich, E. J., D. G. Addiss, J. J. Marx, B. L. P. Ungar, and D. D. Juranek. 1993. Increased exposure of *Cryptosporidia* among dairy farmers in Wisconsin. J. Infect. Dis. 167:1252–1255.

171. Levine, M. M. 1987. *Escherichia coli* that cause: enterotoxigenic, enteropathogenic, enterovisie, enterohemorrhagic and enteroadherent. J. Infect. Dis. 155: 377–389.

172. Levine, O. S., and M. M. Levine. 1991. Houseflies *(Musca domestica)* as mechanical vectors of shigellosis. Rev. Infect. Dis. 13:688–696.

173. Lew, J. F., C. L. Moe, S. S. Monroe, J. R. Allen, B. M. Harrison, B. D. Forrester, S. E. Stine, P. A. Woods, J. C. Hierholzer, J. E. Herrmann, et al. 1991. Astrovirus and adenovirus associated with diarrhea in children in day care settings. J. Infect. Dis. 164:673–678.

174. Lighthart, B., and A. J. Mohr. 1987, Estimating downwind concentrations of viable airborne microorganisms in dynamic atmospheric conditions. Appl. Environ. Microbiol. 53:1580–1583.

175. Little, J. W., R. G. Douglas, and W. J. Hall. 1979. Attenuated influenza produced by experimental intranasal inoculation. J. Med. Virol. 3:177–180.

176. Loyd, J. E., R. M. Des Prez, and R. A. Goodwin, Jr. 1990. *Histoplasma capsulatum,* pp. 1989–1999. *In* G. L. Mandell, R. G. Douglas, Jr., and J. E. Bennett, eds., Principles and practices of infectious diseases, 3rd ed. Churchill, Livingstone, New York.

177. Lucena, F., A. Bosch, J. Jofre, and L. Schwartzbord. 1985. Identification of viruses isolated from sewage, riverwater and coastal seawater in Barcelona. Water Res. 19:1237–1239.

178. MacKenzie, W., M. Neil, N. Hoxie, M. Proctor, M. Gradus, K. Blair, D. Peterson, J. Kazmierczak, D. Addiss, K. Fox, J. Rose, and J. Davis. 1994. A massive outbreak in Milwaukee of *Cryptosporidium* infection transmitted through the public water supply. N. Engl. J. Med. 331:161–167.

179. Macler, B. A. 1993. Acceptable risk and U.S. microbial drinking water standards, pp. 619–626. *In* G. F. Craun, ed., Safety of water disinfection. International Life Sciences Institute Press, Washington, DC.

180. Macler, B. A., and S. Regli. 1993. Use of microbial risk assessment in setting United States drinking water standards. Int. J. Food Microbiol. 18:245–256.

181. Madore, H. P., J. J. Treanor, and R. Dolin. 1986. Characterization of Snow Mountain Agent of viral gastroenteritis. J. Virol. 58:487–492.

182. Mahl, M. C., and C. Sadler. 1975. Virus survival on inanimate surfaces. Can. J. Microbiol. 21:819–823.

183. Mahy, B. W. J. 1988. Classification and general properties, pp. 1–18. *In* M. Bendinelli and H. Friedman, eds., Coxsackieviruses: a general update. Plenum Press, New York.

184. Mallet, R., M. Riberre, F. Bonnenfant, B. Labrune, and L. Reyrole. 1969. Les pneumopathies graves adeno-virus. Arch. Fr. Pediatr. 23:1057–1073.

185. Marshall, M. M., D. Naumovitz, Y. Ortega, and C. R. Sterling. 1997. Waterborne protozoan parasites. Clin. Microbiol. Rev. 10:67–85.

186. Martone, W. J., J. C. Hierholzer, R. A. Keenlyside, D. W. Fraser, L. J. D'Angelo, and W. G. Winkler. 1980. An outbreak of adenovirus type 3 disease at a private recreation center swimming pool. Am. J. Epidemiol. 111:229–237.

187. Mathiesen, L. R., P. Skinholj, J. O. Nielsen, R. H. Purcell, D. C. Wong, and L. Ranek. 1980. Hepatitis A, B, and non-A, non-B in fulminant hepatitis. Gut 21: 72–77.

188. McCullough, N. B., and C. W. Eisele.1951. Experimental human salmonellosis II. Immunity studies following experimental illness with *Salmonella meleagridis* and *Salmonella anatom.* J. Immunol. 66:595–608.

189. McLean, D. M., R. M. Bannatyne, and K. Gibosn. 1967. Myxovirus dissemination by air. Can. Med. Assoc. J. 96:1449–1453.

190. McMillan, N. S., S. A. Martin, M. D. Sobsey, D. A. Wait, R. A. Meriwether, and J. N. MacCormack. 1992. Outbreak of pharyngoconjunctival fever at a summer camp: North Carolina, 1991. Morb. Mortal. Wkly. Rep. 41:342–347.

191. McMinn, P. C., J. Stewart, and C. J. Burrell. 1991. A community outbreak of epidemic keratoconjunctivitis in central Australia due to adenovirus type 8. J. Infect. Dis. 164:1113–1118.

192. Medema, G. J., P. F. M. Teunis, A. H. Havelaar, and C. N. Haas. 1996. Assessment of the dose–response relationship of *Campylobacter jejuni.* Int. J. Food Microbiol. 30:101–111.

193. Meiklejohn, G. 1983. Viral respiratory disease at Lowrey Air Force Base in Denver, 1952–1982. J. Infect. Dis. 148:775–784.

194. Melnick, J. L. 1982. Enteroviruses, pp. 187–251. *In* A. S. Evans, eds., Viral infections of humans: epidemology and control, 2nd ed. Plenum Medical Book Co., New York.

195. Mertens, T., D. Gruneklee, and H. J. Heggers. 1983. Neutralizing antibodies against Coxsackie B viruses in patients with recent onset of type 1 diabetes. Eur. J. Pediatr. 140:293–294.

196. Miller, W. S., and M. S. Artenstein. 1967. Aerosol stability of three active respiratory disease viruses. Proc. Soc. Exp. Biol. Med. 125:222–232.

197. Mills, J., J. E. Vankirk, P. F. Wright, and R. M. Chanock. 1971. Experimental respiratory syncytial virus infection in adults. J. Immunol. 107:123–130.

198. Mishu, B., and M. J. Blazer. 1993. Role of infection due to camplyobacter jejuni in the initation of Guillain–Barré syndrome. Clin. Infect. Dis. 17:104–108.

199. Modlin, J. F. 1990. Coxsackieviruses, echoviruses, and newer enteroviruses, pp. 1367–1379. *In* G. L. Mandell, R. G. Douglas, Jr., and J. E. Bennett, eds., Principles and practice of infectious diseases, 3rd ed. Churchill Livingstone, New York.

200. Moffett, H. L., H. K. Shulenberger, and E. R. Burkholder. 1968. Epidemiology and etiology of severe infantile diarrhea. Pediatrics. 72:1.

201. Mohr, A. J. 1991. Development of models to explain the survival of viruses and bacteria in aerosols, pp. 160–190. *In* C. J. Hurst, ed., Modeling the environmental fate of microorganisms. American Society for Microbiology, Washington, DC.

202. Moler, F. W., A. S. Kahn, A. S. Melinoes, J. Custer, J. Palmisano, and T. C. Shape. 1992. Respiratory syncytial virus morbidity and mortality estimates in congenital heart disease patients: a recent experience. Crit. Care Med. 20:1406–1413.

203. Moore, M., R. C. Baron, M. R. Filtein, J. P. Lofgren, D. L. Rowley, L. B. Schonberger, and M. H. Hatch. 1983. Aseptic meningitis and high school football players, 1978 and 1980. J. Am. Med. Assoc. 149:2039–2042.

204. Moore, M., M. H. Kaplan, J. McPhee, D. J. Bregman, and S. W. Klein. 1984. Epidemiologic, clinical, and laboratory features of Coxsackie B1–BS infections in the United States, 1970–1979. Public Health Rep. 99:515–222.

205. Morens, D. M., M. A. Pallansch, and M. Moore. 1991. Polioviruses and other enteroviruses, pp. 427–497. *In* R. B. Belshe, ed., Textbook of human virology, 2nd ed., Mosby, St. Louis, MO.

206. Morton, S., C. L. R. Bartlett, L. F. Bibby, D. N. Hutchinson, J. V. Dyer, and P. J. Dennis. 1986. Outbreak of Legionnaire's disease from a cooling tower system in a power station. Br. J. Ind. Med. 43:630–635.

207. Muilenberg, M., and H. Burge. 1996. Aerobiology. CRC Press, Boca Raton, FL.

208. Murtagh, P., C. Cerqueiro, A. Halae, M. Avila, and A. Kajon. 1993. Adenovirus type 7h respiratory infections: a report of 29 cases of acute lower respiratory disease. Acta Paediatr. 82:557–561.

209. Myint, H., M. M. Soe, M. Khin, T. M. Myint, and K. M. Tin. 1985. A clinical and epidemiological study of an epidemic of non-A non-B hepatitis in Rangoon. Am. J. Trop. Med. Hyg. 34:1183–1189.

210. National Research Council. 1993. Managing wastewater in coastal urban areas. National Academy Press, Washington, DC.

211. Navas, L., E. Wang, V. De Carralho, and J. Robinson. 1992. Improved outcome of respiratory syncytial virus infection in a high risk hospitalized population of Canadian children. Pediatric Investigations Collaborative Network on Infections in Canada. J. Pediatr. 121:348–354.

212. Omenaas, E., P. Bakke, G. Haukened, R. Hanoa, and A. Gulsvik. 1995. Respiratory virus antibodies in adults of a Norwegian community: prevalences and risk factors. Int. J. Epidemiol. 24:223–231.

213. Ormsby, H. L., and W. S. Aitchison. 1955. The role of the swimming pool in the transmission of pharyngeal–conjunctival fever. Can. Med. Assoc. J. 73:864–866.

214. Ortega, Y. R., C. R. Sterling, and R. H. Gilman. 1998. *Cyclospora cayetanensis.* Adv. Parasitol. 40:399–418.

215. Papapetropoulu, M., and A. Vantarakis. 1996. Detection of an adenovirus outbreak at a municipal swimming pool by nested PCR amplification. Abstracts of the Health-Related Water Microbiology Conference, October 6–10, Mallorca, Spain.

216. Pattison, C. P., K. M. Boyner, J. E. Maynard, and P. C. Kelley. 1974. Epidemic hepatitis in a clinical laboratory: possible association with computer card handling. J. Am. Med. Assoc. 230:854–857.

217. Pavia, A. T., and R. V. Tauxe. 1991. Salmonellosis non-typhoidal. *In* A. S. Evans and P. S. Brachman, eds., Bacterial infections of humans. Plenum Press, New York.

218. Payne, S. B., E. A. Grilli, A. J. Smith, and T. W. Hoskins. 1984. Investigation of an outbreak of adenovirus type 3 infection in a boys' boarding school. J. Hyg. 93:277–283.

219. Petersen, L. R., M. L. Cartter, and J. L. Hadler. 1988. A foodborne outbreak of *Giardia lamblia.* J. Infect. Dis. 157:846–848.

220. Pickering, L. K., W. E. Woodward, H. L. DuPont, et al. 1984. Occurrence of *Giardia lamblia* in children in day-care centers. J. Pediatr. 104:522–526.

221. Porter, I. A., and T. M. S. Reid. 1980. A milk-borne outbreak of *Campylobacter* infection. J. Hyg. 84:415–419.

222. Prusiner, S. B. 1994. Biology and genetics of prion diseases. Annu. Rev. Microbiol. 48:655–686.

223. Rab, M. A., M. K. Bile, M. M. Mubarik, H. Asghar, Z. Sami, S. Siddigi, A. S. Dil, M. A. Barzgar, M. A. Chaudhry, and M. I. Burney. 1997. Water-borne hepatitis E virus epidemic in Islamabad, Pakistan: a common source outbreak traced to the malfunction of a modern water treatment plant. Am. J. Trop. Med. Hyg. 57:151–157.

224. Rabeneck, L., F. Gyorkey, M. Genta, P. Gyorkey, L. W. Foote, and J. M. Risser. 1993. The role of microsporidia in the pathogenesis of HIV-related chronic diarrhea. Ann. Intern. Med. 119:895–899.

225. Ravdin, J. I., and W. A. Pettri. 1990. *Entamoeba histolytica* (amebiasis), pp. 2036–2049. *In* G. L. Mandell, R. G. Douglas, Jr., and J. E. Bennett, eds., Principles and practice of infectious diseases, 3rd ed. Churchill Livingstone, New York.

226. Ray, C. G., C. J. Holberg, L. L. Minnich, Z. M. Shehab, A. L. Wright, L. M. Taussig, and the Group Health Medical Associates. 1993. Acute lower respiratory illnesses during the first three years of life: potential roles for various etiologic agents. Pediatr. Infect. Dis. J. 12:10–14.

227. Ray, C. G., L. L. Minnich, C. J. Holber, Z. M. Shehab, A. L. Wright, L. L. Barton, L. M. Taussig, and the Group Health Medical Associates. 1993. Respiratory syncytial virus-associated lower respiratory illnesses: possible influence of other agents. Pediatr. Infect. Dis. J. 12:15–19.

228. Reed, S. E. 1975. An investigation of possible transmission of rhinovirus colds through contact. J. Hyg. (Camb.) 75:249–258.

229. Reed, S. L., and J. I. Ravdin. 1995. Amebiasis, pp. 1065–1080. *In* M. J. Blazer, P. O. Smith, J. I. Ravdin, H. B. Greenberg, and R. L. Guerrant, eds., Infections of the gastrointestinal tract. Raven Press, New York.

230. Regli, S., J. B. Rose, C. N. Haas, and C. P. Gerba. 1991. Modeling the risk from *Giardia* and viruses in drinking water. Am. Water Works Assoc. 83:76–84.

231. Reyes, M. P., and A. M. Lerner. 1985. Coxsackievirus myocarditis: with special reference to acute and chronic effects. Prog. Cardiovasc. Dis. 27:373–394.

232. Richmond, S. J., S. M. Dunn, E. O. Caul, C. R. Ashley, and S. K. R. Clarke. 1979. An outbreak of gastroenteritis in young children caused by adenoviruses. Lancet (June 2):1178–1180.

233. Rose, J. B., J. T. Lisle, and M. Le Chevallier. 1997. Waterborne cryptosporidiosis: incidence, outbreaks, and treatment strategies, pp. 93–109. R. Fayer, ed., *Cryptosporidium* and cryptosporidosis. CRC Press, Boca Raton, FL.

234. Rusin, P. A., J. B. Rose, C. N. Haas, and C. P. Gerba. 1997. Risk assessment of opportunistic bacterial pathogens in water. Rev. Environ. Contam. Toxicol. 152: 57–83.

235. Sattar, S. A., and V. S. Springthorpe. 1996. Transmission of viral infections through animate and inanimate surfaces and infection control through chemical disinfection, pp. 224–257. *In* C. J. Hurst, ed., Modeling disease transmission and its prevention by disinfection. Cambridge University Press, United Kingdom.

236. Schnurr, D., and M. E. Dondero. 1993. Two new candidate adenovirus serotypes. Int. Virol. 36:79–83.

237. Schwartz, D., R. Bryan, R. Weber, and G. Visvesvara. 1994. Microsporidiosis in HIV positive patients: current methods for diagnosis using biopsy, cytologic, ultrastructural, immunological and tissue culture techniques. Folia Parasitol. 41: 101–109.

238. Scott, E., S. F. Bloomfield, and C. G. Barlow. 1982. An investigation of microbial contamination of the home. J. Hyg. (Camb.) 89:279–293.

239. Sealy, D. P., and S. H. Schuman. 1983. Endemic giardiasis and daycare. Pediatrics. 72:659–662.

240. See, D. M., and J. G. Tiles. 1991. Viral myocarditis. Rev. Infect. Dis. 13:951–956.

241. Shields, A. F., R. C. Hackman, K. H. Fife, L. Corey, and J. D. Meyers. 1985. Adenovirus infections in patients undergoing bone-marrow transplantation. N. Engl. J. Med. 312:529–533.

242. Smith, H. V., and J. B. Rose. 1998. Waterborne cryptosporidiosis: current status. Parasitol. Today 14:14–22.

243. Smith, J. L., S. A. Palumbo, and I. Wallis. 1993. Relationship between foodborne bacterial pathogens and reactive arthritides. J. Food Safety 13:209–236.

244. Smith, M. H. D., K. W. Newell, and J. Sullianti. 1965. Epidemiology of enteropathogenic Escherichia coli infection in non-hospitalized children. Antimicrob. Agents Chemother. 5:77–83.

245. Sparel, J. M., C. Sarfati, O. Liguory, B. Caroff, N. Dumoutier, B. Gueglio, E. Billaud, F. Raffi, J. M. Molina, M. Miegeville, and F. Derovin. 1997. Detection of microsporidia and identification of *Enterocytozoon bieneusi* in surface water by filtration followed by specific PCR. J. Eukaryot. Microbiol. 44:78S.

246. Spendlove, J. C., and K. F. Fannin. 1982. Methods for the characterization of virus aerosols, pp. 261–329. *In* C. P. Gerba and S. M. Goyal, eds., Methods in environmental virology. Marcel Dekker, New York.

247. Spers, R., R. A. Shooter, H. Gaya, N. Putel, and J. H. Hewitt. 1969. Contamination of nurses' uniforms with *Staphylococcus aureus*. Lancet 1:84–88.

248. Springthorpe, V. S., and S. A. Sattar. 1990. Chemical disinfection of virus contaminated surfaces. CRC Crit. Rev. Environ. Control 20:169–229.

249. Stalder, H., J. C. Hierholzer, and M. N. Oxman. 1977. New human adenovirus (candidate adenovirus type 35) causing fatal disseminated infection in a renal transplant recipient. J. Clin. Microbiol. 6:257–265.

250. Steahr, T. E., and T. Roberts. 1993. Microbial Foodborne Disease: Hospitalizations, Medical Costs and Potential Demand for Safer Food. Department of Resource Economics, United States Department of Agriculture, University of Connecticut, Storrs, CT.

251. Storch, G., W. B. Baine, D. W. Fraser, C. V. Broome, H. W. Clegg, M. L. Cohen, S. A. J. Goings, B. D. Politi, W. A. Terranova, T. F. Plikaytis, B. D. Shepard, and J. V. Bennett. 1979. Sporadic community-acquired Legionnaire's disease in the United States: a case-control study. Ann. Intern. Med. 90:596–600.

252. Stout, J. E., V. L. Yu, P. Muraca, J. Jolly, N. Troup, and L. S. Tompkins. 1992. Potable water as a cause of sporadic cases of community-acquired Legionnaire's disease. N. Engl. J. Med. 326:151–155.

253. Tamayo, J. F., W. H. Mosley, M. G. Alvero, R. R. Joseph, C. Z. Gomez, T. Montague, J. J. Dizon, and D. A. Henderson. 1965. Studies of cholera El Tor in the Phillipines. 3. Transmission of infection among household contacts of cholera patients. Bull. WHO 33:645–649.

254. Thompson, M. A., J. W. T. Benson, and P. A. Wright. 1987. Two-year study of *Cryptosporidium* infection. Arch. Dis. Child. 62:559–563.

255. Tiemenn, E. M., P. L. Shipley, R. A. Correia, D. S. Shields, and R. L. Guerrant. 1985. Sulfamethoxazole-trimethoprim-resistant *Shigella flexneri* in northeastern Brazil. Antimicrob. Agents Chemother. 25:653–654.

256. Toniolo, A., G. Federico, F. Basolo, and T. Onodera. 1988. Diabetes mellitus, pp. 351–382. *In* M. Bendinelli and H. Friedman, eds., Coxsackieviruses: a general update. Plenum Press, New York.

257. Turner, M., G. R. Istre, H. Beauchamp, M. Baum, and S. Arnold. 1987. Community outbreak of adenovirus type 7a infections associated with a swimming pool. South. Med. J. 80:712–715.

258. Tyrell, D. A. J., M. Bynoe, K. Birkum, S. Petersen, and M. S. Perura. 1959. Inoculation of human volunteers with parainfluenza viruses 1 and 3 (HA_2 and HA_1). Br. Med. J. 2:909–911.

259. Uhnoo, I., L. Svensson, and G. Wadell. 1990. Enteric adenoviruses, pp. 627–642. *In* M. J. G. Farthing, ed., Baillière's clinical gastroenterology, Vol. 4, No. 3. Baillière Tindall, London.

260. Van, R., C.-C. Wun, M. L. O'Ryan, D. O. Matson, L. Jackson, and L. K. Pickering. 1991. Outbreaks of human enteric adenovirus types 40 and 41 in Houston day-care centers. J. Pediatr. 120:516–521.

261. van Gool, T., J. C. Vetter, B. Weinmayr, A. Van Dam, F. Deroin, and J. Dankert. 1997. High seroprevalance of *Encephalitozoon* species in immunocompetent subjects. J. Infect. Dis. 175:1020–1024.

262. Velazquez, O., H. C. Stetler, C. Avila, G. Ornelas, C. Alvarez, S. C. Hadler, D. Bradley, and J. Sepulveda. 1990. Epidemic transmission of enterically transmitted non-A, non-B hepatitis in Mexico, 1986–1987. J. Am. Med. Assoc. 263:3261–3285.

263. Villarejos, V. M., J. Serra, K. Anderson-Visona, and J. W. Mosely. 1982. Hepatitis A virus infections in households. Am. J. Epidemiol. 116:577–586.

264. Wadell, G. 1984. Molecular epidemiology of human adenoviruses. Curr. Top. Microbiol. Immunol. 110:191–220.

265. Walsh, J. A. 1988. Prevalence of *Entamoeba histolytica* infection, pp. 93–105. *In* J. I. Ravdin, ed., Amebiasis: human infection by *Entamoeba histolytica*. Churchill Livingstone, New York.

266. Ward, R. L., D. I. Bernstein, D. R. Knowlton, J. R. Sherwood, E. C. Young, T. M. Cusack, T. M. Rubino, and G. M. Schiff. 1991. Prevention of surface-to-human transmission of rotavirus by treatment with disinfectant spray. J. Clin. Microbiol. 29:1991–1996.

267. Webb, D. H., A. F. Shields, and K. H. Fife. 1987. Genomic variation of adenovirus type 5 isolates recovered from bone marrow transplant recipients. J. Clin. Microbiol. 25:305–308.

268. Weber, R., R. Bryan, D. Schwartz, and R. Owen. 1994. Human microsporidial infections. Clin. Microbiol. Rev. 7:426–461.

269. Wendt, C. H., D. J. Weisdorf, M. C. Jordon, H. H. Balfour, Jr., and M. I. Hertz. 1992. Parainfluenza virus respiratory infection after bone marrow transplantation. N. Engl. J. Med. 326:921–926.

270. Wood, R. C., K. L. MacDonald, and M. T. Osterholm. 1992. *Campylobacter* enteritis outbreaks associated with drinking raw milk during youth activities. J. Am. Med. Assoc. 268:3228–3230.

271. Woodward, W. E. 1971. Chorea reinfection in man. J. Infect. Dis. 123:61–66.

272. Yolken, R. H., F. Lawrence, F. Leister, H. E. Takiff, and S. E. Strauss. 1982. Gastroenteritis associated with enteric type adenovirus in hospitalized infants. J. Pediatr. 101:21–26.

273. Yoon, J. W., P. R. McClintock, C. J. Bachurski, J. D. Longstreth, and A. L. Notkins. 1985. Virus-induced diabetes mellitus: no evidence for immune mechanisms in the destruction of beta cells by encephalomyocarditis virus. Diabetes 34:922–925.

274. Yu, V. L. 1990. *Legionella pneumophila* (Legionnaire's disease), pp. 1764–1774. *In* G. L. Mandell, R. G. Douglas, Jr., and J. E. Bennett, Eds., Principles and practice of infectious diseases, 3rd ed. Churchill Livingstone, New York.

275. Zahradnik, J. M., M. J. Spencer, and D. D. Porter. 1980. Adenovirus infection in the immunocompromised patient. Am. J. Med. 68:725–732.

CHAPTER 3

RISK ASSESSMENT PARADIGMS

Risk assessment may be viewed by some as a scientific discipline that is interdisciplinary in nature and by others as a professional process that includes the participation of many established scientific disciplines. As defined, risk assessment is the qualitative or quantitative characterization and estimation of potential adverse health effects associated with exposure of individuals or populations to hazards (materials or situations, physical, chemical, and or microbial agents). Risk assessment is not used in isolation but is part of what is known in a broader context as *risk analysis*. Risk analysis includes risk assessment, risk management, and risk communication (Table 3-1).

CHEMICAL RISK ASSESSMENT: NATIONAL ACADEMY OF SCIENCES PARADIGM

Risk assessment has had an interesting and controversial history. While the process attempted to identify the risks and quantify the degree of risk for exposed individuals or populations, the development of this field of study was tied closely to governmental policies that focused on the control of environmental chemical contaminants. For air and drinking water this began in the early 1970s with the congressional mandates of the Clean Air Act and the Safe Drinking Water Act Amendments to these acts required that better estimates of potential hazards be made for risk management purposes. Thus began a series of studies and reports from the National Academy of Sciences–National Research Council (16). The principles, process, and methods were further refined, and in 1983 the National Research Council published the

TABLE 3-1 Definitions Used in Risk Analysis

- *Risk assessment:* the qualitative or quantitative characterization and estimation of potential adverse health effects associated with exposure of individuals or populations to hazards (materials or situations, physical, chemical, and or microbial agents).
- *Risk management:* the process for controlling risks, weighing alternatives, and selecting appropriate action, taking into account risk assessment, values, engineering, economics, and legal and political issues.
- *Risk communication* the communication of risks to managers, stakeholders, public officials, and the public; includes public perception and ability to exchange scientific information

Source: Ref. 13.

"Red Book" (17), which officially recognized the field of risk assessment. The report also recommended that risk assessment and risk management be kept distinct and that uniform guidelines for risk assessment be established for both cancer and noncancer effects.

Four steps in the risk assessment process were defined (Table 3-2). While hazard identification was defined in human and animal studies, dose–response modeling was done primarily in animal models. Exposure involved monitoring of the environment and the transport and fate of the chemicals through the various exposure pathways (17).

Risk assessments have generally followed two approaches: one for cancer endpoints and one for noncancer. Other approaches, for reproductive effects, for example, are now being developed. Chemicals that may be carcinogenic have been evaluated using a weight-of-evidence classification system, which was originally developed from International Agency for Research on Cancer of the World Health Organization. This uses both human and animal data to identify the possible cancer risk. Another approach used by the National Toxicology Program (NTP) is to evaluate each animal study individually without

TABLE 3-2 Risk Assessment Paradigm for Human Health Effects

1. *Hazard identification:* to describe acute and chronic human health effects associated with any particular hazard, including toxicity, carcinogenicity, mutagenicity, developmental toxicity, reproductive toxicity, and neurotoxicity.
2. *Dose–response assessment:* to characterize the relationship between various doses administered and the incidence of the health effect.
3. *Exposure assessment:* to determine the size and nature of the population exposed and the route, amount, and duration of the exposure.
4. *Risk characterization:* to integrate the information from exposure, dose–response, and health steps in order to estimate the magnitude of the public health problem and to evaluate variability and uncertainty.

necessarily offering this as an explanation of a chemical's ability to cause cancer in humans (although generally presumed). Until recently, carcinogenic chemicals have been considered to be what is known as nonthreshold (i.e., any level of the chemical will pose some level of cancer risk). In examining animal dose–response studies, the slope of the line defines the cancer potency factor (CPF), which is the rate of increase in cancer risk as a function of increasing dose (Figure 3-1). Within the Safe Drinking Water Act, for example, this approach was used to set maximum contaminant levels (MCLs) for chemicals that could have the potential to cause human cancer effects at exposures considered to have a *de minimis* health impact [often at 1 in a million (10^{-6})].

Those chemicals demonstrating a threshold, or some dose, below which there is no response in the people exposed have been described for noncancer endpoints. In these cases the levels of no effect known as the *no observable adverse effect level* (NOAEL) and the *lowest observable adverse effect level* (LOAEL) were determined in animals with organ systems similar to humans or animals that may be the most sensitive. Safe levels were defined by the NOAEL and LOAEL, including *acceptable daily intake* (ADI) by the FDA and *reference dose* (RfD), which includes an uncertainty factor by the EPA.

Exposures in the risk assessment process have been defined for chemical risk assessment as an average lifetime daily dose (concentration $\times$ contact rate $\times$ duration/body weight $\times$ lifetime). Yet defining this average has involved different approaches depending on the exposure medium and the legislative mandate. Under Superfund, for example, it was suggested that 30 years of exposure with the *reasonable maximum exposure* be determined, while in the Clean Air Act, a maximally *exposed individual* for 70 years was to be used.

Early risk assessment focusing on chemical contaminants included many other fields of study, such as toxicology, epidemiology, animal bioassays, environmental monitoring (with methods development), and statistical modeling. The controversies surrounding the risk assessment methods were focused on four major areas: (1) sensitivity and limitations of epidemiological

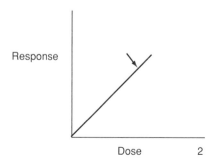

Figure 3-1   Cancer potency factor.

studies, (2) use of animals for extrapolating health effects for humans, (3) type of mathematical models to be used for extrapolation of high dose to low dose, and (4) approaches for handling uncertainty in the estimates. In addition, the use of real-world exposure was often limited, worst-case scenarios, and maximum exposed individuals and vulnerable populations were assessed. As the various governmental agencies took different approaches to address exposure and dose–response, limited uniform methodologies were developed (Table 3-3). The U.S. Department of Energy, Department of Defense, Department of Health and Human Services, and Department of Agriculture have also made extensive use of risk assessments.

Although risk assessment and risk management were, in theory, separate processes, in practice, assumptions and other decisions made in the analysis phase were closely tied to management options. While the risk assessment methodologies may not be prescribed in legislative statutes, the laws may dictate very specific risk management directives or mandates (Table 3-4). These have influenced the methods used for risk assessment, and as other risk paradigms developed it was clear that risk assessment and risk management as well as risk communication in some cases needed to be integrated.

ECOLOGICAL RISK ASSESSMENT

The protection of the environment in addition to human health is clearly a goal shared by most. Increasingly, issues surrounding global climate change, loss of biological diversity (e.g., the rain forest), and sustainability of resources (e.g., fisheries) have focused pollution control on the protection of environmental systems or ecosystems. The framework for ecological risk as-

TABLE 3-3 Some Agencies Involved in Risk Assessment

Agency	Legislative Programs
Consumer Product Safety Commission	Consumer products
Environmental Protection Agency, mandated in 1977 to use better risk assessment for SDWA and in 1990 for CAA	Clean Air Act (CAA), Clean Water Act (CWA), Safe Drinking Water Act (SDWA), Resource Conservation and Recovery Act (RCRA), Superfund, Toxic Substances and Control Act (TOSCA), Pesticide Program
Food and Drug Administration	Food Additive Program
Occupational Safety and Health Administration mandated in 1980 by U.S. Supreme Court to undertake risk assessments for toxic chemicals	Worker exposure and permissible exposure levels

TABLE 3-4 Statutory Mandates on Risk

- *Zero-risk or pure-risk standards* associated with the *Delaney Clause,* are mandated as part of the Federal Food, Drug, and Cosmetic Act, and prohibit any food additive that has been found to "induce cancer"; provisions in the Clean Air Act associated with the national ambient air quality standards call for protection of public health without regard to technology or cost. Within the Safe Drinking Water Act, *Maximum contaminant levels* (MCLs) were set for chemicals based on levels of risk of 10^{-5} to 10^{-6}.

- *Technology-based standards* as part of the Safe Drinking Water Act and the Clean Water Act focus on the cost and effectiveness of alternative control technologies to reduce risks. These include *best-practicable control technology, best conventional technology, best available technology economically achievable,* and *best demonstrated control technology.*

- *No unreasonable risk* requires the balancing of risks against benefits in making risk management decisions. The Federal Insecticide, Fungicide, and Rodenticide Act and the Toxic Substances Control Act require the registration of pesticides that will not cause "unreasonable adverse effects on the environment" and assessment of chemical substances that "present an unreasonable risk of injury to health or the environment" given the benefits of the chemical, magnitude of the exposure, and the possibility of substitutes.

sessment established by the EPA in 1992 (30) was developed conceptually from the National Academy's paradigm for chemicals and human health. However, it was recognized that (1) ecosystems are complex communities and that the complex interactions are site specific, (2) ecosystems are made up of a variety of species which may have different sensitivities to the hazards of concern, (3) exposure pathways are difficult to ascertain, (4) nonchemical hazards (such as sediments effects on seagrasses) need to be considered, and (5) quantitative risk estimates would be difficult to develop. To highlight these differences, the terminology used and the risk assessment process were slightly altered (Figure 3-2).

It was clear that risk management and risk assessment and even risk communication would need to be better integrated. Thus a problem formulation phase was added to the risk paradigm. This took into account an initial evaluation of the site-specific factors, resources, and values to be protected, types of stressors, scope of the evaluation, policy and regulatory issues, and the scientific data needs. The term *stressor* was often used and *stressor–response* (rather than *dose–response*) because *dose* in the previous risk assessment approaches was used within agencies to address chemicals. In addition, the adverse effects were multifaceted and included not only direct effects (increased mortality, reproductive changes, changes in maturation) but also indirect effects (decrease in habitat might mean a decrease in spawning). Ecological effects were often descriptive and reflective of the complexities and the difficulties in modeling these interactions.

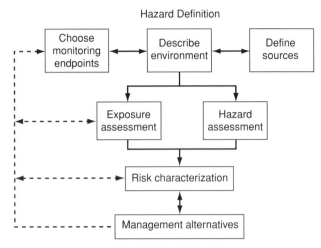

Figure 3-2 Ecological risk.

The EPA's approach to ecological risk assessment may still be seen as controversial in that in some cases it is the policy that drives the need for the scientific data, often developing research that fits the policy rather than developing policy that addresses the scientific assessment. Thus others have described a risk paradigm that more closely resembles the National Academy of Sciences original proposal (27).

Monitoring is still seen as a key to the assessment of risk. In some cases the stressor can be monitored for directly (heavy metals and nutrients) and the response (toxicity and phytoplankton production) can be ascertained in laboratory studies or in field studies. Both types of scientific data are seen as necessary. While dose–response relationships can be developed, for example between nitrogen inputs and chlorophyll a, as a measure of the potential for development of eutrophic conditions, the relationship of this response to *adverse* effects, such as toxic alga blooms, is not so apparent (18). Identification of the transport and fate of the contaminants across media (air, water, and sediments) and the most sensitive species are targeted, as essential scientific data needs.

While human health risk assessment and ecological risk assessment followed similar approaches, they were implemented as distinct fields of scientific study. Endpoints of acceptable contaminant loading to waters, land, or air could differ greatly, and in some cases, standards for discharges, say under the *Clean Water Act,* were more stringent, based on the protection of the ecosystem rather than human health. There is a growing awareness that human health and ecosystem health are related and that a more integrated approach is needed to address both endpoints.

Example 3-1 **Pfiesteria:** *A Case for Ecological and Human Health Risk Assessment.* Pfiesteria spp. are dinoflagellates found in estuarine waters. One of the species emerging is *P. piscicida,* which produces a toxin that potentially can affect both fish and human health, causing ulcers and death in fish and ulcers, headaches, respiratory effects and short-term memory loss in humans (4). It has 24 life stages and is free-living but can act as a predator when fish are abundant and is particularly stimulated by nutrients. Both recreators and fisherman have been affected through direct exposure and inhalation. The toxic dinoflagellate has plagued estuaries and coastal waters of the mid-Atlantic and southeastern United States and has been implicated as a cause of major fish kills, more than 10 in 1995 to 1996, with as many as 1 million fish killed per episode. The fish kills occurred in areas where there were high nutrient inputs; thus partly responsible for these major and unusual blooms of this microorganism may be enhanced eutrophication of these waters. Studies have reported that the zoospore stage increased in numbers as much as 100 times, particularly with increases in phosphates. In environmental surveys, zoospore levels were six times higher at sites affected by wastewater dischargers.

APPROACHES FOR ASSESSING MICROBIAL RISKS

Infectious disease associated with exposure to pathogenic microorganisms is an old and established risk but is also one that is emerging along with antibiotic resistance and newly identified human pathogens (*Cryptosporidium, Cyclospora,* and *E. coli* O157:H7). Infectious disease risk numbers and statistics are for the most part determined by *measured rates.* That is the number of people who have disease *X* per 1000 people per some unit of time (usually over a year's time). This of course is dependent on a disease surveillance program (from clinical laboratories, physicians and health departments) and a reporting system. The better the program and reporting, the more accurate the rates. It is generally accepted that these systems greatly underestimate the level of disease in a community, and while providing a picture of past risk may not accurately reflect future risk. This also becomes problematic for new pathogens for which there is no established procedure for testing of patients and rarely addresses the various exposure or transmission pathways. In addition, outcome may be assessed by mortality in the extreme or by a case without identification of consequence (severity of the illness, number of days sick, medical care, etc.). It is often presumed that infectious disease exposure occurs through person-to-person transmission. However, clearly, waterborne and in some cases foodborne disease are established risks associated with environmental transmission and the contamination of water, air, or the food chain.

Since cholera was first identified and associated with waterborne transmission in the famous Broad Street pump study in London, epidemiology has always been the major science used to study the transmission of infectious disease and the role of the environment. "Epidemiology may be defined as the study of the occurrence and distribution of disease and injury specified by person, place and time" (1). This was traditionally used to study epidemics or excess cases of disease, and therefore the focus for microorganisms was on epidemics or waterborne/foodborne outbreaks of disease. With the shift from infectious agents to chronic diseases such as cancer, environmental epidemiology arose. "Environmental epidemiology is the study of environmental factors that influence the distribution and determinants of disease in human populations" (1). Epidemiology has always been used and will remain an integral part of the risk assessment process for microbials. As epidemiological studies focus on actual human health effects instead of hypothesized outcomes, the data carry with them a great deal of scientific validity.

The use of epidemiology alone without other scientific fields integrated into the process will not fulfill the needs for a complete risk assessment. Epidemiology may be limited by the sensitivity of the study, seeking to associate statistically very small risks relative to the background. Therefore, the study of endemic disease risks becomes much more difficult. In addition, exposure data are lacking, incomplete or imprecise, confounded by the nature of the human subjects. For example, the study by Payment et al. (21) found that 35% of the diarrheal illness was associated with tap water consumption. However, the microbial agents, their concentrations, distributions, and sources, and the potential for other serious or chronic health effects were not elucidated. Similarly, risk management strategies were not developed.

Risk assessment methods following the National Academy paradigm have been used only on a limited scale for judging waterborne pathogenic microorganisms between 1983 and 1991 (10–12,22–24). As this is an estimate of the risk based on models, the risks were criticized as not carrying the validity of the measured rates. A number of conferences and authors attempted to deal with pathogen risks in a qualitative manner, but few quantitative attempts were made until the 1980s. These early attempts lacked an adequate database for dose–response and exposure as well as an understanding of the type of risks associated with microorganisms in water and a framework for analyzing such risks.

Haas (11) was the first to look quantitatively at microbial risks associated with drinking waters based on dose–response modeling. He examined mathematical models which could best estimate the probability of infection from the existing databases associated with human exposure experiments. He found that for viruses a beta-Poisson model best described the probability of infection. This model was used to estimate the risk of infection, clinical disease, and mortality to hypothetical levels of viruses in drinking water, calculating annual and lifetime risks (10,12). Rose et al. (24) then used an exponential model to evaluate daily and annual risks of *Giardia* infections from exposure to contaminated water after various levels of reduction through treatment. This particular study used survey data for assessing the needed treatment for polluted and pristine waters based on *Giardia* cyst occurrence. The use of "probability of infection" models for development of standards for bacteria, viruses, and protozoa in water was suggested (23), and this approach was used in the development of the Surface Water Treatment Rule to address in particular the performance-based standards in the control of *Giardia* (29,31).

In a study for the U.S. Army, Cooper et al. (5) attempted to quantify the risks of water-related infection and illness exposure to Army units in the field. They reviewed the literature for information on infectious dose and clinical illness for potentially waterborne pathogens. Using this information, the probability of infection was assessed using logistic, beta, exponential, and lognormal models. A generalized model was then developed incorporating expected pathogen concentrations, consumption volume, and risk of infection for different military units. The study attempted to incorporate organism concentrations, effective treatment, and risk of infection; however, there was a limited existing database on microbial concentrations and infectious dose. Data used for estimation of subclinical to clinical illness rates for enteroviruses did not incorporate clinical data or reviews of clinical research. In addition, the data on occurrence of the pathogens in source waters and the likely efficiency of removal by treatment processes were limited.

One of the major limitations of the early attempts to use a risk assessment approach for microorganisms was the lack of data and analysis of those data regarding all the steps in the risk assessment process, the health effects, and dose–response and exposure assessment. Sobsey et al. (26) published as part of the report *Drinking Water and Health in the Year 2000* (26) a conceptual framework of data needs for addressing microorganisms. Microbial risk assessment needed better identification of the specific microorganisms, assessment of human health effects, development of dose–response data, understanding of physiological host-microorganism interactions, and incorporation of epidemiological data. With regard to exposure, the authors recommended better data on occurrence, transport and fate, regrowth potential, and susceptibility to water treatment processes.

The National Academy's four-tiered approach can be used for the microbial risk assessment process. However, because microorganisms are living entities, the risk assessment does require modification and the use of terminology that is specific to microbial risks. A combination of scientific fields which also have different languages are needed to form the basis of microbial risk assessment, including but not limited to epidemiology, medicine, clinical and environmental microbiology, and engineering.

Hazard Identification

The hazard identification is both identification of the microbial agent and the spectrum of human illnesses and disease associated with the specific microorganism. The types of clinical outcome range from asymptomatic infections to death (see Chapters 1 and 4). These data come from the clinical literature and studies from clinical microbiologists. The pathogenicity and virulence of the microorganism itself are of great interest as well as the full spectrum of human disease that can result from specific microorganisms. The host response to the microorganisms in regard to immunity and multiple exposures and the adequacy of animal models for studying human impacts should be addressed here. Endemic and epidemic disease investigations, case studies, hospitalization studies, and other epidemiological data are needed to complete this step in the risk assessment. The transmission of disease is often microbial specific (e.g., rabies, vectorborne diseases such as malaria or influenza); therefore, in some cases the transmission (and to some extent the exposure) is tied into hazard identification for microbial risks. The use of new methods, such as molecular epidemiology, that can be used to track specific microorganisms from the patient back to the environment will be of great value for refining the role of the environment in the transmission of various types of microorganisms and diseases.

Dose–Response Assessment

The dose–response is aimed at the mathematical characterization of the relationship between the dose administered and the probability of infection or disease in the exposed population. The microorganisms are measured in doses that are routinely used to count the specific microbe in the laboratory, such as colony counts on agar media for bacteria, plaque counts in cell culture for viruses, and direct microscopic count of cysts/oocysts for the protozoa. However, this means that for the protozoa this results in essentially particle counts (nonviable organisms viewed microscopically could be counted in the dose), while with bacteria and viruses the opposite problem exists, viable but non-culturable organisms are *not* counted (see Chapters 5 and 7). Despite these limitations in estimation of the dose, the methods used are similar to those used to detect these same microorganisms in environment samples. Natural routes of exposure are used: direct ingestion, inhalation, or contact. Both disease and infection can be measured in these studies as the endpoint. In most cases less virulent strains of the microorganisms and healthy human adults were used. Multiple exposures should be evaluated, but in most past studies, they were not.

ISSUES SURROUNDING A THRESHOLD IN MICROBIAL RISK DOSE–RESPONSE MODELING

One of the more controversial areas surrounding microbial modeling is the potential for a single organism to initiate an infection. Current scientific data support the independent-action (or single-organism) hypothesis (25). That a single bacterium or virus or protozoan can reproduce is a known biological phenomenon that has been proven in laboratory studies. This concept has also been suggested as providing the explanation for sporadic cases of infectious disease. Although it is clear that the host defenses (immunity at the cellular and humoral level) do play a critical role in determination of which individuals may develop infection and particularly develop more severe disease, it has also been suggested that these do not provide the complete explanation (25). In the early literature, it was suggested that many microorganisms are needed to act cooperatively to overcome host defenses to initiate infection (3). The independent-action theory, however, suggests that each microorganism alone is capable of initiating the infection, but more than one is needed, as the probability that a single microorganism will evade host defenses successfully is small (25). This is analogous to another biological phenomenon, that of spermatozoa and fertilization (25).

The evaluation of the dose–response data sets also support the independent-action hypothesis, as in almost every case the exponential or beta models provided a statistically significant improvement in fit over the lognormal model, which could be used to predict a threshold (11). Currently, there are no scientific data to support a threshold level for these microorganisms.

It should be kept in mind that healthy human volunteers were used in most of the dose studies, all with normal-functioning immune systems. There was no attempt to screen out those who had antibodies and were previously exposed, except in the case of the *Cryptosporidium* study (8,20). Thus the argument that the immune system would influence these models suggests that the models could actually be less conservative and underestimate the risks associated with the sensitive or vulnerable populations.

Although the human data sets are extensive, they are not exhaustive in terms of answering many of the questions regarding the host–microbe interaction. In the future, more human and animal studies will be needed to further address both hazard and dose–response, including virulence, strain variation and immunity, and multiple exposures.

Exposure Assessment

Exposure assessment is an attempt to determine the size and nature of the population exposed and the route, concentrations, and distribution of the microorganisms and the duration of the exposure. The description of exposure includes not only occurrence based on concentrations but the prevalence (how often the microorganisms are found) or distribution of microorganisms in space and over time. Exposure assessment depends on adequate methods for recovery, detection, quantification, sensitivity, specificity, virulence, and viability, as well as studies and models addressing transport and fate through the environment. For many microorganisms, the methods, studies, and models are not available (see Chapter 5). Often, the concentration in the medium associated with the direct exposure (e.g., drinking water, food) is not known, but must be estimated from other databases. Therefore, knowledge on the ecology of these microorganisms, sources in the environment, transport, and fate are needed, including inactivation rates and survival in the environment, ability to regrow as in the case of some bacteria and resistance to environmental factors (temperatures, humidity, sunlight, etc.), and movement through soil, air and water. Finally, because the current methods for monitoring microorganisms in environmental samples often do not afford the necessary sensitivity to examine treated water or food in most cases to the levels desirable, a greater database is needed on the inactivation/removal of microorganisms through treatment processes. These data can be used to estimate levels in the final treated product.

Risk Characterization

Risk characterization is an integration of the three steps to estimate the magnitude of the public health problem, understand the variability, and uncertainty of the hazard. This definition encompasses essentially four distributions:

1. The spectrum of health outcomes
2. The confidence limits surrounding the dose–response model
3. The distribution of the occurrence of the microorganism
4. The exposure distribution

The occurrence and exposure can be further delineated by distributions surrounding the method recovery and survival (treatment) distributions. It may be possible to group microorganisms rather than assess each individually if all four aspects of the risk characterization are similar. In addition, parts of the risk assessment (health outcomes and dose–response models) may have applicability to many transmission routes and different exposures (shellfish, recreational waters, and drinking waters). The assumptions made at each level influence the outcome and variability of the risk outcome. For example, one may assume that all individuals are equally susceptible, therefore underestimating the risk to immunocompromised individuals in regard to the health outcome for certain pathogens. Where dose–response models are not available for new and emerging pathogens such as *E. coli* O157:H7, because of relatedness, the *Shigella* model may be assumed to be useful for predictions of risks.

The testing of risk characterization or risk assessment plausibility may be evaluated against key epidemiological studies where the microbial hazard, transmission, concentration, and outcome have been established. This opportunity usually occurs during investigations of foodborne or waterborne outbreaks. Thus assumptions and extrapolations can be evaluated and the data that exhibit the greatest variability or uncertainty to the model can be explored.

The marriage of risk assessment methods with epidemiological models that describe the transmission of disease through a population has been suggested as an approach for examining population risks compared to the individual's risk (9). These take into account effects such as incubation times (time from exposure to infection and illness), immunity (protective as well as impaired), and secondary spread in a population, in addition to exposure and dose–response. However, the parameters associated with the mathematical permutation of these models can require as many as 13 model parameters and 22 different pieces of data, most of which are not well known. Therefore, many assumptions are made and the complexities of the models may make them more difficult to evaluate in terms of plausibility and validation.

Risk assessment methodologies are now being accepted for application to microbial contaminants that cause human disease through environmental exposure. It is an approach that attempts to address the uncertainty of environmental contamination, generally at low levels of exposure and the potential resulting health effects. Although the assessments are made through the use of assumptions, resulting in quantitations with a large range of variation and uncertainty, the method is useful for ranking the risks, comparing different environmental problems and different solutions. This is an unusual approach for evaluating risks associated with microorganisms that cause infectious disease, which has relied in the past on outbreak data (epidemics) for assessing human health outcomes and indicator microorganisms or surrogates for assessing the potential for contamination. For the first time, monitoring data that show low levels of pathogenic microorganisms in water, food, air, surfaces, or soil can be interpreted. The role of the environmental contamination contributing to endemic disease as well as epidemic disease can now be studied using this approach.

As stated by Cothern (6): "One of the main impediments to the development and use of quantitative risk assessment is the lack of complete information and data as input to this process." Although it has been possible to evaluate and compile a comprehensive database on microbial dose–response models, there is a limited database on occurrence and this impedes most of the risk characterizations.

DEVELOPMENT OF A MICROBIAL RISK ASSESSMENT FRAMEWORK AND PROCESS

As with other fields using risk assessment, the integration of risk management and risk assessment is seen as a necessary requirement in the development of a workable framework (Figure 3-3). Regulatory agencies are now attempting to develop the best approach for undertaking and using microbial risk assessment for policies that will improve water quality, food safety, and public health. The U.S. EPA first used risk assessment based on dose–response models for the development of the *surface water treatment rule* for *Giardia* (31). Teunis et al. (28) with the National Institute of Public Health and Environmental Protection in the Netherlands have also used formal risk assessment procedures for microorganisms. The *food safety initiative* developed in the United States in 1995 spurred on a great deal of microbial risk assessment

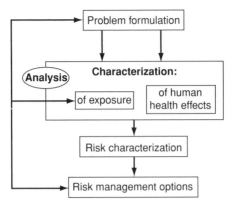

Figure 3-3 Framework for integration of risk management and microbial risk assessment.

work. Risk management involves the industry, state, local as well as federal agencies, and it will be important that the microbial risk assessment process be better understood at all levels so that it will become an effective tool for risk managers.

The EPA established a national committee in the United States in 1995 to assist in development of a framework for pathogen risk assessment. Thus an approach for how to go about a risk assessment, the type of data needed, and the tools available were prepared by a multidisciplinary group of scientists representing epidemiology, medicine, microbiology, water treatment, food safety, chemical risk assessment, and public policy (14). As was suggested in the ecological risk assessment, there was a need to integrate into the process, management, site specificity regarding environmental transmission, and risk communication. Therefore, the initial step in the process was developed as the problem formulation. The data needs of the managers for making a decision and types and priorities of risks to be addressed are to be articulated in this step. The risk assessment itself is defined by an analysis phase involving the exposure characterization and health characterization.

As mentioned previously the analysis phase of the risk assessment includes four distributions (Figure 3-4): human *health effects* (symptomatic and asymptomatic infection; severity, duration, hospitalization, medical care; mortality; host immune status; susceptible populations), *dose–response modeling, exposure analysis* (vehicle, amount, route, single versus multiple, demographics of those exposed), and *occurrence assessment* (methods, concentrations, frequency, spatial and temporal variation, regrowth, die-off, and transport).

Example 3-2 Microbial Risk Assessment for Determining Data Needs for Improving the Safety of Drinking Water. Since the large waterborne outbreak of cryptosporidiosis in 1993 in Milwaukee and reauthorization of the

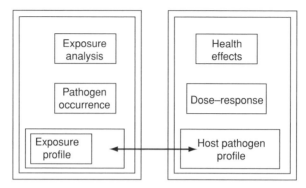

Figure 3-4   Analysis phase for microbial risk assessment. (From Ref. 14.)

Safe Drinking Water Act, in 1996, there has been a new focus on emerging microorganisms that may pose an increased risk to the safety of U.S. drinking water supplies (15). This has been further highlighted worldwide with the introduction and explosive spread of old diseases such as cholera in South and Central America (7). Most fecal–oral microorganisms have the potential to be spread through contaminated water. These agents come from animal and human wastes. Other microorganisms such as *Legionella* and *Mycobacterium avium complex,* seem to be naturally occurring and can grow in the water distribution system. The balancing of chemical risks disinfection, and the potential cancer risks from the disinfection by-products formed with the microbial risks by which the disinfection is a major management option for control, is a worldwide issue in drinking water and cannot be solved without appropriate risk assessment. One of the major areas where microbial risk assessment can be used is the identification of needed databases (Tables 3-5 and 3-6). The approach can be used to analyze existing and new data in a systematic fashion that will aid in the prioritization of the various microbial contaminants and management strategies. Separate from dose–response modeling, the risk and data needs can be divided in two areas: Exposure and Health Outcome.

The focus of risk assessment for waterborne disease is now on defining the specifics of the watershed and community. This means that cities, counties, and states will need to engage scientific teams that can implement risk assessments in a collaborative fashion with the medical, public health, and water utilities, as well as with agencies such as the U.S. Geological Survey, and environmental departments.

Although water treatment such as chlorination may readily control some microbial risks, such as *Shigella* or *Campylobacter,* the reliability of treatment must be included in any risk assessment. For example, if disinfection failure continues to occur in a percentage of facilities that use highly polluted water

TABLE 3-5 Some Health Factors Involved in Assessment of Microbial Risks Associated with Drinking Water[a]

Health Effect	Data Needs
Evaluation of waterborne outbreaks	Magnitude of community impact, attack rates, hospitalization and mortality, demographics, sensitive populations, level of contamination, duration, medical costs, community costs, course of immune response and secondary transmission
Evaluation of endemic disease	Incidence, prevalence, geographic distribution, temporal distribution, percentage associated with various transmission routes (e.g., water versus food), demographics, sensitive populations, hospitalization, individual medical costs, antibody prevalence, infection rates, illness rates
Immune status	Protection versus issues for sensitive populations, lifetime protection versus temporary, age (e.g., elderly) versus disease (e.g., AIDs) impacts
Description of microbial pathogens	Mechanism of pathogenicity (how it causes disease), virulence factors, virulence genes, antibiotic resistance
Disease description	Types of diseases, duration, severity, medical treatment and costs, time loss, chronic sequelae, contributing risks [e.g., pregnancy, nutritional status, lifestyle (e.g., smoking, *Legionella*), immune status]
Methods for diagnosis	Availability, whether in routine use or requiring special requests, ease in use, cost, time

[a]Clinical diagnostic tests must be available before other databases can be adequately established.

supplies, what type of risk will the population have? It is critical that occurrence databases be developed for the microorganisms that may exhibit a high level of virulence and contribute to a greater severity in the population exposed, despite the ease in control because of treatment reliability issues.

In regard to indicators of water quality, water treatment, and microbial safety, this is an issue of some scientific uncertainty. It is clear the coliform bacteria that have been used for over 100 years to evaluate the biological quality and safety of water cannot be relied upon for viruses, parasites, and bacteria in biofilms. This has already been acknowledged in the development of the surface water treatment rule (31). Thus while surrogates for treatment reliability may play a role in microbial risk assessment, it is doubtful that these types of data will be useful for predictions of exposure.

The use of decision trees may be useful in regard to the gathering of critical data to finalize the exposure even without full characterization of health ef-

TABLE 3-6   Some Exposure Factors Involved in Assessment of Microbial Risks Associated with Drinking Water[a]

Exposure Factor	Data Needs
Transmission	Define fecal–oral, respiratory, contact, or multiple routes
Environmental source	Levels found in human waste, animal waste, sediments, and biofilms; potential loading to a water system (identification of animal reservoirs, sewage treatment facilities, stormwater discharges, septic tanks, etc.)
Survival potential	Inactivation in waste, soil, groundwater, surface water sediments, and biofilms; effects of temperature, sunlight, and desiccation
Regrowth potential	Growth in waste, soil, groundwater, surface water, sediments, and biofilms; effects of temperature and nutrients
Occurrence in raw water supplies	Raw water type and level of contamination, temporal and spatial occurrences
Resistance to treatment	Reduction by waste treatment, drinking water treatment, and distribution treatment; resistance to disinfection, removal by filtration, etc; adequacy of surrogates (coliform bacteria, turbidity) to evaluate removal
Environmental transport	In storm events, in solids, in aerosols, to groundwater, in distribution systems
Availability of methods	Methods for assessing source water, identification of environmental sources, quantification, viability, assessing treated water

[a]Analytical methods must be available before other databases can be developed.

fects. These can be used for prioritization and eventually, management decisions in regard to reducing the potential for exposing the population to microorganisms through drinking water.

Example 3-3   Use of Microbial Risk Assessment in Food Safety and the HACCP System. The *hazard analysis critical control point* (HACCP) system is aimed at specific operations whose ultimate goal is to ensure food safety. The physical, chemical, or biological hazard is defined as well as the specification of the control criteria. The point where the hazard can be controlled to acceptable levels or eliminated is then defined and is known as the *critical control point* (CCP) (Table 3-7). In theory the CCP must be a point where the process or operation can be monitored to meet the performance specified to meet the level of hazard reduction, but in practice very little has been done to implement this approach.

HACCP is food specific; therefore, the hazards and CCP may be different for beef, shellfish, or produce. Often, the goal would be to establish a number of CCPs representing the entire food chain from the farm to the table, with multiple barriers for protection. HACCP is a management strategy but has

TABLE 3-7 Examples of Hazards and CCPs for Specific Foods

Food	Hazard(s)	Sources of Hazards	Control Points
Shellfish	*Vibrio vulnificus* and other species, enteric viruses (e.g., hepatitis A and Norwalk virus)	*Vibrio* is naturally occurring growth related to temperature and salinity; viruses associated with human wastewater contamination of shellfish harvesting waters	Protection of shellfish waters from wastewater: depuration[a] cooking[b]
Beef	*E. coli* O157:H7	Infection in herds, food preparers	Reduce infection in herds; reduce fecal contamination during slaughter and processing; store at proper temperatures to avoid regrowth; cooking.
Produce	*Cyclospora*	Human feces	Protection of irrigation waters; increase hygiene for field and packing workers; maintenance of temperatures that deter maturation of oocyst[c]

[a] Depuration is a process whereby the contaminated shellfish are allowed to filter with clean water and eliminate the microorganism.
[b] Virus outbreaks have been associated with cooked (baked, steamed, and fried) shellfish, as these viruses are thermotolerant.
[c] The oocysts are excreted in feces and mature in the environment at optimal temperatures between 28 and 35°C.

elements of the risk assessment process inherent in it. This includes part of a hazard identification and exposure assessment. The system is largely non-quantitative and does not incorporate the dose–response or the risk characterization. It can be argued that better management decisions regarding the controls, performance standards, and monitoring can be made if a more thorough risk assessment is undertaken. When one thinks of the food supply, there are several elements in the chain (farm to table). This includes production, harvest, processing, transport, packaging, storage, and wholesale and retail marketing. Finally, there is preparation (which may take place in the home or restaurants). Several examples can be used to address the risk assessment process and the relationship to HACCP.

Once the hazards to be assessed are determined, clinical data can be used to address the hazard identification step. Both foodborne outbreaks and endemic disease should be analyzed. Some hazards may have been associated

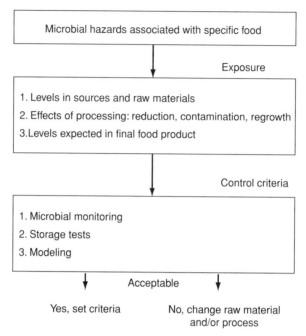

Figure 3-5 Use of risk assessment for setting criteria in HACCP. (From Ref. 19.)

with multiple food items, and others may be associated predominantly with one type of food. Dose–response models can be used, but consideration may need to be given to synergistic effects (e.g., foods with buffering capacity may enhance the infectivity and change the dose–response curve). Exposure assessment may need models that address the potential for bioaccumulation in shellfish, for example, or regrowth of bacteria associated with temperature abuse. Data are needed in the exposure assessment to determine the average size of a serving and the numbers of servings per person per some time frame (e.g., per year). Monitoring of foods to define the exposure can be difficult. Therefore, the source of microorganisms may be a better target for monitoring; followed by modeling based on various environmental conditions that can be defined (Figure 3-5).

The acceptable level of risk per serving at the table needs to be defined. If zero tolerance of the hazard is to be promoted, then only through risk assessment can the costs and benefits be evaluated adequately. Better methods associated with the specific hazard and food will be needed to monitor CCPs in the future to fully implement HACCP.

REFERENCES

1. Aldrich T., and J. Griffith, eds. 1993. Environmental epidemiology and risk assessment. Van Nostrand Reinhold, New York.

2. AWWARF. 1992.

3. Blaser, M. J., and L. S. Newman. 1982. A review of human salmonellosis. I. Infective dose. Rev. Infect. Dis. 4:1096–1106.

4. Burkholder, J. M., and H. B. Glasgow, Jr., 1997. *Pfiesteria piscida* and other *Pfiesteria-like* dinoflagellates: behavior, impacts and environmental control. Limnol. Oceanogr. 42(5, p. 2):1052–1075.

5. Cooper, R. C., A. W. Olivieri, R. E. Danielson, P. G. Badger, R. C. Spear, and S. Selvin. Evaluation of military field-water quality. Vol. 5. Infectious organisms of military concern associated with consumption: assessment of health risks, and recommendation for establishing related standards. Lawrence Livermore National Laboratory UCRL-21008, Vol. 5. U.S. Army Medical Research and Development Command, Fort Detrick, Frederick, MD.

6. Cothern, C. R., ed. 1992. Comparative environmental risk assessment. Lewis Publishers, Boca Raton, FL.

7. Craun, G. F., ed. 1996. Water quality in Latin America. International Life Sciences Institute Press, Washington, DC.

8. Dupont, H., C. Chappell, C. Sterling, P. Okhuysen, J. Rose, and W. Jakubowski. 1995. Infectivity of *Cryptosporidium parvum* in healthy volunteers. New Engl. J. Med. 332:855, 1.

9. Eisenberg, J. N., E. Y. W. Seto, A. W. Olivieri, and R. C. Spear. 1996. Quantifying water pathogen risk in an epidemiological framework. Risk Anal. 16:549–563.

10. Gerba, C. P., and C. N. Haas. 1988. Assessment of risks associated with enteric viruses in contaminated drinking water. ASTM Spec. Tech. Publ. 976:489–494.

11. Haas, C. N. 1983. Estimation of risk due to low doses of microorganisms: a comparison of alternative methodologies. Am. J. Epidemiol. 118:573–582.

12. Haas, C. N, J. B. Rose, C. Gerba, and S. Regli. 1993. Risk assessment of virus in drinking water. Risk Anal. 13:545–552.

13. Hoppin, J. 1993. Risk assessment in the federal government: questions and answers. Center for Risk Analysis, Harvard School of Public Health, Boston.

14. International Life Sciences Institute. 1996. A conceptual framework to assess the risks of human disease following exposure to pathogens. Risk Anal. 16:841–848.

15. Mac Kenzie, W. R., N. J. Hoxie, M. E. Proctor, S. Gradus, K. A. Blair, D. E. Peterson, J. J. Kazmierczak, K. Fox, D. G. Addiss, J. B. Rose, and J. P. Davis. 1994. Massive waterborne outbreak of *Cryptosporidium* infection associated with a filtered public water supply, Milwaukee, Wisconsin, March and April, 1993. N. Engl. J. 331(3):161–167.

16. National Academy of Sciences. 1977. Drinking water and health. Safe Drinking Water Committee. National Academy Press, Washington, DC.

17. National Academy of Sciences, 1983. Risk assessment in the federal government: managing the process. National Academy Press, Washington, DC.

18. National Research Council. 1993. Managing wastewater in coastal urban areas. National Academy Press, Washington, DC.

19. Notermans, S., and G. C. Mead. 1996. Incorporation of elements of quantitative risk analysis in the HACCP system. Int. J. Food Microbiol. 30:157–185

20. Okhuysen, P. C., C. L. Chappell, C. R. Sterling, W. Jakubowski, and H. L. Dupont. 1998. Susceptibility and serological response of healthy adults to reinfection with *Cryptosporidium parvum.* Infect. Immun. 66:441–443.

21. Payment, P., L. Richardson, J. Siemiatycki, R. Dewar, M. Edwardes, and E. Franco. 1991. A randomized trial to evaluate the risk of gastrointestinal disease due to consumption of drinking water meeting current microbiological standards. Am. J. Public Health 81:703–708.

22. Regli, S., J. B. Rose, C. N. Haas, and C. P. Gerba. 1991. Modeling the risk from *Giardia* and viruses in drinking water. J. Am. Water Works Assoc. 83:76–84.

23. Rose, J. B., and C. P. Gerba. 1991. Use of risk assessment for development of microbial standards. Water Sci. Technol. 24:29–34.

24. Rose, J., C. N. Haas, and S. Regli. 1991. Risk assessment and control of waterborne giardiasis. Am. J. Public Health 1:709–713.

25. Rubin, L. G. 1987. Bacterial colonization and infection resulting from multiplication of a single organism. Rev. Infect. Dis. 9(3):488–493.

26. Sobsey, M. D., A. P. Dufour, C. P. Gerba, M. W. LeChevallier, and P. Payment. 1993. Using a conceptual framework for assessing risks to health from microbes in drinking water. J. Am. Water Works Assoc. 85:44–48.

27. Suter, G. W., II. 1993. Predictive risk assessments of chemicals, pp. 49–88. *In* G. W. Suter II, ed., *Ecological risk assessment.* Lewis Publishers, Boca Raton, FL.

28. Teunis, P. F. M., A. H. Havelaar, and G. J. Medema. 1994. A literature survey on the assessment of microbiological risk for drinking water. Report 734301006. National Institute of Public Health and Environmental Protection, Bilthoven, The Netherlands.

29. U.S. Environmental Protection Agency. 1989. Guidance manual for compliance with filtration and disinfection requirements for public water systems using surface water sources. EPA Report 570/9-89/018. U.S. EPA, Washington, DC.

30. U.S. Environmental Protection Agency. 1992. Framework for Ecological Risk Assessment. EPA Report 630/R-92/001. U.S. EPA, Washington DC.

31. U.S. Environmental Protection Agency. 1989. National primary drinking water regulations; filtration and disinfection; turbidity; *Giardia lamblia,* viruses, *Legionella,* and heterotrophic bacteria. Fed. Reg. 54(124):27486–27541.

CHAPTER 4

CONDUCTING THE HAZARD IDENTIFICATION

The understanding and description of the human health hazards associated with microorganisms has primarily been part of three fields of study. Within the medical field this is the specialty in infectious diseases. Within public health, the field of epidemiology has focused on specific transmission routes, such as foodborne, vectorborne, and waterborne microbial agents. Finally, the field of clinical microbiology has examined the nature of the infectious microorganism. The term *infectious* is an adjective that comes from the word *infect,* meaning to contaminate with disease-producing substance, germs, or bacteria and describes those agents that are able to reproduce in association with the disease process and therefore to be transmitted to others. The major groups of infectious microorganisms are described in Chapter 2 and include bacteria, fungi, prions, protozoa, and viruses. Algae and dinoflagellates are also groups of microorganisms that naturally occur in water; however, the resulting human health hazards are associated with exposure to toxins produced by these microorganisms and not through an infectious process. This is similar to some foodborne bacteria that after growth produce a toxin in the food that results in disease (e.g., *Clostridium*). The elements of the hazard identification of microorganisms are descriptive, mechanistic, and in some cases quantitative and involve assessment of the microorganisms, the disease process, and disease surveillance.

The steps in microbial hazard identification are as follows:

1. Identification of the microorganism as a cause of human illness associated with proof using Koch's postulates, which demonstrate that the agent is found and is the cause of specific types of disease and when transmitted causes a similar disease in the person newly exposed.

2. Development of diagnostic tools that identify the symptoms, the infection, and more specifically, the microorganism in host specimens (e.g., sputum, stools, blood).
3. Understanding of the disease process from exposure (e.g., respiratory) to infection (colonization of the human body) to development of pathology, disease, and death.
4. Identification of possible transmission routes.
5. Assessment of virulence factors and components of the microorganism and its life cycle that aid in understanding transmission and the disease process.
6. Use of the diagnostic tools to evaluate the incidence and prevalence of disease in populations (endemic risks) and for investigation of outbreaks (epidemic risks).
7. Development of models (usually, animal models) to study the disease process and approaches for treatment.
8. Evaluation of the role of the host immune system in combating the infection and the possible development of vaccines for prevention.
9. Epidemiological studies associated with various exposures.

IDENTIFYING AND DIAGNOSING INFECTIOUS DISEASE

The identification of disease (or illness) is made by one of several methods (Table 4-1). The difference between disease and illness may be minor in some cases, but from a medical viewpoint and for hazard identification, these do not necessarily mean the same thing. *Disease* is defined as the process or mechanism that ultimately results in an illness or a condition that impairs vital functions. A person could have a disease without initially having any symptoms. *Symptoms* are a state where the effects of the illness (e.g., headache, diarrhea, stomach cramps, vomiting) can be described by the person who is ill. *Clinical assessment of the illness* is generally defined by a measurable description of the illness (e.g., fever, bloody stool). Infection is colonization of the microorganism in the body and may result in disease and symptoms, as this is the initial step in the microbial disease process. However, this may also result in *asymptomatic* or *subclinical* infections. Symptoms and clinical descriptions (e.g., fever, rash, inflammation) can be very specific, such as with measles, which is associated with a specific agent, or they can be generic, such as with diarrhea, which is associated with many different types of microorganisms.

The second means of identification is *clinical diagnosis* by detection of the specific microorganism in a host specimen (laboratory identification in a liquid stool of an enteric pathogen). This requires the collection of a specimen (sputum, feces, blood, biopsy) and a specific diagnostic test (specific growth,

TABLE 4-1 Methods for Diagnosing Infections and Disease

Method	Approach	Advantages/Disadvantages
Symptoms and clinical descriptors	Based on person's feelings (headache) and measurable impacts (fever, rash)	Can easily diagnose or identify those infected; however, is not generally agent specific but more generic (e.g., diarrhea).
Clinical diagnosis	Based on testing specimens (sputum, feces, or blood) for presence of the agent[a]	Can specifically identify agent; however, patient must deliver a specimen and there must exist a test method for the agent
Antibody response	An indirect test (blood or in some cases, saliva) for the presence of antibodies that the body produces as a result of infection[b]	Is specific to the agent; however, may not be able to determine the timing of the exposure and infection

[a] Asymptomatic infections can be detected.
[b] Antibody response may or may not be protective from subsequent exposure and infection and does not usually occur without infection.

biochemical tests, stains, genetic or protein markers, microscopic identification). This also means that there is some understanding of the agents that may be responsible for the disease symptoms and the process of disease, resulting in the infection of specific cells and/or organs in the body affected. Infection without the person reporting symptoms (an asymptomatic infection) can be detected in this manner.

The final method is associated with the response of the host system to infection, which illicits an *antibody response* that can be detected in blood or in some cases, saliva. This antibody response may be associated with past or current exposure. In some cases, depending on the type of antibody and amount, one can determine the approximate timing of the exposure and infection. Exposure without infection rarely causes an antibody response except in the case of repeated exposure to very high concentrations of the agent such as occurs with some vaccinations.

The first method, symptomology (headache, stomach ache, diarrhea), has been used in numerous studies for assessing disease in a population. Example 4-1 demonstrates the use of this approach for identification of the microbial hazards associated with drinking water.

Example 4-1   Study Examining Health Impacts from Drinking Water (35)
The study: An intervention study was undertaken that examined the health of populations drinking tap water compared to a similar population that drank water that was further treated through point-of-use-device reverse osmosis filters.

Health effects: One member of each family was responsible for recording and reporting disease for all members (nausea, vomiting, diarrhea, fever, cramps, muscular pain, cold, flu, sore throat, absence from work or school, visit to a doctor, hospitalization).

Episode definition: One or more days of symptoms (vomiting or liquid diarrhea; or nausea or soft diarrhea combined with abdominal cramps) with at least six consecutive symptom-free days between episodes.

Results: Annual incidence of episodes per person was 0.76 in those drinking tap water and 0.5 in those drinking tap water that was further treated, so it was estimated that 35% of reported gastrointestinal illness was tap water related.

Advantages and disadvantages: Identified the transmission route and some health endpoints that are useful for cost–benefit analysis; however, specific microbial hazards were not identified, chronic health outcomes with more rare occurrences (e.g., viral myocarditis) were not identified, and risk management options were not addressed.

It is generally acknowledged that the identification of the disease burden and the specific microbial hazard in populations are underestimated. This is due primarily to the lack of diagnostic tests and the use of those tests to screen populations.

Example 4-2   Underdiagnosis of **Cryptosporidium** *as a Cause of Disease in Populations*

- There was no recognition of human disease until 1976; intestinal biopsies used for diagnosis, even though the parasite was first described in 1895.
- Diagnostic tests developed in early 1980s, testing feces for oocysts showed 0.5 to 20% prevalence of disease throughout the world. Antibody tests showed 30 to 50% prevalence.
- Most people do not go to a physician or submit an adequate specimen to the laboratory for diagnosis.
- Test for *Cryptosporidium* is not run routinely in the parasite laboratory except by physician's request.
- Not reportable to most state or national statistical databases until the 1990s.

HEALTH OUTCOMES ASSOCIATED WITH MICROBIAL INFECTIONS

After exposure to a microorganism, once infection begins (defined by dose–response or attack rates; see the section "Epidemiological Methods for Undertaking Hazard Identification" and Chapter 5), there are a number of possible outcomes, including asymptomatic illness, various levels of acute and chronic disease (mild illness, to more severe illness, to chronic problems, to that which requires hospitalization) and potentially, death (known as *mortality*). In particular, there has been inadequate recognition and documentation of the chronic diseases and long-term sequelae associated with microorganisms such as degenerative heart disease and insulin-dependent diabetes caused by Coxsackievirus B infections and peptic ulcers and stomach cancer caused by the bacterium *Helicobacter pylori.*

Figure 4-1 demonstrates various outcomes. It has been difficult to predict, based on the current health databases, the quantitative probability of each possible outcome, as it may be microorganism specific, even isolate specific, and can depend on the host status. The goal of hazard identification, however, is to define these outcomes to the extent possible. Each outcome can be described as a ratio or percentage; the numerator and denominator need to be adequately defined as well as the populations associated with the data. Table 4-2 is an example of the level of symptomatic infections for *Salmonella* infections. The term used to address this outcome is *morbidity.* In a total of 12 studies, conducted primarily during outbreak investigations, the percentage of people who exhibited symptoms divided by the total numbers infected (in the case of *Salmonella,* this was measured by detection of the bacterium in fecal specimens, indicating infection) averaged 41%, with a range of 6 to 80%, or stated conversely, an average of 59% were asymptomatic (ranging from 20 to 94%). These data do not indicate the type of disease or the severity, and

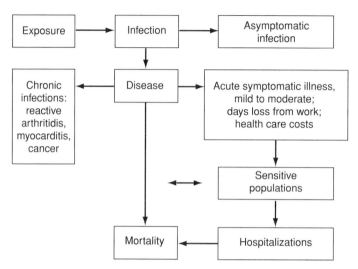

Figure 4-1 Outcomes of the infection process for quantification.

often other databases are needed to address those aspects of the health outcome.

While acute outcomes have been described for most microbial infections, the chronic impacts may be more significant but are rarer (in terms of cases) and generally, not as quantifiable. Table 4-3 lists some of the bacteria, protozoa, and viruses for which both acute and chronic effects have been documented. Some of the chronic impacts are more severe and particularly affect the more sensitive populations. For example, enteropathogenic *E. coli* causes

TABLE 4-2 Morbidity Ratios for *Salmonella* (Nontyphi).

Study	Population/Situation	Morbidity Percentage
1	Children/food handlers	50
2	Restaurant outbreak	55
3	College residence outbreak	69
4	Nursing home employees	7
5	Hospital dietary personnel	8
6	Hospital dietary personnel	6
7	Nosocomial outbreak	27
8	Summer camp outbreak	80
9	Nursing home outbreak	23
10	Nosocomial outbreak	43
11	Foodborne outbreak	54
12	Foodborne outbreak	66
Average		41

Source: Ref. 6.

TABLE 4-3   Acute and Chronic Outcomes Associated with Microbial Infections

Microorganism	Acute Disease Outcomes	Chronic Disease Outcomes
Campylobacter	Diarrhea	Guillain–Barré syndrome
E. coli O15:H7	Diarrhea	Hemolytic uremic syndrome
Helicobacter	Gastritis	Ulcers and stomach cancer
Salmonella, Shigella, and *Yersinia*	Diarrhea	Reactive arthritis
Coxsackievirus B	Encephalitis, aseptic meningitis, diarrhea, respiratory disease	Myocarditis, reactive insulin-dependent diabetes
Giardia	Diarrhea	Failure to thrive, lactose intolerance, chronic joint pain
Toxoplama	Newborn syndrome, hearing and visual loss	Mental retardation, dementia, seizures

mild illness in adults (1 to 2 days of mild diarrhea), while *E. coli* O157:H7 causes hemorrhagic uremia, with death more likely in children and the elderly. Figure 4-2 quantifies some of the health outcomes documented during the outbreak associated with contaminated hamburgers in the Northwest in 1993 (4). *Salmonella, Shigella,* and *Yersinia* have all been shown to initiate a reactive arthritis in about 2.3% of those infected with these bacteria, whether

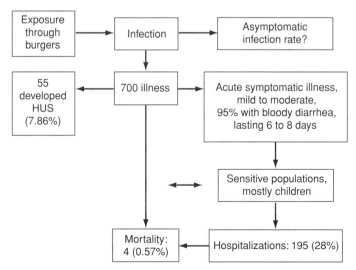

Figure 4-2   Health outcomes associated with *E. coli* O15:H7 infections during a foodborne outbreak.

symptomatic or not. *Toxoplama gondii* is associated with very little disease in adults but causes congenital malformations. It has been estimated that there is a 45% chance (30 to 60% range) of maternal-to-fetal transmission of the infection if the mother is seropositive. The outcome, while normal for 55% of the newborns, is associated with 2% mortality, 11% retardation, and 6% blindness (38). Guillain–Barré syndrome (GBS) is a major cause of neuromuscular paralysis in the United States, causing an estimated 2628 to 9575 cases each year, between 20 and 40% caused by infections with *Campylobacter* (5). Figure 4-3 shows some of the health outcomes associated with *Campylobacter*-associated GBS that have been estimated.

Hospitalization ratios during waterborne disease outbreaks have been shown to be highly dependent on the etiologic agent (Table 4-4 and Figure 4-4). The highest percentages of hospitalizations during waterborne outbreaks in the United States were associated with hepatitis A virus and bacteria such as *Shigella, Salmonella,* and *E. coli.* Levels of 1.0% were found for *Cryptosporidium;* in the outbreak in Texas, the ratio was 0.85%. In the Carrollton, Georgia, outbreak, although no hospitalizations were documented in the literature, the numbers of visits to the hospital emergency room (ER) increased five- to sixfold, giving approximately an 0.8% ratio (ER visits/total cases) (17). In Milwaukee, Wisconsin, 4000 hospitalizations were estimated (32), less than 1% for *Giardia* and Norwalk-like enteric viruses. The undetermined etiologic agents responsible for AGI also resulted in hospitalizations of <1%; however, a large number of cases (874) felt that their illness was serious enough to warrant treatment in the emergency room. Almost four times the

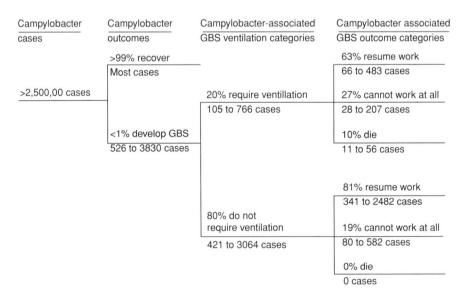

Figure 4-3   Health outcomes associated with *Campylobacter*-associated Guillain–Barré syndrome.

TABLE 4-4 Hospitalizations Associated with Microorganisms Responsible for Waterborne Outbreaks Reported in the United States, 1971–1992

Microorganism	Waterborne Outbreak Data		Hospitalization Data During the Outbreaks		
	Numbers of Outbreaks	Number of Cases	Number of Outbreaks	Ratio of Cases Hospitalized to Total Cases	Percent Hospitalized
Viruses					
Hepatitis A	29	807	9	75/265	28.3
Viral gastroenteritis	30	12,699	4	10/1,154	0.9
Bacteria					
Salmonella	12	2,370	3	12/293	4.1
Shigella	54	9,967	24	339/5,768	5.9
Campylobacter jejuni	13	5,257	5	73/2,152	3.4
Yersenia entercolyitca	2	103	2	20/103	19.4
Escherichia coli[a]	3	1,323	2	41/323	12.7
Typhoid	7	293	4	235/277	84.8
Protozoan parasites					
Giardia lamblia	118	26,733	18	60/13,239	0.5
Cryptosporidium[b]	5	433,517	3	4,105/415,960	0.99
Unknown etiology					
Acute gastrointestinal illness	341	82,486	50	253/40,039	0.6

Source: Data from Refs. 10, 18, and 31.

[a]Two outbreaks O157:H7; one outbreak of O6:H16.

[b]Includes Milwaukee, Wisconsin, outbreak and emergency room visits for Carrollton, Georgia, outbreak (17,26,32).

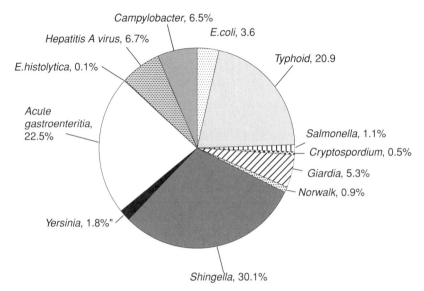

Figure 4-4   Hospitalizations during waterborne outbreaks by etiological agents, 1971–1990.

number of hospitalized cases of AGI were treated in the ER. The only other illness with significant numbers of cases that was treated in the ER was giardiasis; twice the number of hospitalized cases (60) were seen in the ER (120). While the more severe illnesses appear to be caused by bacteria and hepatitis A virus infections, *Giardia* and the etiologic agents of AGI contribute to a significant number of hospitalized cases, representing 5.3% and 22.5% of all hospitalized cases, respectively (Figure 4-4).

Mortality or death is often an endpoint that is most likely to be identified, reported, and used. However, mortality ratios may be overestimated, as the denominator (total cases) is not well defined and is underestimated. See Table 4-6 for the estimated mortality ratios for various microorganisms compared to ratios in the elderly populations. Often, the severity associated with acute outcomes as well as mortality is much more significant in sensitive populations.

SENSITIVE POPULATIONS

Select populations such as pregnant women, the elderly, infants, and the immunocompromised currently represent almost 20% of the general population (Table 4-5), and their numbers are expected to increase in the years ahead (48). The elderly account for almost 67% of the vulnerable group, with a great percentage residing in long-term care facilities. Risks of increased se-

TABLE 4-5   Sensitive Populations in the United States

Population Group	Number	Year Surveyed
Pregnancies	5,657,900	1989
Neonates	4,002,000	1989
Elderly (over 65)	29,400,000	1989
Residences in nursing homes or related care facilities	1,553,000	1986
Cancer patients (non-hospitalized)	2,411,000	1986
Transplant organ patients (1981–1989)	110,270	1981–1989
AIDS patients	142,000	1981–1990

Source: Ref. 48.

verity and mortality are greater for these populations when exposed to many of the waterborne pathogens.

Women During Pregnancy Neonates, and Young Children

Women during pregnancy may be at increased risk from waterborne agents and may also act as a source of infection for neonates. During the past decade at least 30 outbreaks of hepatitis E have been documented in 17 countries due to contaminated water (8). Although outbreaks of hepatitis E have not been reported in the United States, cases do occur among tourists returning from developing countries. Waterborne outbreaks have at times involved thousands of people. Overall, case fatality ratios have ranged from 1 to 2% during outbreaks, which is significantly higher than that for hepatitis A virus. However, for pregnant women, the ratio is generally between 10 and 20%, but can be as high as 40% (8,15). In contrast, hepatitis A does not appear to manifest itself differently in well-nourished pregnant women than in non-pregnant populations (28). However, this may not be true for poorly nourished pregnant women. Numerous reports from developing countries recount disease of great severity, often leading to fulminant hepatitis, particularly in the third trimester (28).

Viral infection during pregnancy may also result in the transmission of infection from the mother to the child in utero, during birth, or shortly thereafter. This appears to be a common mode of transmission of Coxsackie- and echoviruses (22). Neonates are uniquely susceptible to enterovirus infections. This group of viruses are capable of causing severe disease and death when infection occurs within the first 10 to 14 days of life. Acquisition of Coxsackievirus B infections early in life is the most significant risk factor leading to fatal disease. Most fatal cases caused by this virus are probably transmitted transplacentally at term (22). Among 41 documented cases of fatal infection in infants, 24 of their mothers had symptomatic illness consisting of fever, symptoms of upper respiratory tract involvement, pleurodynia, or meningitis.

Symptomatic infection occurred between 10 days antepartum and 5 days post-partum. The observed case fatality ratio for Coxsackievirus B in a New England county was almost 13%, with a morbidity ratio of 50.2 per 100,000 live births. Stillbirth late in pregnancy has been reported for the echoviruses and Coxsackievirus B (29). More recently, Coxsackievirus B has been implicated as a potential important causative agent in spontaneous abortions (11). Anomalies (urogenital system, heart defects, digestive malformation) in children born to mothers infected with Coxsackievirus B have been suggested in several studies (29).

Echoviruses can also be transmitted from the mother to the unborn child or shortly after birth with a potentially serious outcome (30). An average case fatality ratio of 3.4% was observed in 16 documented outbreaks of echovirus in newborn nurseries. In two outbreaks of Coxsackievirus B in nurseries, the infant mortality ratio from myocarditis ranged from 50 to 60% (29). Coxsackievirus and rotavirus have also been associated as a potential cause of sudden infant death syndrome in young children. An outbreak of sudden infant death syndrome associated with rotavirus infection over a 3-week period was observed in the emergency facility of a hospital in which two of five children died (53).

The Elderly

Mortality from diarrhea shows a bimodal curve, with the greatest risk of mortality among the very young and very old. A majority of diarrheal deaths that occur in the United States are among the elderly, especially those older than 74 years of age (51%), followed by adults 55 to 74 years of age (27%), and those younger than 5 years (11%) (25). Most epidemiological studies concerning a specific agent in the elderly are focused around nursing homes since the impact can be more easily observed in a confined group of people. Mortality ratios for specific enteric pathogens are 10 to 100 times greater in this group than in the general population (Table 4-6). One documented outbreak of rotavirus in a nursing home was characterized by high attack rates (66%), with few, if any, asymptomatic cases (16). While the number of days of illness was within the range observed for other age groups (1 to 5 days), the convalescence was prolonged for some people. The mortality ratio was 1%. Gordon reported a mortality ratio of 1.3% among a retirement community during a foodborne outbreak of Snow Mountain agent, a calicivirus (13). They pointed out that several of the residents sustained serious injuries from falling because of near-syncopal episodes due to dehydration from the gastroenteritis. The elderly are expected to be more prone than younger adults to such injuries because of greater illness severity. Outbreaks of Norwalk-like viruses and enteric adenoviruses have been reported in nursing homes and geriatric wards in hospitals (22,33,36,37). Although no mortality was observed during these outbreaks, higher attack rates occurred among the residents, as well as a more severe or protracted illness compared to the staff (33,37). Interestingly, no

TABLE 4-6   Mortality Ratios for Enteric Pathogens in Nursing Homes Versus General Population

Microorganism	Mortality Ratio (%) in:	
	General Population	Nursing Home Population
Camplylobacter jejuni	0.1	1.1
Escherichia coli O157:H7	0.2	11.8
Salmonella	0.1	3.8
Rotavirus	0.01	1.0
Snow Mountain agent (Norwalk-like virus)	[a]	1.3

Source: Data from Refs. 1, 13, 16, 22, 24, 33, 36, 37, and 41.

[a]Only documented deaths have been in the elderly in nursing homes.

increase in illness severity or attack rates has been observed among the elderly during outbreaks of Norwalk virus (22).

Hepatitis A virus usually causes a mild and often asymptomatic infection in children. However, in adults the illness typically produces clinical illness that can lead to death (23). Waterborne outbreaks of hepatitis A are often characterized by high attack rates with all or most of the infected persons exhibiting clinical illness (2). The case/fatality ratio of hepatitis A increases significantly with age. In a recent 8-year review of hepatitis A cases from England, Wales, and Ireland, the case/fatality ratio of hepatitis A for patients less than 55 years of age was 0.02 to 0.03%, 0.9% at 55 to 64 years of age, and 1.5% for older patients. The median age of those dying from hepatitis A is over 60 in the United Kingdom (14).

The elderly also experience higher mortality from enteric bacterial gastro-enteritis (Table 4-6). The overall case/fatality ratio for foodborne outbreaks in nursing homes in the period 1975 to 1987 was 1.0%, compared to 0.1% for outbreaks at other locations (34). For domestically acquired cases of typhoid, the case/fatality ratio is higher among those 55 years or older (41). In a developing country, at highest risk of complications and death were children from birth through 1 year of age and adults greater than 31 years of age (41).

The Immunocompromised

The impact of the acquired immunodeficiency syndrome (AIDS) epidemic has increased the number of diarrheal deaths in the 25- to 54-year-old age group (25), and enteric diseases are among the most common and devastating problems. The majority of AIDS patients (50 to 90%) suffer from chronic diarrheal illnesses, that can be fatal (21). Many studies have shown that the rates of diarrheal disease among HIV-infected persons in developing countries are higher than rates in developed countries and may reflect more frequent exposure to enteric pathogens by contaminated food and water (3). Adenoviruses and rotavirus are the most common enteric viruses isolated in the stools of AIDS-infected persons (21). A comprehensive study of Australian men showed that 54% of diarrheal illnesses in AIDS patients were caused by viruses and that 37% of the viral diarrheas were adenovirus related (11). Overall, it is estimated that 12% of AIDS patients with clinical disease suffer from adenovirus infections and that 45% of these cases result in death within 2 months (19). The other enteric viruses do not appear to be a significant problem in AIDS-associated gastroenteritis. Enteric bacterial infections are more severe in AIDS patients. For example, patients with *Salmonella, Shigella,* and *Campylobacter* often develop bacteremia (12).

Although patients with AIDS may not have more severe illness when infected with *Giardia,* they do exhibit impaired immune response to the parasite (27). *Cryptosporidium* is a serious problem among AIDS patients. A severe and protracted diarrhea results, with fluid losses of several liters per day in some cases. Symptoms may persist for months, resulting in severe weight loss and mortality. *Cryptosporidium* may cause 7 to 38% of the diarrhea in immunocompromised patients (43). Currently, there is no treatment for the disease. Mortality ratios for *Giardia* are low and do not seem to increase in vulnerable populations. Waterborne outbreaks of *Cryptosporidium* in the United Kingdom increased the incidence of disease in the AIDS population, with severe consequences (7). In the two years following the waterborne outbreak in Milwaukee, there were 54 cryptosporidiosis-associated deaths, compared to only 4 deaths in the two years prior to the outbreak (20).

Cryptosporidium may account for 16% of the cases of diarrhea in AIDS patients (and as much as 50% in the developing world), with 87% associated with chronic illness related to CD4 counts ($<180 \times 10^6$ L^{-1}). Three waterborne outbreaks associated with drinking water have found those with AIDS to be at grave risk (Table 4-7). Community-wide exposure did not increase the attack rates in the AIDS patients; however, the outcome of the disease was severe, with 52 to 68% mortality within 6 months to a year after the outbreaks. During the Milwaukee outbreak, in a cohort of 73 AIDS patients (33 with *Cryptosporidium*), morbidity was also much more severe, with 400 of 444 hospital days logged in by those with the protozoal infection and extra medical costs of $795,699 (39).

One year after Milwaukee, a cluster of cases and deaths in AIDS patients in Las Vegas, Nevada, alerted health officials to another outbreak (39). Those who drank any unboiled tap water were four times more likely than those drinking only bottled water to develop cryptosporidiosis. It was hypothesized that contamination of the drinking water had been over an extended time period with intermittent low levels of oocysts as opposed to a massive contamination event as was the case in the Milwaukee and Oxford/Swindon outbreaks, both associated with rainfall events.

Cancer patients undergo intensive chemotherapy with cytotoxic and immunosuppressive drugs and often radiation treatment in attempts to destroy

TABLE 4-7 Impact of Waterborne Outbreaks of Cryptosporidiosis on Patients with AIDS

Outbreak	Attack Rate	Mortality Ratio %	Comments
Oxford/Swindon, U.K., 1989	36	Not reported	3 of 28 renal transplant patients were found to be excreting oocysts without severe symptoms
Milwaukee, Wisconsin, 1993	45	68	17% biliary disease; CD4 counts <50 were associated with high risks
	40		14 of 82 AIDS patients had biliary symptoms
Las Vegas, Nevada, 1994	Not known	52.6	CD4 counts <100 high risk
	21[a]		Bottled water in the case-controls protective

Source: Ref. 39.

[a] Increase in positive stools for *Cryptosporidium* from 4% in 1993 to 21% in the first quarter of 1994.

neoplastic growth. These measures also attack the immune system, leaving the patient with little defense against opportunistic pathogens. For example, in cancer immunosuppressed patients, the mortality ratio for adenovirus infection is 53% (19). Bone-marrow transplantation is an effective therapy in patients with severe aplastic anemia or acute leukemia. However, because of a very weakened immune system they are very susceptible to infection. The mortality ratio among bone-marrow transplants with enteric viral (rotavirus, Coxsackie, adeno) infection was 59% in one study (53). Five of eight patients with rotavirus infections died. The mortality ratios for adenoviruses for bone-marrow patients ranged from 53 to 69%, depending on subgenus (Table 4-8).

TABLE 4-8 Mortality Ratios Among Specific Immunocompromised Patient Groups with Adenovirus Infection

Patient Group	Percent Mortality (Case-Fatality Ratio)	Overall Mean Age of Patient Group (years)
Bone marrow transplants	60	15.6
Liver transplant recipients	53	2.0
Renal transplant recipients	18	35.6
Cancer patients	53	25
AIDS patients	45	31.1

Source: Ref. 19.

Coxsackievirus Al infection, which seldom causes diarrhea in healthy persons, resulted in the deaths of six of seven bone-marrow patients in one outbreak (47).

Hypogammaglobulinemic patients are at increased risk of chronic meningoencephalitis from enteroviruses (40). Chronic meningoencephalitis is most frequently associated with echovirus infection but has occasionally been reported in association with Coxsackievirus B infection.

DATABASES FOR STATISTICAL ASSESSMENT OF DISEASE

In the United States, official statistics on diseases are compiled as part of the National Notifiable Diseases Surveillance System (NNDSS). As a part of a national effort for assessing vital statistics, certain diseases have been reported since 1920; up to 1960 the health statistics were obtained from publications of the National Office of Vital Statistics. Initially part of the Public Health Service, the National Center for Health Statistics is now a part of the Center for Disease Control and Prevention (CDC). The number of reported cases is summarized by type of disease, reported month, state, age, and race in some cases. There is some temporal, spatial, and demographic assessment of the health outcomes. At the state level, the cases are generally reported by county. The data represent only clinically identified cases, and then case ratios (cases to total population) or incidence rates are most often reported annually (numbers of cases/total population per year, often referred to as cases per 100,000 population for U.S. data). It is clear that some diseases are easily identified and reported consistently by all states (e.g., human rabies), while other diseases (e.g., salmonellosis) are underreported. Diseases such as giardiasis are reported at the state level but not at the national level, and some (e.g., infections associated with *Helicobacter* and enteric viruses) are not reported at all. Generic illnesses or symptoms are not reported, and antibody prevalence is not reported.

Many diseases associated with transmission through the environment are not reportable. The hazards and health outcomes are underappreciated and inadequately quantified, because these rates are reported without regard to who is exposed or who may experience the gravest consequence as a result of the risk. Worldwide, the assessment and reporting in most countries of most diseases are poor. Given the inadequacy of clinical tests and people's failure to seek medical attention and even get the proper tests done, it is known that these reports woefully underestimate the burden of disease in most populations. Table 4-9 summarizes the diseases reported at the national level in the United States for select years. Only a few changes have been made over the last 50 years in the diseases reported, despite many new emerging diseases and the use of vaccines that have essentially eliminated or reduced dramatically the level of some diseases in the United States.

TABLE 4-9   Summary of Reported Cases at the National Level in the United States for Selected Years[a]

Disease	1943	1970	1990	1996
U.S. population:	134,245,000	203,805,000	248,710,000	65,284,000
AIDS			41,595	66,885
Amebiasis	3,329	2,888	3,217	
Antrax	72	2		
Aseptc meningitis		6,480	11,852	
Botulism		12	92	119
Brucellosis	3,733	213	95	112
Chancroid	8,354	1,416	4,212	386
Cholera			6	4
Diptheria	14,811	435	4	2
Encephalitis	771	1,580	1,341	
Gonorrhea	275,070	600,072	690,169	325,883
Granuloma inguinale	1,748	124	97	
Hansen disease (leprosy)	35	129	198	112
Haemophilus influenzae		First reported 1991		1,170
Hepatitis A		56,797	31,441	31,032
Hepatitis B		8,310	21,102	10,637
Non-A, non-B hepatitis			2,553	3,716
Unspecified hepatitis			1,671	
Legionellosis			1,370	1,198
Leptospirosis		47	77	
Lyme disease		First reported 1991		16,455
Lymphgranuloma venereum	2,593	612	471	
Malaria	54,554	3,051	1,278	1,800
Measles	633,627	47,351	9,643	508
Mennigococcal	18,223	2,505	2,130	3,437
Mumps		104,953	4,264	751
Murine typhus fever	4,528	27	43	
Pertussis	191,890	4,249	2,719	7,796
Plague	1	13	11	5
Poliomyelitis	12,450	33	8	5
Psittacosis	1	35	94	42
Rabies (animals)	9,649	3,224	6,910	6,982
Rabies (humans)	47	3	3	3
Rheumatic fever		3,227	127	
Rocky mountain spotted fever	473	380	628	831
Rubella (German measles)		56,552	1,401	238
Rubella, congenital		77	11	4
Salmonellosis	731	22,096	48,603	45,471
Shigellosis	31,590	13,845	27,077	25,978
Smallpox	765		Last case 1949	
Syphillis	82,204	21,982	134,255	52,976
Tetnus		91,382	64	36
Toxic-shock syndrome			322	145

TABLE 4-9 (*Continued*)

Disease	1943	1970	1990	1996
Trichnosis		148	129	11
Tuberculosis	120,253	109	25,701	21,337
Tularemia	966	37,137	152	
Typhoid fever	4,690	172	552	396
Varicella (chickenpox)		346	173,099	83,511
Yellow fever	Last indigenous case 1911, last imported cases 1924			1

Source: Data from Refs. 49 and 51.

[a] Blank indicates no data, not identified as a disease, or no requirement for national reporting.

In 1996, *E. coli* O15:H7 was added to the national database, with 2741 cases reported and 102 cases of the hemolytic uremic syndrome (HUS) reported. Cryptosporidiosis was made reportable in 1995 with 2972 cases reported from 27 states and 2426 cases reported in 1996 from 42 states. *Hantavirus* pulmonary syndrome associated with environmental transmission from rodents was reported in 26 states: 22 cases in 1996 and another 138 cases as of May 1997 (the case/fatality mortality ratio was 47.5%; thus although the incidence was low, the consequence or health outcome was significant). Dramatic foodborne, waterborne, and vectorborne outbreaks in each case associated with deaths forced the change in the system.

ICD Codes

Within the National Center for Health Statistics are several other databases that are often used to examine health outcomes and relative risks. In 1990, the National Ambulatory Medical Care Survey provided data from office-based physicians through examination of patient records and gave an indication of the number of persons who seek a physician and are diagnosed. Like the Foodnet program developed in 1996, which enlists state health departments and local network systems, these systems have demonstrated that for most mild illnesses, patients do not seek a physician and do not undergo the correct clinical testing for diagnosing the diseases. Therefore, testing and not reporting appears to be the limiting factor for most diseases. Other significant databases that have been used to examine more severe disease outcomes and death include the National Hospital Discharge Survey (begun in 1988 to assess the number of patients treated in hospitals) and National Mortality Followback Survey (representing about 1% of the U.S. resident deaths). These systems use the International Classification of Disease, which assign codes for specific diseases.

The ninth revision of the Clinical Modification (ICD-9-CM) codes have been used to examine foodborne disease. However, those cases of enteric infectious diseases may or may not have been transmitted by the foodborne disease or may actually have been waterborne; thus the exposure may not be well defined. Nevertheless, medical discharge certificates are useful for assessing severity, costs, and relative health outcomes. Costs have been estimated by the average length of stay in the hospital and the 1990 national average cost per day of $687. The numbers of cases per year over a 4-year period ranged from 5344 for *Shigella* to 530,689 for unspecified acute gastroenteritis at costs between $16 million and $2 billion. These data have also been used to examine the demographics of those affected, including AIDS patients and the elderly (Table 4-10). These data suggest that for HIV patients, protozoa are of greater significance, and for the elderly, other enteric bacteria are of greater concern. In all cases unspecified intestinal illness appears to be a major cause of disease, and thus the hazard remains unidentified.

By comparing Table 4-9 to Table 4-10 for 1990, one can see that in trying to define hospitalization ratios based on these data, the denominator will be greatly underestimated for some diseases. For example, for hepatitis A virus, total hospitalizations/total national reports gives a ratio of 6643/31,441 or 21%, compared to 28.3% for hospitalizations during waterborne outbreaks (see Table 4-4). However, for *Salmonella,* the hospitalizations/total reports was 17,984/48,603 or 37%, compared to 4.1% during waterborne outbreaks. If the cases are divided by total population to get the rates, quantification of the sequence of outcomes from exposure to asymptomatic to symptomatic to severe disease, and therefore the distribution of health outcomes is not assessed.

EPIDEMIOLOGICAL METHODS FOR UNDERTAKING HAZARD IDENTIFICATION

Epidemiology is the study of occurrence and causes of diseases in populations. The field has focused on exposure and the relationship to health outcome using statistical methods to show a significant association between exposure and health. To some extent epidemiologists have attempted to describe the influence of environmental factors. The field has also been described as providing an application of the knowledge to the prevention and control of health problems, so in that vein it is tied to risk management. Epidemiological studies have often been referred to as *risk assessment* because there is an attempt to examine both exposure (or what are often referred to as *risk factors*) and health outcome. However, exposure in these studies is rarely specific or quantitative for microbial contaminants, and in many cases the health hazard is defined by symptomology and does not address the specific hazard as it becomes much more extensive investigation to undertake clinical tests or antibody tests (see Table 4-1).

The health endpoints measured in epidemiological studies of infectious diseases can be divided into several groups: (1) endemic risks are the constant

TABLE 4-10 Summary of Patients Discharged from Hospitals by Category of Disease in the United States, 1990[a]

Disease	ICD-9 Code	Total Mentions	Percent Distribution by Mention with HIV/AIDS	Percent Distribution by Mention with Elderly[b]
Amebiasis	006.0	1,341		0.11
Botulism	005.1	729		
Cholera	001.9	92		0.04
E. coli (part of intestinal due to other)	008.0	2,258		
Giardiasis [part of other protozoa (007)]	007.1	2,967		
Hepatitis A	070.1	6,643	4.1	0.66
Hepatitis all	070	12,810		
Ill-defined intestinal infections	009	28,303	12.1	1.62
Intestinal infections due to other microorganisms	008	176,282	15.1	25.11
Listeriosis	027.0	1,248		0.28
Other noninfectious gastroenteritis and colitis or unspecified	558	562,047	54.1	69.64
Protozoal intestinal disease (excluding amebiasis)	007	4,864	12.1	0.02
Salmonellosis	003	17,984	2.6	1.11
Shigellosis	004	2,113		0.04
Staphylococcal	005.0	308		0.88
Toxic effects of noxious substances (e.g., mushrooms, shellfish)	988	352		
Trichinosis	124	42		
Typhoid fever and paratyphoid fevers	002	2,056		0.49

Source: Data from Refs. 44 to 46.

[a]U.S. population in 1990: 203,805,000.

[b]Elderly defined as 65 years of age and older.

low levels of diseases or infections that are present in a population; (2) epidemic risks are disease cases in excess of the number of cases normally found or expected, constituted as an outbreak if limited to a specific population; and (3) outbreaks are two or more cases associated with a common exposure in time and place or source (see Figure 1-2). In most cases these studies rely on routine health surveillance methods, whereby people seek medical attention, submit laboratory samples, and are diagnosed. Often this is done retrospectively, through the examination of records or through personnel interviews and recall.

Waterborne and Foodborne Outbreaks

Outbreaks are generally associated with a specific population and exposure in time and place. There are community outbreaks that may be associated with the contamination of water or food. There are outbreaks at events (weddings, dinners, generally associated with foods), outbreaks associated with recreational exposure (fecal accidents in swimming pools), and outbreaks associated with place (hospital and day-care outbreaks, associated with contamination of food, water, surfaces, instruments, hands, etc.).

From 1971 to 1992 in the United States, more than 164,000 persons were reported ill during 684 documented waterborne outbreaks caused by bacteria (14.4%), viruses (8.6%), protozoa (18.5%), and chemicals (8.8%) (10,18,31). The etiological agents causing acute gastrointestinal illness in a large percentage of the outbreaks were not identified (49.7%). The numbers of outbreaks and cases associated with microbial agents are shown in Table 4-4 (see also Table 1-1). During the investigation of drinking water outbreaks, the source of the water (groundwater, spring, river) is generally identified as well as the treatment deficiency (e.g., no disinfection).

There is generally widespread community impact. Figure 4-5 displays the average attack rates in drinking water outbreaks by etiological agent. The *attack rate* is defined by the ratio of cases that occur relative to the total population exposed. For example, on average, 22% of the populace in the communities exposed developed illness when the drinking water was tainted with *Campylobacter*, while Norwalk virus outbreaks on average caused 53% of the populace to become ill. The problem with the data are that often the numerator (cases) and denominator (those exposed) are not very well defined. These attack rates are not only subject to the accuracy of the investigation but are subject to the level of the contamination (which is rarely identified), amount of contaminated water consumed over time, and the type of microorganism (dose–response). However, it appears that at least in drinking water outbreaks (perhaps under conditions of high levels of contamination), it is the microbial hazard that influences these attack rates, which correlate well with the dose–response parameters for the individual microorganisms (see Chapter 7).

Many more outbreaks are associated with contaminated food than with contaminated water. A total of 2423 outbreaks were documented and 77,373

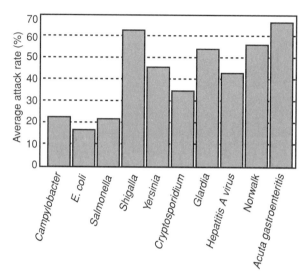

Figure 4-5 Average attack rates by microorganism during waterborne outbreaks.

cases from 1988 to 1992. Like the waterborne investigations however, the hazard went unidentified in 59% of the outbreaks and the remainder were caused by bacteria (33%), chemical agents (5.7%), parasites (0.82%), and viruses (1.6%) (see also Table 1.2). Outbreak data greatly underestimate the disease burden and the hazards, and there have been attempts to estimate the greater extent and magnitude of the risks. Bennett et al. (1) have estimated over 25 million cases of enteric infections, resulting in 10,800 deaths per year in the United States, and of these 940,000 cases, 900 deaths are waterborne. These data suggest that as much as 3.9% of the enteric illnesses experience per year are due to contaminated water, and in regard to the U.S. population the rate of waterborne disease is 4/1000 per year. Foodborne risks have been estimated at 6,496,000 cases and 9100 deaths annually (or a rate of 27/1000 for the U.S. population). Most have concluded that these estimates may be too low, and others have suggested that there may be as much as 33 million cases of foodborne disease per year in the United States.

While the estimation of total disease burden is important (endemic disease), outbreak assessment remains a significant component of the hazard identification. This is due to the extreme costs associated with outbreaks not only in medical care and days lost from work, but costs accrued in assessment of the outbreak, recall of food products, boil orders, communication efforts, remediation, and future safety efforts. The waterborne outbreak in Milwaukee was estimated to cost the community $153 million, not to mention subsequent costs of aversion behavior due to loss of confidence in the water supply (e.g., purchase of bottled water and point-of-use devices to further treat the water).

Controlled Epidemiological Investigations

In controlled epidemiological studies the goals are to identify the health hazards in an exposed population compared to an unexposed population. There are two types of studies used to identify hazards (9):

1. *Cohort studies.* Exposed and nonexposed groups are followed over time while assessing the health outcome (usually prospective, usually numerous possible health outcomes evaluated).
2. *Case-control studies.* Cases with specific illnesses are selected and compared to those without the disease among the population (controls) for relevant risk factors (e.g., exposure) (generally retrospective and one possible health outcome evaluated).

The cohort study is generally seen as more robust. In most of these types of studies, generic symptoms are used to describe the health outcome (diarrhea, stomach cramps, vomiting) and then reported as cases per population exposed (or numbers in the study, assuming equal exposure, referred to again as *attack rates*). This is divided by the cases per population in the control group or nonexposed group. This comparison is reported as the odds ratio (incidence of disease in exposed group/incidence of disease in nonexposed). This odds ratio can also be adjusted for age (sensitive populations) or other types of variables.

Example 4-3 shows how these studies are used to examine microbial hazards associated with one type of environmental exposure (recreational water) (52). As presented in Example 4-1, the health effects are based on symptomology and clinical descriptions and may be due to any number of specific microorganisms, which may or may not have been present during the study period. One of the assumptions in these types of studies is that diarrhea is indicative of other types of health outcomes. However, this may not be true for many types of health effects and microorganisms, in that infection and not necessarily diarrheal illness is associated with outcomes such as viral myocarditis. While it was indicated that viruses were detected in the recreational waters, the viral types were not indicated in the study nor was specific clinical diagnosis or antibody prevalence evaluated. Recreational outbreaks of *E. coli* O157:H7 have been documented (one in 1998 with a total of 12 cases reported) and the potential severity of the outcome has caused much concern regarding water quality and children's safety. These incidents demonstrate that assessing attack rates without consequence based on specific microorganisms underestimates the possible health outcomes. The target population in Example 4-3 was athletes, and therefore the assumption was that exposure and health outcomes would be similar for the general population. The study assumed that the year of study represented all exposures to the various microbial hazards past and present. These types of studies partially address exposure, examine one type of mild acute outcome, but do little to address the remaining boxes shown in Figure 4-1.

In addition, cohort studies are judged to suggest that whatever disease exists in the nonexposed group represents an acceptable low level of background of infections in the community by which to make a comparison. Therefore, as in the case of this example, 3.6% diarrheal attack rates were seen as the acceptable marker by which to make the comparison. However, had this type of study taken place in a community with a higher level of endemic disease, a poorer water quality, supposedly a greater level of microbial hazards, and higher levels of health impacts would have been found to be indistinguishable. This study was designed and used purposely for risk management to set a standard for recreational water quality. The surrogate used to assess risk was based on indicator microorganisms, which have been shown to have severe limitations in regard to the presence of specific microbial hazards (see Chapter 5). However, there usually is a significant database available for indicator microorganisms (or surrogates of risk), and these types of epidemiological studies have a long history of use.

Example 4-3   Study Examining Health Impacts from Recreational Water (52)

Exposure: recreational water exposure by triathletes; many types of microbial hazards possible but no specific identification. Suggestion that indicator bacteria (range between 10 and 1000) are surrogates for relative increased levels of microbial hazards, although no studies have demonstrated such correlations.

Asymptomatic infections: not evaluated.

Acute symptomatic illness: generic diarrhea, vomiting, stomach ache, and so on, used as the health outcome. Attack rates ranged from 0.4 to 5.2%, depending on the symptoms reported (type of episode definition used). Significance of the health outcome compared to background endemic disease of nonexposed of about 3.0 to 3.6% diarrheal (episode definition) cases per year.

Chronic disease, sensitive populations, hospitalization, and mortality: not evaluated; too rare an outcome to be detected (total study population examined was a population of 827).

Note: Recent (1998) recreational outbreaks of *E. coli* O157:H7 have highlighted the importance of the type of microbial hazard in relationship to outcome.

HAZARD IDENTIFICATION DATA USED IN THE RISK ASSESSMENT PROCESS

Many health outcomes are microbial specific, and while it may not be possible to evaluate all possible outcomes and data sets for every microorganism, high-priority risks can be assessed in greater detail. One such risk is *Cryptospor-*

idium in drinking water. This protozoan has been associated with recent waterborne outbreaks worldwide (in the United States, Canada, United Kingdom, and Japan). The oocyst is extremely resistant to routine water treatment and is found routinely in water supplies. Finally, the risk to sensitive populations, primarily immunocompromised, is extreme (death) (20,26,39).

Work by Perz et al. (35) presents a risk assessment using in part specific hazard identification data and compares this to the current disease surveillance system for reporting cases of *Cryptosporidium.* This study makes an interesting example.

Example 4-4   **Cryptosporidium** *Risks in Drinking Water and Comparative Disease Surveillance (35)*

Exposure: Drinking water; average level of contamination was assumed to be 1 oocyst per 1000 L in the New York City population.

Sensitive populations: Adults and children with and without AIDS were evaluated.

Asymptomatic illness: At least one feeding study showed that 39% healthy adults had no symptoms; on average for non-AIDS patients, 70% [range of 35 to 100% (range represents the 95% confidence limits)] symptomatic rates were used, while for AIDS patients, 95% were assumed symptomatic.

Acute symptomology: Moderate illness was defined as an average of 4 days duration; for non-AIDS the percentage of adults that experienced this level of disease was 15% [range of 8 to 30% (range represents the 95% confidence limits)], for children it was estimated at 20% (range of 10 to 30%), and for AIDS patients it was estimated at 95% (range of 80 to 100%).

Detection:

Physician's visit
    Adult non-AIDS (33%; 17–66%)
    Children non-AIDS (50%; 25–75%)
    Adult AIDS (90%; 80–100%)
    Children AIDS (95%; 70–100%)
Clinical diagnosis: ova and parasite examination
    Adult non-AIDS (25%; 10–63%)
    Children non-AIDS (50%; 25–75%)
    Adult AIDS (90%; 80–100%)
    Children AIDS (95%; 70–100%)
Clinical diagnosis: *Cryptosporidium* examination
    Adult non-AIDS (10%; 5–20%)
    Children non-AIDS (15%; 8–30%)
    AIDS (95%;80–100%)
Reporting

Non-AIDS (60%, 40–80%)
AIDS (95%, 80–100%)

Table 4-11 shows the results of the Perz et al. (35) analysis. The exposed population was divided by age and AIDS status, and the *Cryptosporidium* hazard associated with drinking water was assessed given the current surveillance system. Overall, the analysis predicted that infection would result in approximately 3 reported cases out of every 10,000, in children the rate was ten times higher, while in AIDS patients diagnosis and reporting were much more diligent, due to the significant mortality in the health outcome. The analysis also predicted that tap water would contribute to between 5 and 10% of the cases at exposures of 1 oocyst per 1000 L.

Many different data sets (outbreak investigations, disease surveillance, and epidemiological data) are available and all should be used and evaluated when undertaking the hazard identification. Until the microbial hazards are identified by the medical community as causing disease and diagnostic tools are available and used, there will remain significant gaps in this portion of the risk assessment process. Modeling approaches with the best available data may provide useful insights into these microbial agents, their transmission through the environment, and the health risks associated with exposure. Adequate cost–benefit analysis cannot be undertaken under various risk management scenarios until the health outcomes have been assessed; otherwise, the benefits will always remain underappreciated and underestimated.

TABLE 4-11 Prediction of Waterborne Cryptosporidiosis in New York City in AIDS Patients Compared to the General Population

	Adults	Children	Adults with AIDS	Pediatric AIDS
Total NYC population	6,080,000	1,360,000	30,000	1,200
Reported cases (1995)	40	30	390	10
Predicted tapwater-related reported cases (% of total actually reported)	2 (5%)	3 (10%)	33 (8.5%)	1 (10%)
Predicted annual risk from tapwater unreported (% of those predicted to be reported)	5,400 (0.03%)	940 (0.3%)	56 (59%)	1 (100%)

Source: Ref. 35.

REFERENCES

1. Bennett, J. V., S. D. Holmberg, M. F. Rogers, and S. L. Solomon. 1987. Infectious and parasitic diseases. Am. J. Prev. Med. 3(5 Suppl.):102–114.

2. Bowen, G. S., and M. A., McCarthy. 1983. Hepatitis A associated with a hardware store water fountain and a contaminated well in Lancaster County, Pennsylvania, 1980. Am. J. Epidemol. 117:695–705.

3. Bulter, T., A. Islam, I. Kabir, and P. K. Jones. 1991. Patterns of morbidity in typhoid fever dependent on age and gender: review of 55 hospitalized patients with diarrhea. Rev. Infect. Dis. 13:85–90.

4. Buzby, J. C., T. Roberts, C. T. J. Lin, and J. M. MacDonald. 1996. Bacterial foodborne disease: medical costs and productivity losses. Economic Research Division, U.S. Department of Agriculture, Washington, DC, August.

5. Buzby, J. C., T. Roberts, and B. M. Allos. 1997. Estimated annual costs of *Campylobacter*-associated GBS. Economic Research Service, U.S. Department of Agriculture, Washington, DC, July.

6. Chalker, R. B., and M. J. Blaser. 1988. A review of human salmonellosis. III. Magnitude of *Salmonella* infection in the United States. Rev. Infect. Dis. 10:111–124.

7. Clifford, C. P., D. W. M. Crook, C. P. Conlon, A. P. Fraise, and D. G. Day, et al. 1990. Impact of waterborne outbreak of cryptosporidiosis on AIDS and renal transplant patients. Lancet 335:1455–1456.

8. Craske, J. 1990. Hepatitis C and non-A non-B hepatitis revisited: hepatitis E, F, and G. J. Infect. 25:243–250.

9. Craun, G. F., R. L. Calderon, and F. J. Frost. 1996. An introduction to epidemiology. J. Am. Water Works Assoc. 88(9):54–65.

10. Craun, G. F. 1991. Causes of waterborne outbreaks in the United States. Water Sci. Technol. 24:17–20.

11. Cunningham A. L., G. S. Grohman, J. Harkness, C. Law, D. Marriott, B. Tindall, and D. A. Cooper. 1988. Gastrointestinal viral infections in homosexual men who were symptomatic and seropositive for immunodeficiency virus. J. Infect. Dis. 158:386–391.

12. Gorbach, S. L., J. G. Bartlett, and N. R. Backlow. 1992. Infectious disease. W.B. Saunders, Philadelphia.

13. Gordon, S. M., L. S. Oshiro, W. R. Jarvis, D. Donenfeld, M. Ho, F. Taylor, H. B. Greenberg, R. Glass, H. P. Madore, R. Dolin, and O. Tablan. 1990. Foodborne Snow Mountain agent gastroenteritis with secondary person-to-person spread in a retirement community. Am. J. Epidemol. 131:702–710.

14. Gust, I. D., and R. H. Purcell. 1987. Report of a workshop: waterborne non-A, non-B hepatitis. J. Infect. Dis. 156:630–635.

15. Gust, I. 1990. Design of hepatitis A vaccines. Br. Med. Bull. 46:319–328.

16. Halvorsrud, J., and I. Orstavik. 1980. An epidemic of rotavirus-associated gastroenteritis in nursing home for the elderly. Scand. J. Infect. Dis. 12:161–164.

17. Hayes, E. B., T. D. Matte, T. R. O'Brien, T. W. McKinley, G. S. Logsdon, J. B. Rose, B. L. P. Ungar, D. M. Word, P. F. Pinsky, M. L. Cummings, M. A. Wilson,

E. G. Long, E. S. Hurwitz, and D. D. Juranek. 1989. Large community outbreak of cryptosporidiosis due to contamination of filtered public water supply. N Engl. J. Med. 320:1372–1376.

18. Herwaldt, B. L., G. F. Craun, S. L. Stokes, et al. 1992. Outbreaks of waterborne disease in the United States, 1989–90. J. Am. Water Works Assoc. 84:129–135.

19. Hierholzer, J. C. 1992. Adenoviruses in the immunocompromised host. Clin. Microbiol. Rev. 5:262–274.

20. Hoxie, N. J., J. P. Davis, J. M. Vergeront, R. D. Nashold, and K. A. Blair. 1997. Cryptosporidiosis-associated mortality following a massive waterborne outbreak in Milwaukee, Wisconsin. Am. J. Public Health 87(12):2032–2035.

21. Janoff, E. D., and P. D. Smith. 1988. Perspectives on gastrointestinal infections in AIDS. Gastroenterol. Clin. N. Am. 17:451–463.

22. Kaplan, M. H., S. W. Klein, J. McPhee, and R. G. Harper. 1983. Group B Coxsackievirus infections in infants younger than three months of age: a serious childhood illness. Rev. Infect. Dis. 5:1019–1032.

23. Ledner, W. M., S. M. Lemon, J. W. Kirkpatrick, R. R. Redfield, M. L. Fields, and P. W. Kelly. 1985. Frequency of illness associated with epidemic hepatitis A virus infections in adults. Am. J. Epidemiol. 122:226–233.

24. Levine, W. C., J. F. Smart, D. L. Archer, N. H. Bean, and R. V. Tauxe. 1991. Foodborne disease outbreaks in nursing homes, 1975 through 1987. J. Am. Med. Assoc. 266:2105–2109.

25. Lew, J. F., D. L. Swerdlow, M. E. Dance, P. M. Griffin, C. A. Bopp, M. J. Gillenwater, T. M. Mercatente, and R. I. Glass. 1991. An outbreak of shigellosis aboard a cruise ship caused by a multiple-antibiotic-resistant strain of *Shigella flexneri*. Am. J. Epidemiol. 134:413–420.

26. Lisle, J. T. and J. B. Rose. 1995. *Cryptosporidium* contamination of water in the USA and UK: a mini review. J. Water SRT-Aqua 44(3):103.

27. Mandell, G. L., R. G. Douglas, and J. E. Bennett. eds. 1990. Principles and practice of infectious diseases, 3rd ed. Churchill Livingstone, New York.

28. Mishra, L., and L. B. Seeff. 1992. Viral hepatitis, A through E, complicating pregnancy. Gastroenterol. Clin. N. Am. 21:873–882.

29. Modlin, J. F., and J. S. Kinney. 1987. Prenatal enterovirus infections. Adv. Pediatr. Infect. Dis. 2:57–78.

30. Modlin, J. F. 1986. Perinatal echovirus infection: insights from a literature review of 61 cases of serious illness and 16 outbreaks in nurseries. Rev. Infect. Dis. 8:918–926.

31. Moore, A. C., B. L. Herwaldt, G. F. Craun, R. L. Calderon, A. K. Highsmith, and D. D. Juranek. 1993. Surveillance for waterborne disease outbreaks: United States, 1991–1992. Morb. Mortal. Wkly. Rev. 42(SS-5):1–22.

32. MacKenzie, W. R., N. J. Hoxie, M. E. Proctor, S. Gradus, K. A. Blair, D. E. Peterson, J. J. Kazmierczak, K. Fox, D. G. Addiss, J. B. Rose, J. P. Davis. 1994. Massive waterborne outbreak of *Cryptosporidium* infection associated with a filtered public water supply, Milwaukee, Wisconsin, March and April, 1993. N. Engl. J. Med. 331(3):161–167.

33. Oshiro, L. S., C. E. Halet, R. R. Roberto, J. L. Riggs, M. Croughan, H. Greenberg, and A. Kapikian. 1981. A 27-nm virus isolated during an outbreak of acute in-

fectious nonbacterial gastroenteritis in a convalescent hospital: a possible new serotype. J. Infect. Dis. 143:791–795

34. Payment, P., L. Richardson, J. Siemiatycki, R. Dewar, M. Edwardes, and E. Franco. 1991. A randomized trial to evaluate the risk of gastrointestinal disease due to consumption of drinking water meeting current microbiological standards. Am. J. Public Health 81:703–708.

35. Perz, J. F., F. K. Ennever, and S. M. Le Blancq. 1998. *Cryptosporidium* in tap water. Am. J. Epidemiol. 147(3):289–301.

36. Pether, J. V. S., and E. O. Caul. 1983. An outbreak of foodborne gastroenteritis in two hospitals associated with a Norwalk-like virus. J. Hyg. 91:343–350.

37. Reid, J. A., D. Breckon, and P. R. Hunter. 1990. Infection of staff during an outbreak of viral gastroenteritis in an elderly person's home. J. Hosp. Infect. 16: 81–85.

38. Roberts, T., and J. K. Frenklel. 1990. Estimating income losses and other preventable costs caused by congenital toxoplasmosis in people in the United States. J. Am. Vet. Med. Assoc. 196(2):249–256.

39. Rose, J. B. 1997. Environmental ecology of *Cryptosporidium* and public health implications. Annu. Rev. Public Health 18:135–161.

40. Rubin, R. H., and L. S. Young. 1988. Clinical approach to infection in the compromised host, 2nd ed. Plenum Medical Book Co., New York.

41. Ryan, C. A., R. V. Tauxe, G. W. Hosek, J. G. Wells, P. A. Stoesz, H. W. McFadden, P. W. Smith, G. F. Wright, and P. A. Blake. 1986. *Escherichia coli* O157:H7 diarrhea in a nursing home: clinical, epidemiological, and pathological findings. J. Infect. Dis. 154:631–638.

42. Ryan, C. A., N. T. Hargrett-Bean, and P. A. Blake. 1989. *Salmonella typhi* infections in the United States, 1975–1984: increasing role of foreign travel. Rev. Infect. Dis. 11:1–8.

43. Selik, R. M., E. T. Starcher, and J. W. Curran. 1987. Opportunistic diseases reported in AIDS patients: frequencies, associations and trends. AIDS 1:175–182.

44. Steahr, T. E. 1996. An estimation of foodborne illness in populations with HIV/ AIDS infection, United States. Int. J. Environ. Health Res. 6(2):77–92.

45. Steahr, T. E. 1994. Food-borne illness in the United States: geographic and demographic patterns. Int. J. Environ. Health Res. 4(4):183–195.

46. Steahr, T. E. 1998. An estimate of foodborne illness in the elderly population of the United States. Int. J. Environ. Health Res. 8(1):23–34.

47. Townsend, T. R., R. H. Yolken, C. A. Bishop, G. W. Santos, E. A. Bolyard, W. E. Beschorner, W. H. Burns, and R. Saral. 1982. Outbreak of Coxsackie Al gastroenteritis: a complication of bone-marrow transplantation. Lancet 1:820–823.

48. U.S. Department of Commerce. 1991. Statistical abstract of the United States, 1991: the national data book. Bureau of the Census, Washington, DC.

49. U.S. Department of Health and Human Services. 1996. Summary of notifiable diseases, United States. Morb. Mortal. Wkly. Rep. 45(53).

50. U.S. Department of Health and Human Services. 1996. Surveillance for foodborne-disease outbreaks: United States, 1988–1992. Morb. Mortal. Wkly. Rep. 45 (SS-5).

51. U.S. Department of Health and Human Services. 1992. Summary of notifiable diseases, United States. Morb. Mortal. Wkly. Rep. 41(55).

52. Van Asperen, I. A., G. Medema, M. W. Borgdorff, M. J. W. Sprenger, and A. H. Havelaar. 1998. Risk of Gastroenteritis among triathlethes in relation to faecal pollution of fresh waters. Int. J. Epidemiol. 27:309–315.

53. Yolken, R. H., and M. Murphy. 1982. Sudden infant death syndrome associated with rotavirus infection. J. Med. Virol. 10:291–296.

CHAPTER 5

ANALYTICAL METHODS FOR DEVELOPING OCCURRENCE AND EXPOSURE DATABASES

Critical to the risk assessment processes is the ability to quantify exposure to pathogens. Methods are available to isolate and identify bacteria, fungi, protozoa, and viruses as well as microbiological toxins from environmental samples (24). Standard methods continue to be used such as those published in *Standard Methods for the Examination of Water and Wastewater* (2). However, newer methods using immunomagnetic capture systems and molecular techniques are now being applied to foods: for example, for detection of *E. coli* O157:H7 in hamburger and to water, soil, and air. New methods should be developed, tested, and applied for application to quantitative assessment of exposure and exposure pathways as well as control of exposure. Occurrence databases will need to be developed with final application of risk assessment in mind.

Much of the past microbial occurrence data are nonquantitative, reported as presence/absence, developed with very different protocols and monitoring approaches. Thus often the issue is not the detection method per se but the sampling protocols and schemes and the interpretation of the data. These data have limited application for quantitative risk assessment. It is now recognized that quantitative, statistically evaluated databases must be developed, as often this is the major data gap for adequate risk assessments. These must be combined with models for prediction of transport and fate of microorganisms through the environment and through processes. Thus the field of predictive microbiology is a rapidly developing area that will be able to fill some of the data gaps on exposure.

APPROACHES FOR DEVELOPING OCCURRENCE AND EXPOSURE DATABASES

Exposure could be defined as the monitoring of the source of the exposure over time, that is, the final food product prior to consumption: the glass of water from the tap or the aerosol that is inhaled. This is a difficult and impossible task in most cases. Unlike chemicals, microorganisms act as particles and their concentrations in water, soil, air, food, and on surfaces are not normally nor homogeneously distributed. These microorganisms can change concentrations through die-off or regrowth over time. The sources of the microorganisms (e.g., animal wastes or sewage, etc.) are also diverse in concentrations over time (e.g., seasonal and climatic influences). Finally, many controls have already been implemented (disinfection) to reduce the concentrations and the exposure. Therefore, other strategies have developed for assessing exposure and developing occurrence databases for microorganisms. These include the monitoring of indicators as well as pathogens for:

• Assessing the sources of microorganisms
• Assessing the transport and fate of microorganisms
• Assessing the reduction through the use of treatment/process controls of microorganisms

These approaches include field data as well as laboratory-based data and the use of models for evaluating transport (e.g., subsurface migration) and fate (e.g., inactivation rates). Ecosystem studies are necessary for most microorganisms (e.g., *Legionella* in biofilms and release during aerosolization) and more ecosystem modeling is needed. In the area of food safety the concept of farm to table is being used to follow the microbial contaminants from their source on the farm through harvest and production to the final packing of the food product. For drinking water a similar system based on watershed assessment, drinking water treatment efficacy, and distribution system integrity is being promoted. In some of these cases (e.g., an understanding of infections in the animal or human populations, waste disposal practices, and the transport patterns), survival and regrowth of the microorganisms must be gained and monitoring data must be developed to support the likelihood of exposure through the various pathways. Therefore, the evaluation of exposure will require extensive development of a variety of databases and models (Table 5-1).

METHODOLOGICAL ISSUES

Methods for the detection of bacterial pathogens in clinical specimens developed near the turn of the last century using liquid broth or solid agar to grow and isolate the organisms from clinical specimens. Selective media and bio-

TABLE 5-1   Databases Needed for Monitoring Drinking Water Contaminants

- *Sites:* directly at discharge sites in the watershed, assessment of loading, temporal and spatial differences (sewage treatment plants, animal waste lagoons, stormwater discharges). Inflows and outflows of key water bodies, sediments, intake to the water treatment plant, filter backwash, effluent leaving the treatment plant, distribution samples, tap samples.
- *Occurrence data:* identification of peaks, average levels, frequency of detection, distributions, seasonal variation, association with events (e.g., storms), and other temporal and spatial changes.
- *Hydrologic transport:* association with flows, runoff, precipitation, decreases through lakes or reservoirs, sediment deposition/resuspension, transport through soil.
- *Changes in Concentrations and Exposure:* environmental factors influencing dilution, regrowth, and die-off, and control factors influencing removal by filtration and inactivation through disinfection.
- *Specialized Studies:* may be needed to evaluate viability; species identification (e.g., *Cryptosporidium* from birds or animals), transport, and fate, controls (e.g., seeded pilot studies); and best use of indicators, surrogates, and treatment indices.

chemical methods were developed to differentiate the specific pathogen, allowing for identification. However, development of methods for detection of specific microbial hazards in environmental samples, such as food and water, has proven to be more problematic. Pathogens are present in much lower numbers (usually, orders of magnitude less), and not in as a robust state of growth as that found in clinical specimens from infected persons. Thus methods for assessing environmental microbial quality had been focused on indicator systems. The sanitary quality of food, water, and surfaces, for example, have been assessed for many years using an indicator concept in relation to fecal contamination. It was found that coliform bacteria were always present in the feces of warm-blooded animals and were less common in other places in nature. Thus the presence of coliform bacteria in water indicated the possible presence of fecal contamination and enteric pathogens. Coliform bacteria could be detected in water within 24 hours with a simple test that determined the ability of the organism to utilize the sugar lactose. Application of this indicator played a major role in the reduction of waterborne disease (particularly enteric bacteria) in the United States. To this day the coliform test is used as an indicator of drinking water quality.

The rapid development of this simple test resulted in an attitude that the best way to assess and control exposure to waterborne and foodborne pathogens was the use of indicator organisms that could be detected rapidly (24 to 48 hours) at low cost. Little effort went into the development of methods for the detection of enteric pathogens in the environment until the 1960s, when it became evident that enteric viruses were more resistant than coliform bacteria to disinfection. Waterborne outbreaks of *Giardia* and *Cryptosporidium* which followed, and later foodborne outbreaks of *Salmonella, Escherichia coli* O157:H7, and *Cyclospora*, emphasized the need for direct pathogen detection in water and food.

Traditionally, methods for the isolation of microorganism have been by cultivation of the organisms in the laboratory. Today, most pathogenic bacteria transmitted through the environment can be cultivated on solid agar media. In contrast, viruses and protozoan parasites usually require laboratory animals or animal cell culture. Not all pathogens have been cultivated in the laboratory (e.g., Norwalk virus) and other methods have to be used. Methods using antibodies labeled that specifically react with the pathogen or molecular probes (that react with the genetic material) are used to identify the organism. All methods for the detection and quantification of pathogens have limitations. This is especially true when attempting to assess the exposure to pathogens in the environment. In any risk assessment for pathogens it is important to understand these limitations since the greatest amount of uncertainty in a risk assessment may be assessing the variability in exposure. Methods for the detection of pathogens in the environment often involve the series of steps shown in Figure 5-1.

Because of the number of steps and variable characteristics of the individual microorganisms, methods are not 100% efficient. As an example, most methods for the detection of some of the enteric viruses from tapwater are capable of detecting only 30 to 50% of the enteroviruses (37,56). It is probably unrealistic to expect methods for microbial detection to be highly efficient from environmental samples given their wide variety of composition, volumes that must often be processed, and the genetic variability within the populations of microorganisms. The initial sampling often influences the over-

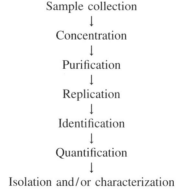

Figure 5-1 Steps involved in the detection and quantification of pathogens.

all methods. For example, water quality will change the efficiency of concentration and detection of microorganisms. Highly turbid organically laden river water can interfere with microbial adsorption techniques, cultivation techniques, and microscopic techniques. Even given these limitations, methods have been developed that are capable of detecting these organisms in thousands of liters of water. On a weight-to-weight basis with water, the methods for viruses have a sensitivity of detection of 10^{-18}. This extreme sensitivity [most chemical methods are only capable of detecting parts per million (10^{-6}) or billion (10^{-9})] is also another probable reason why variability in the efficiency of detection is observed with these methods.

It is desirable in most cases to sample large volumes. For air and water this is often hundreds of liters, and such sampling is done on site by passing the water or air through a device that concentrates the organism. The microorganisms are most commonly concentrated onto a filter by size exclusion (e.g., pore size smaller than the organism) or by adsorption (24), and with air they can also be trapped in a fluid by bubbling of the air through the liquid or can be impacted directly on solid agar (57). In the case of solids such as food, soil, biosolids, and clothing, the organisms are either assayed directly (the case most common with bacteria) or extracted. Because of the difficulty with working with large amounts of solids, most methods cannot process greater than 100 g. Inanimate surfaces (fomites) are usually sampled with a small swab or sponge which is wetted and then placed in contact with the surface. The swab/sponge is then placed in a fluid for extraction of the organisms (13).

Example 5-1   Impact of Sampling Strategies. In the food industry two sampling methods had been developed and used for detection of *Salmonella/ Campylobacter* on chickens:

1. Sampling of the wash water after whole-carcass wash, using concentration and selective cultivation techniques
2. Sampling of a square area of surface from the individual carcass using a swab technique and selective cultivation techniques

Although both approaches could be expressed as colony-forming units (CFU)/cm^2, for example, more chickens and more surface area would be sampled with approach 1. Collection and isolation of bacteria from the wash water could be more efficient than with a swab method; on the other hand, regrowth of bacteria in wash water is a possibility. The use of approach 1 gave much greater results, indicating much greater levels of contamination.

Another step is usually required to further concentrate the organisms to a reasonable volume for assay or purification to remove substances that may interfere with the assay procedure. Traditionally, microbiologists have relied on the growth or replication of the organism that is followed by an identification procedure usually involving growth on selective media, serology, or visual identification by characteristic features. Quantification takes place by direct counting of colonies on an agar plate, a dilution series [e.g., most probable number (MPN)], plaques produced by viruses in a monolayer of animal cells, or direct counting under a microscope (parasites). In more recent years methods using antibodies linked to enzymes or fluorescent dyes, gene probes, or amplification of the genetic material (DNA or RNA) have been used. Finally, further isolation of the specific microorganism may be performed to better characterize the pathogen (e.g., type or strain of the particular pathogen) or to confirm its identity.

There is no one ideal method for the detection and enumeration of microorganisms for all applications and all databases needed for risk assessment. Each method has certain advantages and limitations. Many of the more sophisticated and specific methods require more time at a greater cost to obtain the results. Also, methods are not available for the detection of many pathogens or emerging microbial contaminants. Thus methods development is a continuing process. The ideal method for risk assessment should have the attributes shown in Table 5-2.

The ability to quantify the viable numbers of organisms is desirable when assessing exposure. This may not be possible because many microorganisms may not be culturable by current methods or methods are not available. Culturable methods have been used for the most part for bacteria and viruses. A number of environmental conditions, such as low levels of nutrients or toxic substances (e.g., disinfectants, metals, heat, ultraviolet light, etc.) place stresses on many persistent bacteria and may make it difficult to isolate the cells using cultivation techniques on selective or other media (7). Sometimes, only a small percentage of the total viable organisms may be detected using standard procedures for bacterial detection (7). In the case of viruses, not all of the virons observed under an electron microscope appear capable of infecting cells used in the laboratory. The ratio of virons to tissue culture infective cells may range from 1 : 50,000 in the case of rotavirus in children's stools to 1 : 100 in laboratory-adapted strains of poliovirus (50,61). This may be due to several factors: (1) the genome not containing all or only part of the nucleic acid of the virus, (2) lack of receptors on the host cell for the virus, and (3) inability of all the virons to find receptors during exposure to the cells. The work of Ward et al. (63) suggests that virons present in the stools of infected persons (which are released into the environment via sewage) are likely to be underestimated by current cell culture methods. In summary, all existing methods for pathogen detection in the environment will probably underestimate the true exposure due to inability to recover and detect the microorganisms efficiently.

The speed at which a pathogen is identified can be important for certain applications. For risk assessment the speed of the method is not critical; however, it is significant during extreme events (e.g., floods or hurricanes) for water or food and within the plant treatment or processing when assessment

TABLE 5-2 Method Attributes for Developing Databases for Risk Assessment

Attribute	Useful For:	Limitations
Quantitative	Quantifying relative exposures, regrowth, die-off, change in levels, measuring controls of exposure	Distributions are variable in time and place; nondetects may predominate; precision and accuracy variable because of genetic variability among the same organisms and because of sample matrix effects
Speed	Assessing extreme events: storms, treatment failure, contamination, and prevention of outbreaks; corrective action	Real time (within 1 hour) ideal; sensitivity is not critical
Sensitivity	Low-level exposure and endemic risks post controls and treatments	Time and effort required to assay large volumes of samples
Viability	Evaluation of survival through the environment and/or treatment	Often must be done with models and seeded laboratory or pilot studies
Specificity	Improved hazard identification	Generally, false negatives greater problem than false positives

of exposure is critical in operational malfunctions or public health decisions. The health impact of microbial pathogens is immediate, unlike many chemical contaminates, which may require a long-term or lifetime exposure before there is an impact on health (e.g., trihalomethanes produced by chlorination). Ideally, in these situations methods are needed that can produce results in real time (immediately or within 1 to 2 hours) before exposure can occur and corrective action can be taken. In reality, the most rapid methods require at least several hours to 24 hours to complete. In these situations, sensitivity of the method is not necessarily as important as speed. For example, if it is determined that if 1000 *Cryptosporidium* oocysts per 10 L when present in the raw intake water of a drinking water treatment plant will exceed its ability to prevent an outbreak or will produce an unacceptable risk to the population consuming the water, a method of greater sensitivity is not needed. Speed becomes critical in this situation since it allows an operator of the plant to take corrective action to reduce the level of oocysts or notify public health officials that a "boil water" order may be needed.

The level of sensitivity needed for any given method depends on the expected concentration of organisms at the sampling site and the expected exposure. Highly sensitive methods are not necessary for the detection of enteric pathogens in sewage; only a few liters need to be sampled. Exposure directly to sewage would be expected to be incidental and accidental. In treated drinking water, methods capable of detecting pathogens in hundreds or thousands of liters are needed. Exposure to drinking water is direct and has a daily frequency.

Because the health effects are pathogen specific, it is necessary to have methods that are both specific and/or selective. Methods are required that can identify the genus, species, serological type, strain, and virulence of pathogens. Only certain strains of *E. coli* are capable of producing human illness. These strains represent only a small proportion of all the *E. coli* found in the environment. Over 140 enteric viruses are excreted in human feces, but some have never been identified with illness (e.g., reovirus), while others cause a very serious illness (hepatitis). Not all strains of waterborne protozoan parasites may be capable of producing illness in humans (e.g., *Giardia*).

Assessment of an organism's viability is essential in determining its ability to cause infection. As pointed out previously, not even cultural methods are totally dependable in determining if the organism is viable. Viability of an organism has traditionally been assessed on its ability to grow or reproduce. Bacterial methods typically assess the increase in numbers of the organisms by forming colonies on agar media or broth. Virus viability is addressed by replication in living cells (cell culture) (16,23), and the viability of protozoa by their ability to excyst (a necessary step in replication), infect animal models, or grow in cell culture (53–55). All of these methods require time, which becomes significant particularly for viruses, since it may require several weeks by cell culture methods to establish viability.

Methods are constantly under development to reduce or eliminate many of the problems with pathogen detection. The methods for bacteria, viruses, and protozoa are discussed briefly in the next section.

CULTIVATION TECHNIQUES

Bacteria

While cultivation techniques are well developed for enteric bacterial indicators such as coliform and fecal coliform bacteria from water, little attention has been paid to the development of methods for enteric bacterial pathogens. This is due in part to the success of these indicators in preventing the occurrence of bacterial waterborne disease outbreaks. More interest has been paid to the development of methods for bacterial pathogens in foods because the indicator bacteria are capable of growth in foods and the relationship with bacterial pathogens is not as reliable. If bacteria are in great enough concentrations, they can be assayed directly on agar media and counting the colonies that appear. This technique is useful for highly contaminated foods, water, or surfaces where the pathogens have been able to grow. More often, concentration by filtration is needed or enrichment where low numbers are allowed to replicate in a recovery medium prior to identification.

Three basic methods are used for detection and enumeration of bacteria in environmental samples (58): (1) most probable number (MPN), (2) membrane filter (MF), and (3) presence/absence (PA). The MPN method depends on addition of sample or diluted sample to a series of tubes of liquid media, usually three to five tubes per dilution containing growth medium for the organisms (Figure 5-2). Usually, the growth medium is selective, allowing only the growth of certain types of bacteria. Growth of the organisms is then observed by the production turbidity, acid, or gas. Once the positive tubes have been identified and recorded, it is possible to estimate the total number of organisms in the original sample by using an MPN table or more often a computer program that gives the number of organisms per certain volume. The dilution series usually starts with an enrichment medium. Presence of the pathogen is then confirmed by assaying media from the tubes onto selective media. Agar media can be used for isolation of colonies for further biochemical tests (8). MPN methods are very labor intense and require large amounts of media and glassware, and in the case of pathogens may require several days to complete.

The membrane filter test can be used with water and many liquids. In this procedure a given volume of liquid is passed through a filter with a pore size less than that of the bacteria, and then placed onto a growth medium (agar or a pad soaked in broth). The bacteria then grow on the surface of the membrane as individual colonies (Figure 5-3). This method is more precise,

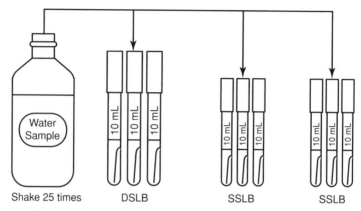

Figure 5-2 Most probable number method (MPN), used to detect and enumerate bacteria.

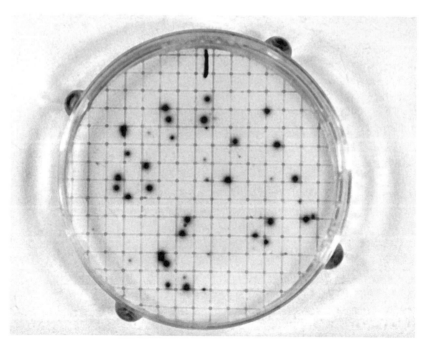

Figure 5-3 Colonies of enterococci bacteria growing on a membrane filter. (Photo by C. P. Gerba.)

less time consuming, and more rapid than the MPN method. Unfortunately, the small pore size of the filter limits the volume that can be assayed (usually, 100 to 1000 mL, depending on the amount of suspended matter in the liquid) and it cannot be used with solids such as food or soil.

Presence/absence tests are not quantitative tests—rather, they answer the simple question of whether or not the target organism is present in a sample. These types of methods have been established according to zero-tolerance standards in risk management approaches. That implies that no detectable pathogens would be present in some predetermined sample size. An example of this is the absence of coliforms per 100 mL of drinking water or the absence of *Listeria* in 10 g of a food product. The presence/absence method can be used as a pass/fail test and is often used in evaluation of processes or treatments aimed at exposure reduction and control.

Viruses

Cultivation methods for viruses began in the 1950s. The first environmental work focused on enteroviruses (such as polioviruses) and recovery and detection in wastewater. Methods for virus detection in water depend on the concentration from volumes of water ranging from 10 to 2000 L (Figure 5-4). This is accomplished by the adsorption of the viruses to positively charged filters that adsorb the negatively charged viruses from water (54). As with the membrane filter test for bacteria, these filters may become clogged from suspended matter in water, or organic matter such as humic acid may interfere with attachment of the virus to the filter. The viruses are then eluted from the filters with a protein solution (beef extract) of about 1 L and then further concentrated to a final volume by precipitation of the proteins to a final volume of 20 to 30 mL before assay (2).

Methods for the detection of viruses in other media, such as air, food, fomites, and biosolids, also require that the viruses are in a liquid medium before assay (24). This requires the elution of viruses or washing the viruses off a surface of a solid. For samples such as shellfish, meat, soil, or biosolids (sludge), the material is usually homogenized with a beef extract or other proteinaceous solution to elute the viruses. For samples such as lettuce, various approaches have been used, such as washing 10 heads of lettuce in a trash barrel, then filtering the wash water as described above.

Concentrates, whether from foods, soil, or water, are assayed on animal cells of a culture of human or primate origin. Because of the expense of producing animal cell culture, it is usually not practical to assay volumes larger than 50 mL. Selection of the cell line depends on the type of virus, cell lines being very specific in the types of viruses that will support growth. The most commonly used cell line, BGM cells, are very selective for the growth of enteroviruses. Thus most available information on viruses in water and food is for enteroviruses. The presence of viruses is indicated by the production of cytopathogenic effects (CPEs) (Figure 5-5): that is, destruction

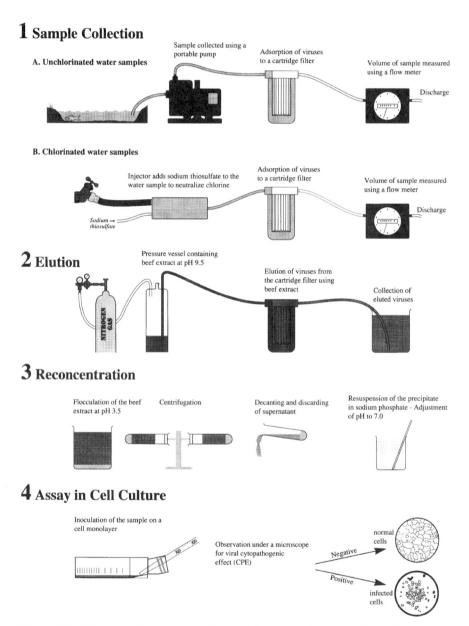

1 Sample Collection

A. Unchlorinated water samples

Sample collected using a portable pump

Adsorption of viruses to a cartridge filter

Volume of sample measured using a flow meter

Discharge

B. Chlorinated water samples

Injector adds sodium thiosulfate to the water sample to neutralize chlorine

Sodium → *thiosulfate*

Adsorption of viruses to a cartridge filter

Volume of sample measured using a flow meter

Discharge

2 Elution

Pressure vessel containing beef extract at pH 9.5

NITROGEN GAS

Elution of viruses from the cartridge filter using beef extract

Collection of eluted viruses

3 Reconcentration

Flocculation of the beef extract at pH 3.5

Centrifugation

Decanting and discarding of supernatant

Resuspension of the precipitate in sodium phosphate - Adjustment of pH to 7.0

4 Assay in Cell Culture

Inoculation of the sample on a cell monolayer

Observation under a microscope for viral cytopathogenic effect (CPE)

Negative

normal cells

Positive

infected cells

Figure 5-4   Procedure for concentration and detection of enteric viruses from water. (From I. L. Pepper, C. P. Gerba, and J. W. Brendecke, *Environmental Microbiology: A Laboratory Manual*, Academic Press, San Diego, Calif., 1995; with permission.)

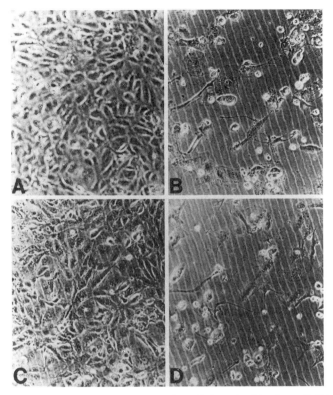

Figure 5-5   (A,C) Monolayer of kidney cells uninfected with virus; (B,D) monolayer of cells infected with poliovirus. Destruction of the cells by the virus results in cytopathogenic effects (CPEs). (Photo by Abid Nasser.)

of the individual cells. This can also be viewed by formation of plaques or clear zones [plaque-forming unit (PFU) method] produced by the destruction of cells under an agar overlay with a viable stain (only living cells are stained). The isolated viruses are then identified by serological neutralization tests.

These methods have largely been optimized for the detection of enteroviruses, and little information is available on other types of enteric viruses that may be present in equal or greater concentrations. The filters used to concentrate viruses from water do not concentrate all types of viruses with equal efficiency because of differences in charge on the different types of viruses (44). Several studies have reported greater concentrations of adenoviruses in sewage and sewage-polluted waters (18) than enteroviruses. Also, many viruses may grow in cell culture without the production of CPEs. Hepatitis A virus may grow quite well in cells without the production of CPEs (35). Cultivation methods for virus assays can take several days to many weeks before the virus produces CPEs and sometimes, substances concentrated from the samples are toxic to the cell culture.

Protozoa

To address infectivity, cell culture methods have been developed for *Cryptosporidium* (53). The method utilizes human Ileocecal adenocarcinoma cells (HCT-8) and detection of the infection process by the various life stages of the parasite after growth by immunofluorescence. The method, known as the *foci detection method* (FDM), is quantifiable by an MPN approach (55) and compares favorably to animal infectivity. Water filter concentrates can be processed with an immunomagnetic capture technique, and infectious oocysts can be detected in as little as 72 hours. The method has been used to examine survival of the parasite in water and inactivation by disinfectants.

MICROSCOPIC TECHNIQUES

Bacteria and protozoan parasites are large enough that they can be observed under a normal light microscope, allowing for detection and quantification. In the case of protozoa and other parasites, this has been the traditional method even to this day.

Bacteria

Microscopic techniques for bacteria through the use of antibodies, genetic probes, image analysis, and flow cytometry have become highly sophisticated, specific, and rapid (28). Although simple Gram stains and description of cell morphology are still used, these have little value for providing exposure data. Staining with specific genetic probes can address not only the total bacterial numbers but the genetic composition of populations and taxonomic status. Thus the state of the microorganism and its identification can now be ascertained. Applications for digital microscopy include quantification, viability, metabolic condition, and the structure of the microenvironment.

Sample concentration and purification are still necessary in some cases. Staining with probes and antibodies may be made on a solid matrix such as a glass slide or in suspension. For example, immunofluorescent microscopy with the development of specific monoclonal antibodies has been used to study the viable nonculturable state of *Vibrio cholera* (22). Generic nucleic acid stains and flow cytometry have followed the fate of *Salmonella typhymurium* in river and marine waters (29). For studies on *Legionella* in biofllms, cultural techniques have proven to be insensitive compared to direct microscopic detection methods (15).

Protozoa

Contaminated water has been the primary focus of methods developed for protozoa, primarily *Cryptosporidium* and *Giardia*. The various adaptations of the methods have been very successful at documenting the widespread occurrence of *Cryptosporidium* and *Giardia* in surface and drinking water in the United States (30–32,45). The cysts and oocysts in water are concentrated by size exclusion through 10-inch cartridge spun filters with a nominal pore size of 1 nm. These filters are also efficient in the collection of suspended matter in the water, and since they are extracted with the parasites, makes microscopic visualization of the parasite cysts or oocyst difficult (33,44). To extract the parasites the filters are cut apart and washed with an eluting solution of detergent. This eluate containing the cysts/oocysts and debris are further concentrated by centrifugation into a single pellet, and an aliquot of the concentrated pellet is layered onto a density gradient of Percoll and sucrose, where centrifugation separates the cysts/oocysts from much of the debris. This semipurified sample is then collected from the gradient and labeled with monoclonal antibodies specific to the cyst or oocyst cell wall using a specific fluorescent antibody (IFA) procedure. The sample can then be examined by epifluorescent microscopy for fluorescence (brilliant apple green), shape (ovoid or spherical), size, and by phase contrast or Nomarski differential interference contrast (DIC) microscopy for internal features (31).

The efficiency of recovery for cysts/oocyst for this process has been investigated in detail (32,40,45,46) with the overall recovery varying from 9 to 59%. The current methods for recovering and detecting parasites always underestimate the true concentration in environmental samples. While the use of fluorescein-labeled antibodies greatly aids the detection of cysts/oocyst background fluorescence, due to naturally fluorescing organisms, and nonspecific binding of the antibody may decrease their accurate identification. Although false positives can be problematic, Clancy et al. (6) found that the larger problem was false negatives. Another limitation is that no single antibody has been found to bind specifically only to species that cause infection in humans; thus those infecting only lower animals may also be detected. The viability of the cysts/oocysts cannot be assessed by IFA. LeChevallier et al. (32) reported that 10 to 30% of the organisms found in water samples were empty without internal features, suggesting that they were not viable. However, it is not clear whether this is an artifact to sample processing. With new cell culture systems the viability question can now be addressed (53).

While density gradients such as sucrose-percoll have been applied to the clarification of the sample, a more promising technique is immunomagnetic separation (IMS) (3,25). IMS uses antibodies tagged to iron beads and a magnetic system to pull the target oocysts and cysts from the suspension. This has applications to both microscopic detection and PCR (9,25). Several IMS kits are now available for *Cryptosporidium* (Dynal, Lake Success, New York; ImmuCell, Crypto-Scan, Portland, Maine).

Other enteric protozoa, such as *Cyclospora,* have had a greater association with foods (21). Thus there has been a need to develop methods to recover the oocysts from the surfaces of fruits and vegetables. Methods for recovery from produce have been relatively simplistic and involve washing the oocysts and cysts from the surfaces of fruits and vegetables with a detergent solution. These methods do not seem to be very efficient for several reasons. The first is the inability to wash the cysts and oocysts off, as they appear to have some stickiness. Second, only a small percentage of the crop or batch can be sampled. It may be more appropriate to sample irrigation water or develop a large batch rinse technique and sample the rinse waters. Better methods are needed for recovering oocysts and cysts from strawberries, raspberries, lettuce, and other produce.

No antibody to date is available for *Cyclospora.* Light microscopy, acid-fast staining, morphology (oocyst wall appearance, size, and shape) and the ability to sporulate have been used for identification. However, this has limited application to environmental samples. The oocyst of *Cyclospora* does have one unique feature and that is the ability to autofluoresce (12). The oocyst produces a distinct blue fluorescence around the outside wall and this aids in its detection. For *Microsporidia,* antibodies tagged similarly to FITC have been used for detection of the spores using immunofluorescent microscopy, but cross-reactivity of the antibody has been problematic for application to water samples (14).

Viruses

Viruses can be visualized with the use of the electron microscope. However, concentrations of 10^5 virons must be present to be seen. This technique is used primarily for clinical specimens and morphological description and identification of viruses after cultivation in cells. Some viruses, such as types of coliphage associated with human and animal fecal pollution, can be identified under the electron microscope.

MOLECULAR TECHNIQUES

Advances in molecular biology have allowed for the development of more rapid, sensitive, and low-cost approaches for detection of pathogens in the environment. These methods are designed to detect and analyze the genetic material of the organisms. Since each organism has a unique genetic code this can be used not only to identify specific species but also to "fingerprint" the strain and clone. Once a new pathogen has been identified, isolated, and its nucleic acid analyzed, these methods can rapidly be adapted and developed for application to risk assessment. These methods also offer the potential ability to detect microorganisms without the need for cultivation.

Probes (FISH)

Fluorescent in situ hybridization (FISH) is a technique that uses genetic probes that bind (hybridize) to the specific target sequences of a specific microorganism. The probe is tagged with fluorescent dyes and the binding is done with the cell intact: thus the name *in situ*. This has proven particularly useful for bacteria, whereby specific probes can be developed against ribosomal RNA sequences (50). FISH has also been used to identify *Cryptosporidium* (36). FISH techniques have great application for detection because the probe binds to the nucleic acid within the protozoan or bacterial cell, and both microscopy and the specificity of the probe can be used for identification and detection. This also allows for instrumentation to be used, such as flow cytometry and digital microscopy, which can greatly reduce the analytical time.

PCR

The polymerase chain reaction (PCR) has offered the most promise for the rapid detection of pathogens in the environment (59). This method involves the specific amplification of the DNA in the genome of the microorganism with the aid of primers. Primers are fragments of DNA that are complementary to the DNA strain to be amplified (sequences specific to the region of the genome to be amplified). Within a matter of a few hours, millions of copies of the genome are produced. The principle of the method involves the repetitive enzymatic synthesis of DNA. In the case of some viruses which only contain RNA as their genome, it is necessary first to convert the RNA to DNA with the use of a reverse transcriptase enzyme. This method is called *RT-PCR*. Amplification takes place only if the specific nucleic acid of the target organism is present (Figure 5-6). PCR as well as RT-PCR have also been used to identify the enteric protozoa, targeting the DNA or message RNA as a viability marker for *Cryptosporidium* and *Cyclospora* (25,27,43).

PCR has a number of issues that must be addressed when used for environmental samples:

- Small assay volumes (volumes must be further concentrated; immunomagnetic separation techniques have proven successful) (25)
- Inhibition by interfering substances in environmental samples (samples must be pretreated)
- Inability to determine live from dead organisms (can be combined with cell culture or potentially messenger RNA) (27,43)
- Nonquantitative assay (may be used with an MPN approach) (49)

Currently, the maximum volume that can be assayed is 100 μL (0.1 mL). Extracts or concentrates from environmental samples for enteric viruses and protozoa range from 2 to 30 mL or more. Thus further concentration is needed. Environmental samples and concentrates usually contain substances that interfere with detection by masking the target DNA or inhibiting the enzyme reaction. This results in often laborious and time-consuming proc-

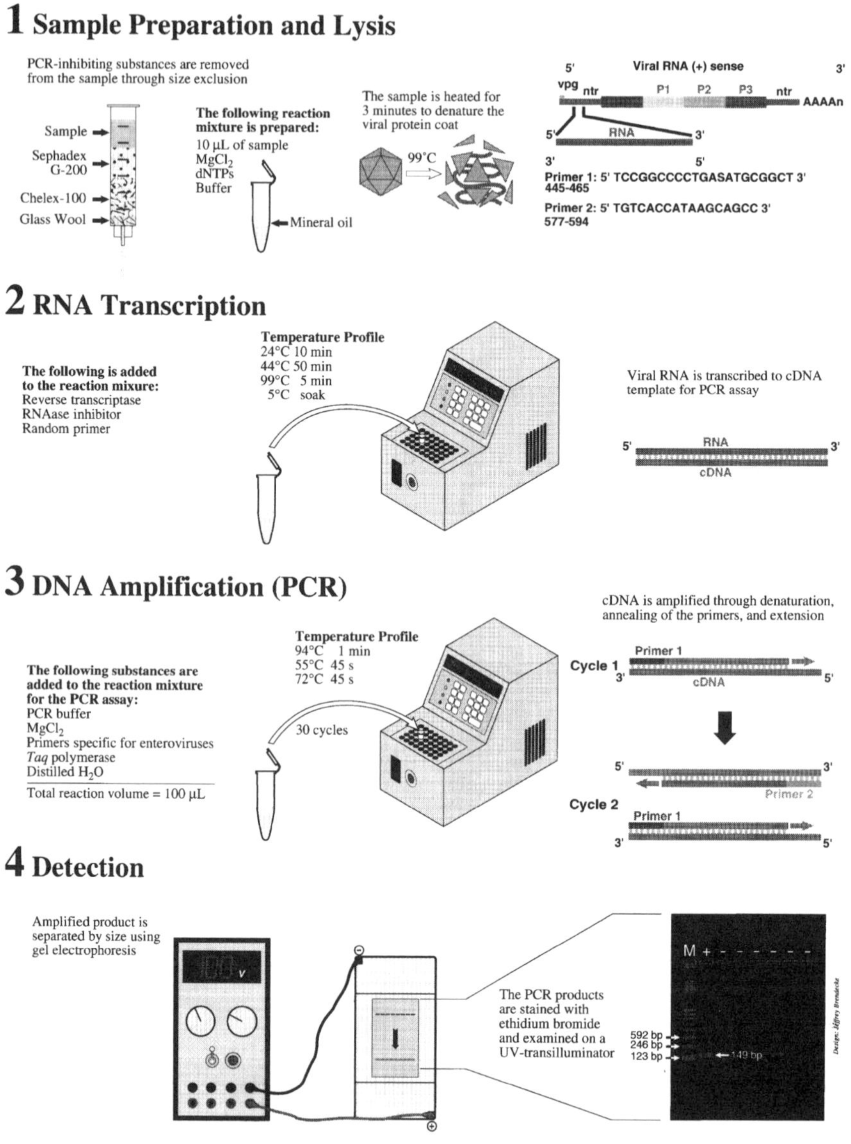

1 Sample Preparation and Lysis

PCR-inhibiting substances are removed
from the sample through size exclusion

Sample →
Sephadex → G-200
Chelex-100 →
Glass Wool →

**The following reaction
mixture is prepared:**
10 μL of sample
MgCl₂
dNTPs
Buffer

← Mineral oil

The sample is heated for
3 minutes to denature the
viral protein coat

99°C

5'   **Viral RNA (+) sense**   3'
vpg ntr   P1   P2   P3   ntr
                                        AAAAn
5'   RNA   3'
3'   5'
**Primer 1: 5' TCCGGCCCCTGASATGCGGCT 3'
445-465**
**Primer 2: 5' TGTCACCATAAGCAGCC 3'
577-594**

2 RNA Transcription

**The following is added
to the reaction mixure:**
Reverse transcriptase
RNAase inhibitor
Random primer

Temperature Profile
24°C 10 min
44°C 50 min
99°C 5 min
5°C   soak

Viral RNA is transcribed to cDNA
template for PCR assay

5'   RNA   3'
        cDNA

3 DNA Amplification (PCR)

**The following substances are
added to the reaction mixture
for the PCR assay:**
PCR buffer
MgCl₂
Primers specific for enteroviruses
Taq polymerase
Distilled H₂O
—————————————
Total reaction volume = 100 μL

30 cycles

Temperature Profile
94°C 1 min
55°C 45 s
72°C 45 s

cDNA is amplified through denaturation,
annealing of the primers, and extension

Primer 1
Cycle 1
3'   cDNA   5'

5'   3'
                    Primer 2
Cycle 2   Primer 1
3'   5'

4 Detection

Amplified product is
separated by size using
gel electrophoresis

The PCR products
are stained with
ethidium bromide
and examined on a
UV-transilluminator

M + - - - - - -

592 bp
246 bp   ← 149 bp
123 bp

Figure 5-6 Detection of enteroviruses by the polymerase chain reaction (PCR).
(From I. L. Pepper, C. P. Gerba, and J. W. Brendecke, *Environmental Microbiology:
A Laboratory Manual*, Academic Press, San Diego, Calif., 1995; with permission.

essing of samples (1,51,59). When these substances are removed it is possible to detect as few as one to two organisms; however, the processes of removing the interfering substances may result in significant loss of the number of target microorganisms. Studies have shown that use of antibody-capture techniques using magnetic beads not only concentrates the sample but removes interfering substances for PCR detection (25) and may also be useful prior to excystation/PCR to assess the potential for viability (9).

PCR will detect dead or inactivated microorganisms. Therefore, without cultivation procedures it is not possible to assess viability. Thus it could not be used to assess disinfection processes but is still useful for assessing occurrence where viability may not be an immediate need. Also, this is the only method available for the detection of some pathogens, such as Norwalk virus, which has not yet been cultivated. Thus, in outbreak investigations PCR has been used to identify the contaminated food (34). PCR, however, has been used successfully to reduce not only the time necessary for virus detection but to address viability when integrated with cell culture (43).

Finally, the test remains qualitative, with results present as either positive or negative. The best approach for developing a quantitative PCR may be a most probable number (49).

Typing

Identification and typing of pathogens has several applications:

- Studies that attempt to trace the sources and routes of transmission
- The assessment of virulence and antibiotic resistance of strains
- Study of phylogenetics or taxonomy, which attempts to clarify genetic relationships among serotypes and strains

Typing is actually an attempt to fingerprint an organism. Molecular methods do this by identifying differences in the nucleic acid of the genome or plasmids (extra chromosomal elements in bacteria). Every organism has a characteristic sequence of nucleotides that define the individual. This can be used to identify the organism and how closely organisms in the same genus or species are related. Three major methods are used (4): (1) restriction fragment length polymorphism (RFLP), (2) plasmid analysis, and (3) polymerase chain reaction (PCR).

RFLP and plasmid fingerprinting utilize restriction enzymes (endonuclease) to digest genomic DNA at specific recognition sites. The location of the restriction sites varies among the bacteria and with the use of gel electrophoresis of the products, bands or patterns are produced that are characteristic of the organism. Plasmid fingerprints are obtained. By selection of primers PCR results in certain sequences to be copied selectively. The primary PCR fingerprinting methods are arbitrarily primed PCR (AR PCR), repetitive se-

TABLE 5-3 Microorganisms and Methods Used for Various Data Sets

Hazards	Application	References
Cryptosporidium and *Giardia*	IFA microscopy for occurrence databases in surface and drinking water	19, 30–32, 47
	IFA microscopy and pilot and field tests for exposure reduction through water treatment (filtration)	38
Microsporidia	PCR detection of human types in water	10
Rotavirus	Cell culture and immunofluorescence for detection in various water types: sewage to drinking water	17
Enteroviruses and adenoviruses	PCR for detection in coastal water for recreational exposure	62
Astroviruses and HAV	PCR detection in mussels	61
Enteric viruses and enteric protozoa	IFA microscopy and cell culture for occurrence in reclaimed water	48
E. coli O157:H7	Immunoassays for rapid detection in beef	65
Vibrio vulnificus	Microscopy for detection of viable nonculturable states in marine waters	64
Bacteria associated with foodborne disease	Molecular subtyping for PulseNet program	58

quence PCR (Rep PCR), and PCR of ribosomal DNA followed by restriction digestion of the products. The last approach has been called PCR ribotyping.

Fingerprinting techniques have been used to determine the source of pathogens during an outbreak. One may examine if the source is animal or human, trace the spread of new strains of pathogens and the virulence of strains detected in food, and link environmental sources with hospitalized cases (5,11,20,26,39).

EXAMPLES OF METHODS AND DATA SETS FOR VARIOUS APPLICATIONS TO RISK ASSESSMENT

The methods cited in Table 5-3 and others are providing new insights into relationships among organisms and aiding in the identification of microorganisms and their sources, transport, and fate. Thus methods will continue to be used for hazard identification and exposure assessment. Interestingly, these methods have been used for assessment of the doses during development of dose–response studies. Therefore, the extrapolation of what is found in the environment and what may cause an infection can be made with greater confidence.

REFERENCES

1. Abbaszadegan, M., M. S. Huber, C. P. Gerba, and I. L. Pepper. 1993. Detection of enteroviruses in groundwater with the polymerase chain reaction. Appl. Environ. Microbiol. 59:1318–1324.

2. American Public Health Association, American Water Works Association, and the Water Environment Federation. 1992. A. E. Greenberg, L. S. Clesceri, and A. D. Eaton (eds.), *Standard methods for the examination of water and wastewater,* 18th ed. APHA, AWWA, and WEF, Baltimore.

3. Bilfulco, J. M., and F. W. Schaeffer III. 1993. Antibody-magnetite method for selective concentration of *Giardia lamblia* cysts from water samples. Appl. Environ. Microbiol. 59:(3):772.

4. Burr, M. D., and I. L. Pepper. 1998. A review of DNA fingerprinting methods for subtyping *Salmonella*. Crit. Rev. Environ. Sci. 28(3):283

5. Champliaud, D., P. Gobet, M. Naciri, O. Vagner, J. Lopez, J. C. Buisson, I. Varga, G. Harly, R. Mancassola, and A. Bonnin. 1998. Failure to differentiate *Cryptosporidium parvum* from *C. melegridis* based on PCR amplification of eight DNA sequences. Appl. Environ. Microbiol. 64:1454–1458.

6. Clancy, J. L., W. D. Gollnitz, and Z. Tabib. 1994. Commercial labs: how accurate are they? J. Am. Water Works Assoc. 86:89–97.

7. Colwell, R. R., P. Brayton, A. Huq, B. Tall, P. Harrington, and M. Levine. 1996. Viable but nonculturable *Vibrio cholera* 1 revert to a cultivable state in the human intestine. World J. Microbiol. Biotechnol. 12:28–31.

8. Cooper, R. C., and R. E. Danielson. 1997. Detection of bacterial pathogens in wastewater and sludge, pp. 222–230. *In* C. J. Hurst, G. R. Knudsen, M. J. McInerney, L. D. Stetsenbach, and M. V. Walter, eds. Manual of environmental microbiology. ASM Press, Washington, DC.

9. Deng, M. Q., D. O. Cliver, and T. W. Mariam. 1997. Immunomagnetic capture PCR to detect viable *Cryptosporidium parvum* oocysts from environmental samples. Appl. Environ. Microbiol. 63(8):3134–3138.

10. Dowd, S. E., C. P. Gerba, and I. Pepper. 1998. Confirmation of the human-pathogenic microsporidia *Enterocytozoon bieneusi, Encephalitozoon intestinalis* and *Vittaforma corneae* in water. Appl. Environ. Microbiol. 64(9). 3333–3335.

11. Dubois, E., F. Le Gutader, L. Haugarreau, H. Kopecka, M. Cormier, and M. Pommepuy. 1997. Molecular epidemiological survey of rotavirus in sewage by reverse transcriptase seminested PCR and restriction fragment length polymorphism. Appl. Environ. Microbiol. 63:1794–1800.

12. Dytrych, J. K., and R. P. D. Cooke. 1995. Autofluorescence of *Cyclospora*. Br. J. Biomed. Sci. 52(1):76.

13. England, B. L. 1982. Detection of viruses on fomites, pp. 179–220. *In* C. P. Gerba and S. M. Goyal, eds., Methods in Environmental virology. Marcel Dekker, New York.

14. Enriquez, F. J., O. Ditrich, J. D. Paiting, and K. Smith. 1997. Simple diagnosis of *Encephalitozoon* sp. Microsporidian infections by using a pan-specific antiexospore monoclonal antibody. J. Clin Microbiol. 35:724–729.

15. Fields, B. S. 1997. *Legionellae* and Legionnaires' disease. Pp. 666–675. *In* C. J. Hurst, G. R. Knudsen, M. J. McInerney, L. D. Stetzenbach, and M. V. Walter, eds., *Manual of environmental microbiology.* ASM Press, Washington, DC.

16. Gerba, C. P. 1987. Recovering viruses from sewage, effluents, and water. *In* G. Berg, ed., Methods for recovering viruses from the environment. CRC Press, Boca Raton, FL.

17. Gerba, C. P., J. B. Rose, C. N. Haas, and K. D. Crabtree. 1996. Waterborne rotavirus: a risk assessment. Water Res. 30(12):2929–2940.

18. Grohmann, G. S., N. J. Ashbolt, M. S. Genova, G. Logan, P. Cox, and C. S. W. Kueh. 1993. Detection of viruses in coastal and river water systems in Sydney, Australia. Water Sci. Technol. 27:457–461

19. Haas, C. N., and J. B. Rose. 1996. Distribution of *Cryptosporidium* oocysts in a water supply. Water Res. 30(10):2251–2254.

20. Herwaldt, B. L., J. F. Lew, C. L. Moe, D. C. Lewis, C. D. Humphery, S. S. Monoe, E. W. Pon, and R. L. Glass. 1994. Characterization of a variant strain of Norwalk virus from a food-borne outbreak of gastroenteritis on a cruise ship in Hawaii. J. Clin. Microbiol. 32:861–866.

21. Herwaldt, B. L., M.-L. Ackers, and T. C. W. Group. 1997. An outbreak in 1996 of cyclosporiasis associated with imported raspberries. N. Engl. J. Med. 336(22): 1548–1556.

22. Huq, A., R. R. Colwell, and R. Rahaman. 1990. Detection of *Vibrio cholerae* 01 in the aquatic environment by fluorescent-monoclonal antibody and culture methods. Appl. Environ. Microbiol. 56:2370–2373.

23. Hurst, C. J. 1997. Detection of viruses in environmental waters, sewage, and sewage sludge. Pp. 168–175. *In* C. J. Hurst, G. R. Knudsen, M. J. McInerney, L. D.

Stetzenbach, and M. V. Walter, eds., Manual of environmental microbiology. ASM Press, Washington, DC.

24. Hurst, C. J., G. R. Knudsen, M. J. McInerney, L. D. Stetzenbach, and M. V. Walter, eds. 1997. Manual of environmental microbiology. ASM Press, Washington, DC.

25. Johnson, D. W., N. J. Pieniazek, D. W. Griffin, L. Misener, and J. B. Rose. 1995. Development of a PCR protocol for sensitive detection of *Cryptosporidium* oocysts in water samples. Appl. Environ. Microbiol. 61(11):3849–3855.

26. Kapperud, G., L. M. Rorvik, V. Hasseltvedt, E. A. Hoiby, B. G. Iversen, K. Staveland, G. Johnsen, J. Leitao, H. Herikstad, Y. Andersson, G. Langeland, B. Gondrosen, and J. Lassen. 1995. Outbreak of *Shigella sonnei* infection traced to imported iceberg lettuce. J. Clin. Microbiol. 33:609–614.

27. Kaucner, C., and T. Stinear. 1998. Sensitive and rapid detection of viable *giardia* cysts and *Cryptosporidium parvum* oocysts in large-volume water samples with wound fiberglass cartridge filters and reverse transcription-PCR. Appl. Environ. Microbiol. 64(5):1743–1749.

28. Lawrence, J. R., J. McInerney, and D. A. Stahl. 1998. Analytical imaging and microscopy techniques, pp. 29–51. *In* C. J. Hurst, G. R. Knudsen, M. J. McInerney, L. D. Stetzenbach, and M. V. Walter, eds., Manual of Environmental Microbiology. ASM Press, Washington, DC.

29. Lebaron, P., N. Parthuisot, and P. Catala. 1998. Comparison of blue nucleic acid dyes for flow cytometric enumeration of bacteria in aquatic systems. Appl. Environ. Microbiol. 64(5):1725–1730.

30. LeChevallier, M. W., and W. D. Norton. 1995. *Giardia* and *Cryptosporidium* in raw and finished drinking water. *J. Am. Water Works Assoc.* 87(9):54.

31. LeChevallier, M. W., W. D. Norton, and R. G. Lee. 1991. Occurrence of *Cryptosporidium* and *Giardia* spp. in surface water supplies. Appl. Environ. Microbiol. 57:2610–2616.

32. LeChevallier, M. W., W. D. Norton, and R. G. Lee. 1991. *Giardia* and *Cryptosporidium* in filtered drinking water supplies. Appl. Environ. Microbiol. 57:2617–2621.

33. LeChevallier, M. W., and T. M. Trok. 1990. Comparison of the zinc sulfate and immunofluorescence techniques for detecting *Giardia* and *Cryptosporidium.* J. Am. Water Works Assoc. 82:75–82.

34. Lees, D. N., K. Henshilwood, J. Green, C. I. Gallomore, and D. W. G. Brown. 1995. Detection of small round structured viruses in shellfish by reverse transcription-PCR. Appl. Environ. Microbiol. 61:4418–4424.

35. Lemon, S. M., L. N. Binn, and R. H. Marchwicki. 1983. Radioimmunofocus assay for quantitation of hepatitis A virus in cell cultures. J. Clin. Microbiol. 17:834–839.

36. Lindquist, J. A. D. 1997. Probes for the specific detection of *Cryptosporidium parvum.* Water Res. 31(10):2668–2671.

37. Melnick, J. L., R. Safferman, V. C. Rao, S. Goyal, G. Berg, D. R. Dahling, B. A. Wright, E. Akin, R. Setler, C. Sorber, B. Moore, M. D. Sobsey, R. Moore, A. L. Lewis, and F. M. Wellings. 1984. Round robin investigation of methods for the recovery of poliovirus from drinking water. Appl. Environ. Microbiol. 47:144–150.

38. Nieminski, E. C., and J. E. Ongerth. 1995. Removing *Giardia* and *Cryptosporidium* by conventional and direct filtration. J. Am. Water Works Assoc. 87:96–106.

39. Niu, M. T., L. B. Plish, B. H. Robertson, B. K. Khanna, B. A. Woodruff, C. N. Shapiro, M. A. Miller, J. D. Smith, J. K. Gedrose, M. J. Alter and H. S. Margolis. 1992. Multistate outbreak of hepatitis A associated with frozen strawberries. J. Infect. Dis. 166:518–524.

40. Ongerth, J. E., and H. H. Stibbs. 1987. Identification of *Cryptosporidium* oocysts in river water. Appl. Environ. Microbiol. 53:672–679.

41. Relman, D. A., T. M. Schmidt, A. Jajadhar, M. Sogin, J. Cross, K. Yoder, O. Sethabutr, and P. Echeverria. 1995. Molecular phylogenetic analysis of *Cyclospora,* the human intestinal pathogen, suggests that it is closely related to *Eimeria* species. J. Infect. Dis. 173:440–445.

42. Reynolds, K. A., C. P. Gerba, and I. L. Pepper. 1991. Detection of infectious enteroviruses by integrated cell culture-PCR procedure. Appl. Environ. Microbiol. 62:1424–1427.

43. Reynolds, K. A., C. P. Gerba, and I. L. Pepper. 1996. Detection of infectious enteroviruses by an integrated cell culture-PCR procedure. Appl. Environ. Microbiol. 62(4):1424–1427.

44. Rose, J. B., S. N. Singh, C. P. Gerba, and L. M. Kelly. 1984. Comparison of microporous filters for concentration of viruses from wastewater. Appl. Environ. Microbiol. 47:989–992.

45. Rose, J. B., H. Darbin, and C. P. Gerba. 1988. Correlations of the protozoa, *Cryptosporidium* and *Giardia* with water quality variables in a watershed. Water Sci. Technol. 20:271–276.

46. Rose, J. B., L. K. Landeen, R. K. Riley, and C. P. Gerba. 1989. Evaluation of immunofluorescence techniques for detection of *Cryptosporidium* oocysts and *Giardia* cysts from environmental samples. Appl. Environ. Microbiol. 55:3189–3195.

47. Rose, J. B., C. P. Gerba, and W. Jakubowski. 1991. Survey of potable water supplies for *Cryptosporidium* and *Giardia.* Environ. Sci. Technol. 25:1393–1400.

48. Rose, J. B., L. J. Dickson, S. R. Farrah, and R. P. Carnahan. 1996. Removal of pathogenic and indicator microorganisms by a full scale water reclamation facility. Water Res. 30(11):2785–2797.

49. Rose J. B., X. Zhou, D. W. Griffin, and J. H. Paul. 1997. Comparison of PCR and plague assay for detection and enumeration of coliphage in polluted marine waters. Appl. Environ. Microbiol. 63(11):4564–4566.

50. Sayler, G. S., and A. C. Layton. 1990. Environmental application of nucleic acid hybridization. Annu. Rev. Microbiol. 44:625–648.

51. Schwab, K. J., R. DeLeon, and M. D. Sobsey. 1995. Concentration and purification of beef extract mock eluates from water samples for the detection of enteroviruses, hepatitis A virus, and Norwalk virus by RT-PCR. Appl. Environ. Microbiol. 61:531–537.

52. Sharp, D. G. 1965. Electron microscopy and viral particle function. Pp. 193–217. *In* G. Berg, ed., Transmission of viruses by the water route. Wiley, New York.

53. Slifko, T. R., D. E. Friedman, J. B. Rose, S. Upton, and W. Jakubowski. 1997. An in-vitro method for detection of infectious *Cryptosporidium* oocysts using cell culture. Appl. Environ. Microbiol. 63(9):3669–3675.

54. Slifko, T. R., D. E. Friedman, J. B. Rose, S. J. Upton, and W. Jakubowski. 1997. Unique cultural methods used to detect viable *Cryptosporidium parvum* oocysts in environmental samples. Water Sci. Technol. 35(11–12):363–368.

55. Slifko, T. R., D. E. Friedman, and J. B. Rose, 1998. Comparison of 4 *Cryptosporidium parvum* viability assays: DAPI/PI, excystation, cell culture and animal infectivity. *In* Proceedings of the Water Quality Technology Conference, San Diego, CA, November, American Water Works Association, Denver, CO.

56. Sobsey, M. D., and J. S. Glass. 1980. Poliovirus concentration from tap water with electropositive adsorbent filters. Appl. Environ. Microbiol. 40:201–210.

57. Spendlove, J. C., and K. F. Fannin. 1982. Methods for the characterization of virus aerosols. *In* C. P. Gerba and S. M. Goyal., eds., Methods in environmental virology. Marcel Dekker, New York.

58. Tauxe, R. V. 1998. New approaches to surveillance and control of emerging food-borne infectious diseases. Emerg. Infect. Dis. 4(3):455–456.

59. Toranzos, G. A. 1997. Environmental applications of nucleic acid amplification techniques. Technomic Publishers, Lancaster, PA.

60. Toranzos, G. A., and G. A. McFeters. 1997. Detection of indicator microorganisms in environmental freshwaters and drinking waters. Pp. 184–194. *In* C. J. Hurst, G. R. Knudsen, M. J. Mclnerney, L. D. Stetsenbach, and M. V. Water, eds., Manual of environmental microbiology. ASM Press, Washington, DC.

61. Traore, O., C. Arnal, B. Mignotte, A. Maul, H. Laveran, S. Billaudel, and L. Schwartzbrod. 1998. Reverse transcriptase PCR detection of astrovirus, hepatitis A virus and poliovirus in experimentally contaminated mussels: comparisons of several extraction and concentration methods. Appl. Environ. Microbiol. 64(8):3118–3122.

62. Vantarkis, A. C., and M. Papapetropoulou. 1998. Detection of enteroviruses and adenoviruses in coastal water of SW Greece by nested polymerase chain reaction. Water Res. 32(8):2365–2372.

63. Ward, R. L., D. R. Knowlton, and M. J. Perce. 1984. Efficiency of human rotavirus propagation in cell culture. J. Clin. Microbiol. 19:748–753.

64. Warner, J. M., and J. D. Oliver. 1998. Randomly amplified polymorphic DNA analysis of starved and viable but nonculturable *Vibrio vunificus* cells. Appl. Environ. Microbiol. 64(8):3025–3028.

65. Woody, J.-M., J. A. Stevenson, R. A. Wilson, and S. J. Knabel. 1998. Comparison of the Difco EZ Coli℠ Rapid Detection System and Petrifilm℠ Test Kit-HEC for Detection of *Escherichia coli* O157:H7 in fresh and frozen ground beef J. Food Prot. 61(1)110–112.

CHAPTER 6

EXPOSURE ASSESSMENT

CONDUCTING THE EXPOSURE ASSESSMENT

The purpose of the exposure assessment is to determine the amount, or number, of organisms that correspond to a single exposure (termed the *dose*), or the total amount or number of organisms that constitute a set of exposures. We are interested in both the expected dose and the distribution of doses. If it is imagined that a large number of exposures occur e.g., either by having a large number of persons exposed, by having a few persons exposed many times, or by some combination), the expected dose would be the average dose among all those exposed, and the dose distribution would be the probability distribution of doses (organisms/exposure).

The problem of ascertaining exposure thus can be divided into one of ascertaining microorganism concentration in a medium (water, air, food) and the consumption amount of the medium. If μ is the concentration and m is the consumption per exposure, the expected dose ($\bar{d}$) would be given by

$$\bar{d} = E(\mu m) \tag{6-1}$$

Furthermore, if μ and m are statistically independent (in other words, if there is no correlation between the amount consumed in a single exposure and the concentration of organisms in that exposure), by the behavior of expectations, the dose may be computed via

$$\bar{d} = \overline{\mu m} \tag{6-2}$$

where the overbar symbol denotes the operation of taking the arithmetic mean.

For example, if the average concentration of a given bacterial pathogen in a ready-to-eat food is 1 organism per gram, and the average portion size is 25 g, the average dose per exposure is 25 organisms.

The focus of this chapter is on estimating the mean and distribution of exposure to microorganisms. The key differentiating characteristic of microorganisms in this regard, compared to chemical agents, is that microorganisms are discrete "particles" at a sufficiently low density that the statistics of their distribution must be considered. In the example above, for instance, exposure to an average of 25 organisms is anticipated, and therefore in repeated doses we might reasonably expect low (e.g., 5 or 10 organisms) and high doses (e.g., 50 organisms). If the risk is substantially nonlinear (i.e., if the risk for a person who is known to ingest 50 organisms is very different than 10 times the risk for a person who is known to ingest 5 organisms), it becomes important to characterize this distribution with some precision and accuracy.

CHARACTERIZING CONCENTRATION/DURATION DISTRIBUTIONS

What are the average and distribution of microorganism concentrations in a medium such as food or water? Of necessity, experimental samples must be obtained, and either the number of organisms in each of a set of samples or the presence or absence of organisms in each of a set of samples must be determined. From this information we wish to determine the average and/or distribution of microorganism densities.

Random (Poisson) Distributions of Organisms

Conceptually, the baseline against which all microorganism occurrence distributions are measured is the *Poisson distribution*. If organisms are distributed "randomly," then in a volume V, the probability that a sample (x) will contain N organisms (including $N = 0$) will be found is given by the Poisson distribution (24,35):

$$P(x = N) = \frac{(\overline{\mu} V)^N}{N!} \exp(-\overline{\mu} V) \qquad (6\text{-}3)$$

where $\overline{\mu}$ is the mean density, which is assumed to be constant among all samples. Since this distribution has only one parameter, $\overline{\mu}$, once the mean density is known, the distribution is completely specified. In particular, the average number of organisms expected to be found in a set of samples each of volume V is equal to $\overline{\mu} V$, and the variance in the number of organisms among replicates of equal volume is also equal to $\overline{\mu} V$ (54).

Equation 6-3 can be generalized to yield a result that we will employ later. The probability (given a Poisson distribution) that a sample (x) will have between N_L and N_U organisms can be related to a sum of Poisson terms as follows:

$$P(N_L \leq x \leq N_U) = \sum_{N=N_L}^{N=N_U} \frac{(\overline{\mu} V)^N}{N!} \exp(-\overline{\mu} V) \tag{6-4}$$

Also, we note that if the upper limit is infinity, we can obtain a complementary cumulative distribution as

$$P(N_L \leq x \leq \infty) = 1 - P(0 \leq x \leq (N_L - 1))$$

$$= 1 - \sum_{N=0}^{N=(N_L-1)} \frac{(\overline{\mu} V)^N}{N!} \exp(-\overline{\mu} V) \tag{6-5}$$

The complementary cumulative Poisson distribution is related to the mathematical incomplete gamma function[1] by the following relationship (79):

$$P(N_L \leq x \leq \infty) = 1 - P(0 \leq x \leq (N_L - 1)) = 1 - \Gamma(N_L, \overline{\mu} V) \tag{6-6}$$

Since a growing number of available software programs have access to the incomplete gamma function, equation 6-6 may be easier to use than equation 6-5.

   If a set of samples from a large body of material (a lake, a finished drinking water, a lot of hamburger, etc.) are taken, in which the actual number of organisms are measured (e.g., 0, 1, 2, etc.) in each sample, we would like to use these data to estimate the mean density $(\overline{\mu})$ in the large body of material from which the samples were taken. The technique may involve plating bacteria on solid medium (and counting colonies), inoculating virus on a "lawn" of cell culture, and counting plaques or infectious foci, or directly counting organisms by microscopy or particle counting instruments. It is presumed that the samples have been taken in a random fashion from the large body of material. Under these circumstances, the *principle of maximum likelihood* (ML) can be used to estimate the mean *if we assume a particular distribution.* The principle states that the best estimate of a set of parameters from a set of data is obtained by maximizing the probability that the particular sample would have been obtained. We will use these maximum likelihood estimators (MLEs) widely in a number of contexts. The MLE process has a number of advantages, in particular (74,75):

· It is asymptotically (for large samples) unbiased and often performs very well for small samples.
· It asymptotically yields the estimator of minimum variance.

---

[1]Here $\Gamma(a,x)$ is defined as $[1/\Gamma(a)] \int_0^x e^{-t} t^{a-1} \, dt$.

· Confidence limits and goodness of fit can easily be computed.

For a Poisson distribution, if a number of samples ($i = 1, 2, \ldots, k$) are taken potentially with different volumes, we measure the number of organisms in each sample (as N_i), and if all samples are independent, the likelihood function (probability of obtaining the results given unknown parameters) can be written as

$$L = \prod_{i=1}^{k} \frac{(\overline{\mu} V_i)^{N_i}}{N_i!} \exp(-\overline{\mu} V_i) \qquad (6\text{-}7)$$

where the symbol Π refers to the product of subscripted terms (analogous to Σ for summation). Note that since the value of $\overline{\mu}$ that maximized 6-7 is independent of the $N_i!$ term, and with logarithms of both sides taken, the MLE can also be obtained by minimizing the following quantity:

$$-\ln(L') = \overline{\mu} \sum_{i=1}^{k} V_i - \sum_{i=1}^{k} N_i \ln(\overline{\mu} V_i)$$

$$= \overline{\mu} \sum_{i=1}^{k} V_i - \ln(\overline{\mu}) \sum_{i=1}^{k} N_i - \sum_{i=1}^{k} N_i \ln(V_i) \qquad (6\text{-}8)$$

Estimation of Poisson Mean in Count Assay (Constant and Variable Volumes). Equation 6-8 can be applied directly to a body of data, and by trial and error the value of $\overline{\mu}$ that minimizes $-\ln(L')$ (negative log-likelihood) can be found. However, for the Poisson distribution, this is one of the few cases in which the optimum likelihood can be found analytically. Equation 6-8 is differentiated with respect to $\overline{\mu}$, and the result set equal to zero, in order to determine the extrema. The process is as follows:

$$\frac{d(-\ln(L'))}{d(\overline{\mu})} = \sum_{i=1}^{k} V_i - \frac{1}{\overline{\mu}} \sum_{i=1}^{k} N_i = 0 \qquad (6\text{-}9a)$$

Therefore,

$$\overline{\mu}_{\mathrm{ML}} = \frac{\sum_{i=1}^{k} N_i}{\sum_{i=1}^{k} V_i} \qquad (6\text{-}9b)$$

By computation of the second derivative, which is strictly positive (not shown), it can be verified that this is indeed a minimum. The subscript ML is used to denote that the quantity $\overline{\mu}_{\mathrm{ML}}$ is the maximum likelihood estimator of the quantity $\overline{\mu}$ (which, unsubscripted, is the "true" value presupposed in the population from which samples were taken). Application of this result is shown in Example 6-1.

Example 6-1.   Two samples of water were analyzed. Counts of organisms were obtained, and results are given in Table 6-1. Different volumes were used for each set. Assume a Poisson distribution and compute the estimated mean.

SOLUTION.   Adding up each column, we get the total volume analyzed (ΣV_i) and the total number of organisms (ΣN_i) for sample A and sample B as 25,27, and 27L, respectively. Hence, in both cases, the ML estimate of mean density is 27/25 or 1.08 L^{-1}.

TABLE 6-1   Count Data for Example 6-1

Volume (L)	Organisms Found	
	Sample A	Sample B
1	1	1
1	2	0
1	1	5
1	0	1
1	0	0
2.5	1	5
2.5	1	0
2.5	3	1
2.5	6	5
5	8	1
5	4	8

Count Assay with Upper Limits.   In some cases, particularly at very high microorganism concentrations, it may be impossible to exactly count the number of organisms in a particular sample. For example, in bacterial counts on agar plates, a result of TNTC (too numerous to count) is recorded if more than 100 colonies occur on a small plate or more than 300 occur on a large plate (1). It is possible to estimate a Poisson mean with these data by a modification of the method described above. If we have a set of analyses consisting of j determinations ($i = 1, \ldots, j$) for which exact counts are available, and k determinations ($i = j + 1, \ldots, j + k$) for which only a lower limit is available (which may be the same or different for each determination), we can use a likelihood function based on equation 6-7 for the former subset of data and a likelihood function based on equation 6-5 for the latter data. This yields

$$L = \left[\prod_{i=1}^{j} \frac{(\overline{\mu} V_i)^{N_i}}{N_i!} \exp(-\overline{\mu} V_i) \right] \left\{ \prod_{i=j+1}^{k+j} [1 - \Gamma((N_{L,i} - 1), \overline{\mu} V_i)] \right\} \quad (6\text{-}10)$$

where $N_{L,i}$ is the lower limit to the uncountable range in sample i (e.g., 100 or 300 colonies). If we take the negative logarithm of this equation, we can obtain (neglecting logarithms of terms not depending on $\overline{\mu}$)

$$-\ln(L') = \overline{\mu} \sum_{i=1}^{j} V_i - \sum_{i=1}^{j} N_i \ln(\overline{\mu} V_i)$$

$$- \sum_{i=j+1}^{j+k} \ln[1 - \Gamma((N_{L,i} - 1), \overline{\mu} V_i)] \qquad (6\text{-}11)$$

Example 6-2 illustrates the use of this equation.

Example 6-2. Table 6-2 shows count data taken on a water. As noted, one of the 10-mL samples and all of the 100-mL samples had TNTC counts (defined as ≥ 100 organisms). Estimate the mean density if a Poisson distribution is assumed.

TABLE 6-2 Count Data for Example 6-2

Volume (mL)	Count
1	12
1	8
1	15
10	40
10	58
10	≥ 100
100	≥ 100
100	≥ 100
100	≥ 100

SOLUTION. Equation 6-11 is used. A range of values of the mean density $(\overline{\mu})$ is assumed and the value of $-\ln(L')$ is tabulated. The results are shown in Figure 6-1. The minimum of the function $-\ln(L')$ is obtained where $\overline{\mu}$ is about 7 mL^{-1}, and this is selected as the value of $\overline{\mu}_{ML}$.

The precise location of the minimum can be found by numerical optimization. EXCEL with the SOLVER add-in function can be used (36), as can general-purpose mathematical software such as MATLAB (106). In the appendix to this chapter we describe the use of the SOLVER in performing an unconstrained optimization.

Estimation with Quantal Assay. In quantal assays of microbial densities, a known amount (mass or volume) of sample is analyzed for the presence or absence of organisms. The prototypical assay is the most probable number (MPN) assay for coliforms (16,35,41), in which an amount of sample is added to a nutrient medium and the presence or absence of growth is taken as

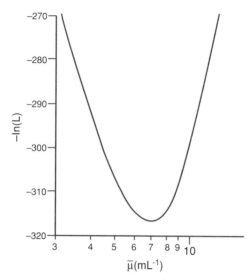

Figure 6-1   Likelihood function versus mean density for example 6-2.

signifying the presence or absence of organisms. In addition, viral and other nonbacterial infectious agents may be analyzed by the ability of a given volume of sample to produce a positive response in a tissue culture system (13,69,70,116). Results from virus monitoring in which negative samples are found (and only negative samples) can also be interpreted using the framework of quantal assays (19).

In a dilution assay, r aliquots of volumes[2] (r may be one or some number greater than 1) $V_1, V_2, \ldots, V_r$ (e.g., 10 mL, 1 mL, 0.1 mL, etc.) of a single sample are each inoculated into $n_1, n_2, \ldots, n_r$ systems (e.g., dilution tubes). At the conclusion of the assay it is determined that $p_1, p_2, \ldots, p_r$ of the systems are "positive" (e.g., show growth). The experimental layout is illustrated schematically in Table 6-3.

Given the results of such an experiment, we wish to determine the best estimate of the density (number/volume) of organisms from which the sample was drawn.

The formal likelihood function for analysis of this experiment can be developed beginning with the theory of the binomial distribution. For any one set i, the probability of observing p_i positive replicates out of n_i trials may be written as

---

[2]For this discussion it is assumed that volumes of sample are used. However, the experiment can be conducted in which different *masses* of sample are used. In this case the density would be determined on a number/mass basis, and the mass inoculated into each system, rather than the volume, would form the basis of the computation.

TABLE 6-3 Schematic Layout of Dilution Assay

Set	Volume per Assay Replicate	Number of Replicates in the Set	Positive Replicates
1	V_1	n_1	p_1
2	V_2	n_2	p_2
3	V_3	n_3	p_3
4	V_4	n_4	p_4

$$P(p_i) = \frac{n_i!}{p_i! \, (n_i - p_i)!} \, \pi_i^{p_i} (1 - \pi_i)^{n_i - p_i} \tag{6-12}$$

where π_i is the probability that a single replicate will have one or more organisms in it. It is assumed that if an organism is inoculated into a particular replicate, a positive response will certainly follow. Under these circumstances, if the organisms are distributed randomly (according to the Poisson distribution), then by application of equation 6-5 (with $N_L = 1$), we can compute

$$\pi_i = 1 - \exp(-\overline{\mu} V_i) \tag{6-13}$$

Now, equation 6-13 is substituted into equation 6-12, and terms for all r (i.e., all volumes) are multiplied, to get the following likelihood function:

$$L = \prod_{i=1}^{r} \frac{n_i!}{p_i! \, (n_i - p_i)!} \, [1 - \exp(-\overline{\mu} V_i)]^{p_i} [\exp(-\overline{\mu} V_i)]^{n_i - p_i} \tag{6-14}$$

By the principle of maximum likelihood, the MPN estimate of $\overline{\mu}$, which is also the MLE estimate of $\overline{\mu}$, is given by finding the value of $\overline{\mu}$ that maximizes L in equation 6-11, or equivalently that minimizes $-\ln(L')$:

$$-\ln(L') = \sum_{i=1}^{r} \{-p_i \ln[1 - \exp(-\overline{\mu} V_i)] + (n_i - p_i)(\overline{\mu} V_i)\} \tag{6-15}$$

If there is only one sample volume used (if $r = 1$), equation 6-12 may be differentiated with respect to $\overline{\mu}$ to get the ML estimate directly. The following process results:

$$\frac{d(-\ln(L'))}{d\overline{\mu}} = \left[\frac{pV \exp(-\overline{\mu} V)}{1 - \exp(-\overline{\mu} V)} + (n - p)V \right] = 0$$

$$\mu = -\frac{1}{V} \ln \frac{n - p}{n} \tag{6-16}$$

Use of this relationship is illustrated below.

Example 6-3. Samples (25 g) of frozen chickens were analyzed for the presence of *Salmonella* (78). Of 31 samples, 10 had detectable *Salmonella*. Assuming that the distribution of organisms was Poisson, and that the assay was capable of detecting one or more organisms in each sample, if they were present, estimate the mean *Salmonella* level.

SOLUTION. Applying equation 6-13 directly leads to the following result:

$$\bar{\mu} = -\frac{1}{25 \text{ g}} \ln \frac{31 - 10}{31} = 0.0155 \text{ g}^{-1}$$

Therefore, we estimate the mean *Salmonella* level as 1.55 per 100 g of frozen chicken.

If multiple volumes are used, the ML estimate must be obtained by solving equation 6-15 for the value of μ that minimizes $-\ln(L')$. The process may be conducted in a manner similar to that in Example 6-2. Example 6-4 illustrates the result for a typical dilution assay used in the estimation of bacterial levels in a water supply.

Example 6-4. Analysis of a water supply by the MPN coliform assay was performed by a three-decimal dilution (10 mL, 1 mL, 0.1 mL) assay using five tubes at each dilution. Five, three, and zero positives, respectively, were found. Estimate the mean concentration in the water that was sampled.

SOLUTION. From equation 6-12, the log-likelihood equation may be written as

$$-\ln(L') = \{-5 \ln[1 - \exp(-\bar{\mu} \times 10)] + (5 - 5)(\bar{\mu} \times 10)\}$$
$$+\{-3 \ln[1 - \exp(-\bar{\mu} \times 1)] + (5 - 3)(\bar{\mu} \times 1)\}$$
$$+\{-0 \ln[1 - \exp(-\bar{\mu} \times 0.1)] + (5 - 0)(\bar{\mu} \times 0.1)\}$$

This can be evaluated at various values of $\bar{\mu}$ in a manner similar to Example 6-2. The value that minimizes this expression is $\bar{\mu} = 0.792$ mL^{-1}, which is estimated at the MLE (or MPN).

In many analyses, the same set of dilutions and number of tubes are used repetitively. For these applications, standard tables reporting the solution of the likelihood equation for the most frequent, and probable, combinations of responses have been tabulated. For example, *Standard Methods* (1) contains tables for experiments based on three- and four-decimal dilution and five tubes per dilution, which are commonly used designs for water coliform determinations.

The precision and accuracy of estimation using the MPN assay are functions of the design of the experiment (number of dilutions, aliquots, dilution volumes). Unlike methods in which actual counts of organisms are made, the MLE estimate of $\overline{\mu}$, like many other ML measurements (44,74,75,96), shows increasing bias as the number of dilutions or aliquots are reduced.

Salama et al. (86) developed a correction to the MPN estimates which appears to reduce the bias by about 90% for typical dilution series used (37). Figure 6-2 presents the bias estimate for a three-dilution (10, 1, and 0.1 mL) five-tube (per dilution) experiment, for a four-dilution (10, 1, 0.1, and 0.01 mL) five-tube experiment, and for a four-dilution 50-tube experiment. In this procedure, if $\overline{\mu}_{ML}$ is the MLE estimate of the mean, a "corrected" estimate is obtained as $\overline{\mu}_B$ by the following correction derived from a Taylor series expansion out to second order:

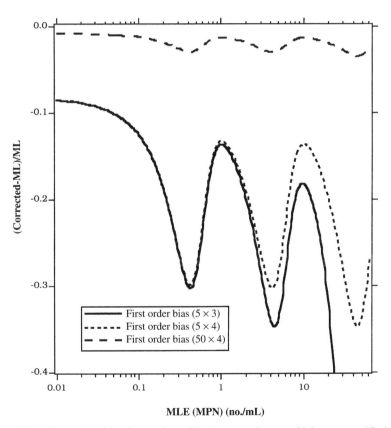

Figure 6-2   First-order bias for various dilution experiments. Volumes are 10, 1, 0.1, and (for the four-dilution experiment) 0.01 mL, five or 50 tubes per dilution. Bias is estimated by the method of Salama.

$$\overline{\mu}_B = \overline{\mu}_{ML} - \frac{1}{2} \sum_{i=1}^{r} w_i z_i \exp(-\overline{\mu}_{ML} V_i)$$

where

$$z_i = n_i[1 - \exp(-\overline{\mu}_{ML})]$$

$$w_i = \frac{V_i^2}{2[1 - \exp(-\overline{\mu}_{ML} V_i)]^2 D^3} \left\{ \sum_{j=1}^{r} \frac{V_j^3 z_j \sinh(\overline{\mu}_{ML} V_j)}{[\cosh(\overline{\mu}_{ML} V_j) - 1]^2} \right\}$$

$$- \frac{V_i^3}{[1 - \exp(-\overline{\mu}_{ML} V_i)][\cosh(\overline{\mu}_{ML} V_i) - 1]D^2}$$

$$D = \sum_{i=1}^{r} \frac{V_i^2 z_i}{2[\cosh(\overline{\mu}_{ML} V_i) - 1]} \tag{6-17}$$

As shown in Figure 6-2, the relative difference in the corrected estimate is substantial for the five-tube experiments, averaging about -20 to -25%; in other words, the value of $\overline{\mu}_{ML}$ (in five-tube experiments) generally overestimates the true mean microorganism density by about 20 to 25%. Although the derivation is obscure, early workers used empirical MPN correction factors of 85% (108), which is consistent with this finding. Also note from Figure 6-2 that by using a large number of replicates per dilution, the bias can be reduced substantially, although even at 50 replicates per dilution, it is still on the order of several percent. Simulation studies have shown that the correction in equations 6-17 can eliminate the bias substantially as well as reducing the overall mean square error of density estimation in MPN experiments (67).

Goodness of Fit to Poisson: Plate Assay. In the analysis presented above, it is assumed that the distribution of organisms is Poisson. This has the advantage that a single mean value (μ) is sufficient to characterize microbial distributions in the population from which samples were taken. However, the validity of this assumption must be tested. Hence we consider in this section how to determine if a set of counts is consistent with a Poisson distribution. In the following section we consider how to ascertain if the results of a dilution series (MPN) experiment are consistent with a Poisson assumption. If the Poisson distribution is not sufficient to describe the results, the depiction of microbial distributions must be approached using alternative models (37).

The approach to analysis of this question depends on the nature of the data set at hand. Therefore, the question is: Given a set of counts, are they consistent with having arisen from a homogeneous Poisson distribution? The general framework of testing and the types of samples are depicted in Table 6-4.

TABLE 6-4   Structure of Count Data Experiments from A Goodness-of-Fit Point of View

Assumed Mean Between Samples	Number of Replicates Within Each Sample	Testing Approach[a]	
		Within Sample	Between Samples[b]
Only one sample	$j > 1$	Simple D^2	NA
Multiple samples with possibly different means	$j = 1$	NA	Likelihood ratio
Multiple samples with possibly different means	$j > 1$	Modified D^2	Likelihood ratio

[a] D^2 index of dispersion; NA, not applicable.
[b] Provided that within-sample error is found to be Poisson.

In the simplest case we may have a number of replicate counts from a single sample. In the second case we have multiple samples (with possibly multiple means), and so our test is whether *if a Poisson distribution is assumed,* it can be characterized by a single mean encompassing all samples. In the third case we have multiple samples and replicates within each sample, allowing a test of Poisson replication error and homogeneity of means across samples. These three situations will be discussed in order.

In the first type of data set, the adherence to a Poisson distribution may most readily be answered by the simple index of dispersion test (39). If $\overline{N}$ is the mean and s_N^2 is the variance of replicate counts, and if there are j replicates, the index of dispersion (D^2) is computed as

$$D^2 = \frac{(j - 1)s_N^2}{\overline{N}} \tag{6-18}$$

If D is greater than the upper $1 - \alpha$ quantile (e.g., $\alpha = 0.05, 0.01$, etc.) of the chi-squared distribution with $(j - 1)$ degrees of freedom, the data are rejected as being inconsistent with a Poisson distribution. Example 6-5 illustrates this test.

Example 6-5.   The following data give replicate counts of heterotrophic plate count bacteria on "washes" from water pipe interior surfaces (in colonies/ mL). Are these consistent with having come from a single Poisson population?

```
27   30   60    60    70    70
74   80   81    82    84    84
93   98   98   101   105   110
```

SOLUTION. The data give a mean colony count $\overline{N} = 78.17$ and a variance $s_N^2 = 116.47$. Hence, from equation 6-15, since $j = 18$, the computed value of D^2 is 116.5. For 17 ($j - 1$) degrees of freedom, the upper 95th percentile of the chi-squared distribution is 28.86; the computed value of 116.5 is at an α value of 2×10^{-17}. Hence we can reject a null hypothesis that the data arise from a single Poisson population.

In the second type of data set, a series of samples (counts) are taken over time. There is no replication—each time a sample is taken, it is assayed only once. In the most general circumstance, each sample may have contained a different volume. In this case, a likelihood ratio (LR) test of the goodness of fit of a single Poisson distribution is employed.

In descriptive terms, a likelihood ratio test computes the ratio, Λ, of the optimum likelihood function under an alternative hypothesis to that under an optimum null hypothesis (103). A likelihood ratio goodness-of-fit test takes as a null hypothesis that all data are drawn from a single Poisson distribution with constant $\overline{\mu}$ (i.e., that the fit is acceptable). The alternative hypothesis is that each sample has a different mean density at the time of sampling; therefore, it has more parameters to be determined.

The likelihood function for the null hypothesis that the data are represented by a Poisson distribution with a constant mean density, equal to that estimated by the maximum likelihood method, $\overline{\mu}_{ML}$ (as in equation 6-9), is given by

$$L^0 = \prod_{i=1}^{r} \frac{(\overline{\mu}_{ML} V_i)^{N_i}}{N_i!} \exp(-\overline{\mu}_{ML} V_i) \tag{6-19}$$

For the alternative hypothesis, we presume that a separate value of $\overline{\mu}_i V_i$ characterizes each sample. The ML estimator of this (for each sample) is simply N_i. Therefore, a likelihood function may be written as

$$L^A = \prod_{i=1}^{r} \frac{(N_i)^{N_i}}{N_i!} \exp(-N_i) \tag{6-20}$$

The likelihood ratio (Λ) is constructed by dividing equation 6-19 by equation 6-20. Since L^A should be greater than L^0 (it fits the data better, since we are allowing for r individual values of $\overline{\mu}_i$ rather than a single value of $\overline{\mu}_{ML}$), Λ is therefore less than 1. We can also construct $-\ln(\Lambda)$, which must therefore be positive. The results after algebraic cancellation and simplification are as follows:

$$\Lambda = \frac{\Pi_{i=1}^{r}[(\overline{\mu}_{ML}V_i)^{N_i}/N_i!] \exp(-\overline{\mu}_{ML}V_i)}{\Pi_{i=1}^{r}[(N_i)^{N_i}/N_i!] \exp(-N_i)}$$

$$= \prod_{i=1}^{r} \left(\frac{\overline{\mu}_{ML}V_i}{N_i}\right)^{N_i} \exp[-(\overline{\mu}_{ML}V_i - N_i)] \tag{6-21}$$

$$-\ln(\Lambda) = \sum_{i=1}^{r} \left[(\overline{\mu}_{ML}V_i - N_i) - N_i \ln \frac{\overline{\mu}_{ML}V_i}{N_i}\right] \tag{6-22}$$

Statistical theory rejects the null hypothesis (that the data come from a single Poisson distribution with constant mean) if the value of $-2\ln(\Lambda)$ exceeds the upper $(1 - \alpha)$th percentile of the chi-squared distribution with $(r - 1)$ degrees of freedom. Note that $(r - 1)$ is the difference between the number of parameters in the alternative hypothesis (r) minus the number of parameters in the null hypothesis (1). The following example is illustrative.

Example 6-6. Repeated (weekly) sampling of a water supply for *Cryptosporidium* oocysts was performed (42). The nature of the assay resulted in variable volumes for each sample. The number of oocysts found and the corresponding volume for each sample are summarized in Table 6-5. Given this information, are the data consistent with having been sampled from a water with a constant Poisson distribution having a constant mean density?

SOLUTION. From the table the total number of oocysts found (ΣN_i) is 29 and the total volume (ΣV_i) (of all samples) is 5621.9 L. Hence, from equation 6-6, the ML estimate of the mean density $(\overline{\mu}_{ML})$ is 0.00516 L^{-1}. From the mean density, the log-likelihood ratio $[-\ln(\Lambda)]$ can be computed from equation 6-22. Note that for volumes whose entries are repeated, there must be a single term for each entry; in other words, this data set contains a total of 52 observations (although some have repeated values of volumes and counts). From equation 6-22 it is determined that $-\ln(\Lambda)$ equals 30.93, and therefore $-2\ln(\Lambda)$ equals 61.85. Since the upper 95th percentile of the chi-squared distribution at 51 degrees of freedom is 68.7, we conclude that the null hypothesis (that the data are from a homogeneous Poisson distribution cannot be rejected). The exact significance level is computed as 14%—in other words, if the Poisson distribution were in fact true, the statistic would be in excess of that computed from our actual data 14% of the time (we would reject the null hypothesis if the computed significance level was less than 5%).

A somewhat different approach to analysis of this second type of data set must occur if the actual counts of all observations have not been determined but if (at least in some cases) only the tally in a certain range is reported. Table 6-6 displays membrane filter total coliform counts obtained November 4–10, 1968 on a cruise in Lake Erie. Here some of the observations are

TABLE 6-5   Sample Volumes (L) in Which Various Numbers of Oocysts Were Found[a]

0 oocysts	1 oocyst	2 oocysts	3 oocysts
48	18.4	95.8	89.9
51	74.1	223.7 (2)	98.4
52	99.9	227.1	100
54.9	100 (5)		
55 (3)	101.1		
57	101.3		
59 (2)	183.5		
85.2	193		
100 (7)			
100.4 (2)			
100.6			
100.7			
101.7			
102 (2)			
102.2 (2)			
103.3			
185.4			
189.3 (2)			
190			
191.4			

[a] Numbers in parentheses indicate the number of occasions in which that combination of oocysts and sample volume was observed (no number indicates only one occasion).

TABLE 6-6   Frequency Distribution of Coliform Counts in 1-mL Samples in Lake Erie

Count	Number of Samples
0	28
1	6
2	1
3	2
4	4
5	2
6	1
9	1
11–14	2
15–18	4
>19	4

Source: Ref. 26.

"binned" into ranges (the last three rows of Table 6-6, for example); in computing the null likelihood, the total expected frequency of all counts in the interval (computed in the case of the Poisson distribution by use of equations 6-4 and 6-5) are used. A slightly different method of writing the likelihood functions is performed. The method described for this type of data is valid only where all the volumes are the same (hence we use an unsubscripted V to denote the common volume to all samples).

The data table has a number of rows, designated as n. Note that we exclude rows in which no observations occur. The symbol f_i is the number of observations in row i. Each row has a lower limit ($N_{L,i}$) and an upper limit ($N_{U,i}$). These limits may be the same in the case of rows whose counts are known precisely (the first 11 rows in Table 6-6), or they may be different (for the remaining rows) where only intervals are known. Some rows may have $N_{U,i}$ as infinity. The total number of observations remains r ($r = \Sigma_{i=1}^{n} f_i$). Given a constant Poisson MLE, the expected frequency (number of observations) for each row in the table can be written as (following equation 6-4)

$$f_{\text{ML},i} = r \left[\sum_{N=N_{L,i}}^{N=N_{U,i}} \frac{(\overline{\mu}_{\text{ML}} V)^N}{N!} \exp(-\overline{\mu}_{\text{ML}} V) \right] \qquad (6\text{-}23)$$

Where the count is known precisely ($N_{L,i} = N_{U,i}$), the summation consists of only a single term. The null hypothesis likelihood function is computed from these frequencies as follows:

$$L^0 = \prod_{i=1}^{n} \left(\frac{f_{\text{ML},i}}{r} \right)^{f_i} \qquad (6\text{-}24)$$

As an alternative hypothesis likelihood function, the observed tallies in each cell are used as the best estimators:

$$L^A = \prod_{i=1}^{n} \left(\frac{f_i}{r} \right)^{f_i} \qquad (6\text{-}25)$$

Dividing L^0 by L^A, we obtain as a likelihood ratio

$$\Lambda = \prod_{i=1}^{n} \left(\frac{f_{\text{ML},i}}{f_i} \right)^{f_i} \qquad (6\text{-}26)$$

and then also

$$-\ln(\Lambda) = -\sum_{i=1}^{n} f_i \ln \frac{f_{\text{ML},i}}{f_i} \qquad (6\text{-}27)$$

The best estimate (MLE) of the density for this type of data is obtained

TABLE 6-7   Computations for Example 6-7

Count	Number of Samples, f_i	$f_{ML,i}$	$-2 \ln(\Lambda)$
0	28	1.08	182.13
1	6	4.25	4.13
2	1	8.35	−4.25
3	2	10.94	−6.8
4	4	10.74	−7.9
5	2	8.43	−5.76
6	1	5.52	−3.42
7–8	0	4.62	0
9	1	0.66	0.82
10	0	0.26	0
11–14	2	0.14	10.77
15–18	4	8.88×10^4	67.3
>19	4	2.15×10^{-6}	115.51

by numerically minimizing -ln(Λ). A goodness-of-fit determination is made by comparing the optimum value of -2 ln(Λ) against a chi-squared distribution with $n - i$ degrees of freedom (*Not r − 1 degrees of freedom*, as in the prior cases without grouping).[3]

The optimization of -ln(Λ) must be made using a numerical procedure, or using graphical trial and error in a manner similar to Example 6-2. Analysis of the data from Table 6-6 is shown in the following example.

Example 6-7.   Given the data in Table 6-6, estimate the Poisson mean density and the goodness of fit to the Poisson distribution.

SOLUTION. By trial and error, $\overline{\mu}_{ML}$ is found to be 3.93 mL^{-1}. Table 6-7 shows the intermediate computations at this optimum value.

Summing up the last column of the table yields the overall value of -2 ln(Λ), which is computed as 252.08. For 12 degrees of freedom (13 rows - 1), the upper 5% of the chi-squared distribution is 21.03, so the null hypothesis of the Poisson distribution can be rejected (the exact significance level is computed as less than 10^{-10}—there is this probability that a result as

---

[3]The test of =2 ln(Λ) computed from equation 6-27 can be shown to be asymptotically identical to the other commonly employed test, the classical Pearson chi-squared statistic, which can be defined using the nomenclature of this section as

$$\chi^2 = \frac{(f_{ML,i} - f_i)^2}{f_{ML,i}}$$

This becomes numerically equal to -2 ln(Λ) as the sample size increases; however, there is not a persuasive argument for the advantage of the χ^2 form versus the statistic in equation 6-27 (103).

poorly fitting or worse would be found if the Poisson distribution were in fact correct). Qualitatively, note that there are far more zero and high counts than would be predicted from the Poisson distribution, and far fewer at the middle ranges.

Given the fact that the Poisson distribution is rejected, the computed value of $\overline{\mu}_{ML}$ is suspect. We will return to this in Example 6-14.

A third type of data set consists of multiple samples in which each sample has been subject to several replicate count determinations. An extended form of the D^2 test can be used to ascertain whether the within-sample replication is Poisson (regardless of whether or not the samples themselves arise from a system with varying means) (39). If the within-sample replication is Poisson, a subsequent likelihood ratio test is computed to ascertain significance of differences between samples.

In this application, the D^2 is computed for each set of replicates (with identical volumes) in a sample. We will characterize the design as having r sets. For each set, $i = 1$ to r, there are n_i replicates (e.g., 2, 3, 4, etc.), with the n_i values being all identical or disparate. The significance level (p_i value) for that set is then computed from a chi-squared distribution at $(n_i - 1)$ degrees of freedom.[4] This yields a set of p_i values (the proportion of area under the chi-squared distribution less than or equal to the computed D^2 values). If the sample has Poisson error between replicates, the set of p_i values should be distributed according to a uniform distribution between 0 and 1. By testing whether the set of p_i values is thus distributed (we will use a Kolmogorov–Smirnov test), the underlying hypothesis (of Poisson errors between replicates) is tested.

The Kolmogorov–Smirnov test (103) can be used to determine the agreement between an experimental distribution of p_i values and the uniform frequency distribution. The test is conducted as follows:

1. The experimental data (in the particular case at hand, the p_i values) are ranked (from low to high) and given ranks 1 to r. In the case of ties, the ranks are averaged among the observations that are tied. The rank of the ith observation is denoted as R_i.
2. For each observation, a deviation[5] is computed from $\delta_i = (R_i/r) - p_i$.
3. The maximum absolute value of the deviations is used as a test statistic, KS:

---

[4]In EXCEL, for example, this can be done by using the formula

$$1\text{-CHIDIST}(x, n - 1)$$

where x is the computed D^2 value and n (i.e., n_i) is the number of replicates in that set. This would give the value of p_i.

[5]Note that R_i/r is the expected value from the uniform distribution of the R_ith of r observations.

$$KS = \max |\delta_i|$$

4. To adjust for the number of observations, an adjusted KS statistic is computed from

$$KS^* = KS \left(\sqrt{r} + 0.12 + \frac{0.11}{\sqrt{r}} \right) \tag{6-28}$$

5. If KS* is greater than a critical value shown in Table 6-8, the null hypothesis (that the underlying distribution is uniform) is rejected, and hence the hypothesis that the replicates are Poisson is rejected. At the usual 5% level, the quantile (0.95) of 1.358 would be required for rejection of the null hypothesis.

TABLE 6-8 Significance Level for the Kilmogorov–Smirnov Test

	Quantile				
	0.85	0.90	0.95	0.975	0.99
$KS\left(\sqrt{r} + 0.12 + \frac{0.11}{\sqrt{r}} \right)$	1.138	1.224	1.358	1.480	1.628

Source: Ref. 58. r is the number of observations in the empirical distribution.

Example 6-8. An experiment was conducted in which three samples (A, B, and C) were analyzed for counts of a particular organism. Replicates at three volumes (100, 10, and 1 mL) were used. The results are shown in Table 6-9. Determine if the distribution between replicates is Poisson, and whether the overall distribution of sample means is Poisson.

SOLUTION. Table 6-10 summarizes the computations. The rows correspond to the rows (sets) in the original data table. For each row, the number of

TABLE 6-9 Count Data for Example 6-8

Sample	Volume (mL)	Counts			
A	100	22	31		
	10	5	3	5	
	1	1	0	0	0
B	100	94	111		
	10	6	4	3	
	1	0	0	1	0
C	100	17	31		
	10	6	7	5	
	1	1	0	1	0

TABLE 6-10   Computations for Example 6-8

n_i	Variance	Mean	D^2	p_i	R_i	$|\delta_i|$
2	40.5	26.5	1.528	0.784	7	0.00585
3	2	4.333	0.923	0.370	3.5	0.0192
4	0.5	0.25	6	0.888	8	0.000499
2	144.5	102.5	1.410	0.765	6	0.0982
3	2	4.333	0.923	0.370	3.5	0.0192
4	0	0.25	0	0	1	0.111
2	98	24	4.083	0.957	9	0.0433
3	0.5	6	0.167	0.080	2	0.1423
4	0.5	0.5	3	0.608	5	0.0528

replicates (n_i), variance in count within each set, and mean count for the set are tabulated.

From this information, the value of D^2 corresponding to that row can be computed by use of equation 6-18. From the D^2 value and the degrees of freedom ($n_i - 1$), the p_i value is computed from the cumulative χ^2 distribution. The column R_i shows the rank of the computed p_i values. The final column gives the absolute value of the deviations.

The maximum value in the final column is 0.1423, which is taken as the KS statistic. KS* is then computed by use of equation 6-28 (with $r = 9$ for the number of samples); it has a value of 0.449. This is less than the 85% quantile in Table 6-8. Since there is more than a 15% chance that a fit as poor (or poorer) would be obtained if the null hypothesis were true, we cannot reject the null hypothesis. Therefore, there is insufficient evidence to reject the hypothesis of random (Poisson) variability among replicates of a single sample.

To determine whether or not the mean density between samples is constant, we approach the problem in a manner similar to Example 6-5. First, for each set, we compute the ML density for that set by

$$\overline{\mu}^i_{\mathrm{ML}} = \frac{\sum_{j=1}^{n_i} N_{ij}}{\sum_{j=1}^{n_i} V_{ij}}$$

It is acceptable to use this relationship since we have already determined that the fit to a within-set Poisson distribution could not be rejected. This notation is doubly subscripted, such that N_{ij} and V_{ij} represent the count and volume in the jth replicate (column) of the ith set (row). This gives the set density in the first column of Table 6-11. Similarly, the overall ML estimate of the mean density, $\overline{\mu}_{\mathrm{ML}}$, is obtained by pooling all observations according to equations 6–9; it is computed to be 0.504 mL^{-1}.

The null hypothesis is that all rows (sets) are characterized by the same Poisson mean. Therefore, the null hypothesis likelihood can be formulated as follows:

**TABLE 6-11   Set Densities and Corresponding
Likelihoods for Example 6-8**

Set Density (No./mL)	L^0	L^A
0.265	2.985×10^{-9}	0.00281
0.4333	0.0042467	0.00497
0.25	0.0670893	0.09197
1.025	9.19×10^{-22}	0.000768
0.4333	0.0035389	0.004142
0.25	0.0670893	0.09197
0.24	2.892×10^{-11}	0.000867
0.6	0.0027476	0.003552
0.5	0.0338314	0.033834

$$L^0 = \prod_{i=1}^{r} \left[\prod_{j=1}^{n_i} \frac{(\bar{\mu}_{ML} V_{ij})^{N_j}}{N_{ij}!} \exp(-\bar{\mu}_{ML} V_{ij}) \right]$$

The second column of the table indicates this value for the respective row (i.e., the term in brackets). Multiplying all values in the second column yields L^0, which has a value of 4.988×10^{-52}.

The alternative hypothesis is that the individual set means are necessary to describe the data. The alternative likelihood is then formulated as

$$L^A = \prod_{i=1}^{r} \left[\prod_{j=1}^{n_i} \frac{(\mu_{ML}^i V_{ij})^{N_j}}{N_{ij}!} \exp(-\bar{\mu}_{ML}^i V_{ij}) \right]$$

The third column of the table provides the alternative likelihood for the respective rows (within the brackets in the preceding equation), which are then multiplied together to get the overall L^A, computed as 3.91×10^{-20}.

Finally, we compute -2 ln(Λ) as -2 ln L^0/L^A. This yields a value of 146.88. This is compared to a χ^2 distribution with 8 degrees of freedom ($r - 1$), and it is found to be highly significant ($p = 8 \times 10^{-28}$). Therefore, the null hypothesis that the data can be represented by a single common density must be rejected.

So the end result is that the within-replicate distribution is Poisson; however, there may be additional factors that cause the density to differ between samples. Further analysis of this data set is suggested in the problems at the end of the chapter.

Goodness of Fit: MPN. Given a set of measurements made by the MPN procedure (dilution series), it is also desirable to determine the consistency with a between-replicate Poisson distribution. Perhaps the first to recognize the need to examine this consistency was Savage and Halvorson (87). The presence of sets of positive tubes that are unlikely was attributed by these authors to deficiencies in incubation medium. The existence of errors in performing dilutions, as well as in precise measurement of samples inoculated into replicate tubes may also increase the variability of an MPN assay (14).

The nature of deviations from a Poisson assumption will be illustrated by two examples. Then some procedures for assessing goodness of fit will be reviewed. Unfortunately, easily implemented fit tests are not as well developed for MPN experiments as for (plate) count experiments.

Descriptively, we may imagine the lack of fit to a Poisson assumption to occur if the fraction of positive responses at the various dilutions in an MPN assay is in some sense unlikely. For example, consider a two-dilution experiment in which 10 tubes at each of four volumes (10, 1, 0.1, and 0.01 mL) are used. An experiment that yields 6, 2, 1, 0 positives (from high to low volume) and an experiment that yields 4, 3, 1, 3 positives have computed MPN values of 0.1215 and 0.1289 mL^{-1}, respectively (by minimizing equation 6-15). However, in some sense the first data set would be more expected then the second set, since the latter combination has a nonmonotonic sequence of positives (as well as neither 0 nor 100% response over a four-order-of-magnitude volume). To develop a criterion for goodness of fit of the within-sample (i.e., between dilution) goodness of fit, we need to quantify the degree of anticipation of individual tube combinations.

There have been a variety of proposals for such tests. Perhaps the first proposal, which we shall show in detail below, relies on a likelihood ratio formulation[6] (44). More rapid tests were proposed by Moran (69,70) and Stevens (100). However, these rapid tests suffer difficulties in the application to the frequent decimal dilution series due to their dependency on the underlying mean density (which results from the periodic behavior of the MPN estimator in decimal dilutions, as evidenced by Figure 6-2). Alternative approaches (which shall be considered with respect to the analogous problem of interpreting dose–response curves) involve testing the adequacy of the simple relationship in equation 6-13 with more complex relationships (again by a likelihood-ratio type of approach).

In the Haldane approach, the null hypothesis (that the result of a dilution experiment on a single sample is fit by a common $\overline{\mu}_{\mathrm{ML}}$) is tested against an alternative hypothesis that each volume in the experiment has a distinct value for $\overline{\mu}_{\mathrm{ML},i}$. Following the nomenclature of equation 6-14, the null likelihood can be written as

---

[6]The astute student, in going back to the original reference, will note that the test of homogeneity presented by Haldane is in the form of the Pearson chi-squared test. Here we present a likelihood-ratio variant. As noted earlier, a likelihood-ratio test of fit is equivalent to a Pearson chi-squared test as sample sizes become sufficiently large. Since dilution series in widespread use tend to employ small numbers of replicates per dilution, it seems more practical to present a likelihood ratio variant of this test.

$$L^0 = \prod_{i=1}^{r} \frac{n_i!}{p_i!(n_i - p_i)!} [1 - \exp(-\overline{\mu}_{ML} V_i)]^{p_i} [\exp(-\overline{\mu}_{ML} V_i)]^{n_i - p_i} \quad (6\text{-}29)$$

The alternative likelihood function can be written as

$$L^A = \prod_{i=1}^{r} \frac{n_i!}{p_i!(n_i - p_i)!} [1 - \exp(-\overline{\mu}_{mL,i} V_i)]^{p_i} [\exp(-\overline{\mu}_{ML,i} V_i)]^{n_i - p_i} \quad (6\text{-}30)$$

The likelihood ratio is constructed by dividing L^0 by L^A, and after algebraic rearrangement, we can obtain

$$-\ln(\Lambda) = \sum_{i=1}^{r} \left[V_i (\overline{\mu}_{ML} - \overline{\mu}_{ML,i})(n_i - p_i) - p_i \ln \frac{1 - \exp(-\overline{\mu}_{ML} V_i)}{(-\overline{\mu}_{ML,i} V_i)} \right] \quad (6\text{-}31)$$

A null hypothesis (of adequacy of fit) is rejected if $-2 \ln(\Lambda)$ exceeds the $1 - \alpha$ (α equals 5%, for example) percentile of the chi-squared distribution. Application is illustrated in Example 6-9.

Example 6-9.   Chang et al. (13) reported the results of a tissue infectivity assay for Coxsackievirus B1. In this assay, aliquots of animal cells (monkey kidney cells) are exposed to known volumes of a suspension containing virus. Given the data in Table 6-12, compute the MPN of the virus in the original suspension, as well as the goodness of fit of the assay.

SOLUTION. For the entire data set, the overall MPN $(\overline{\mu}_{ML})$ can be computed as in earlier problems (by numerical optimization); it is found to be 6.949 mL^{-1}. The values of the row-wise MPNs can be computed by use of equation 6-13 in the following form:

TABLE 6-12   Assay Data for Example 6-9

ML per Replicate	Total Cultures	Positive Cultures
0.032	29	7
0.026	30	4
0.021	28	3
0.018	30	4
0.016	29	2
0.011	25	2
0.005	30	2

$$\overline{\mu}_{ML,j} = -\frac{1}{V_i} \ln \frac{n_i - p_i}{n_i}$$

Table 6-13 gives the results of the row-wise means and the individual terms in the log likelihood. The value of $-2 \ln(\Lambda)$ is thus computed as 2.046. This is compared against the chi-squared distribution at 6 ($= 7 - 1$) degrees of freedom and found to be below the critical value (for 5% α, 12.59), and hence the goodness of fit of the experiment cannot be rejected.

A closely related question with respect to MPN samples is whether multiple samples (each analyzed by an MPN assay) can as an *ensemble* be characterized by a constant mean density, or whether there is some distribution of densities between replicate samples. This can be handled by a simple extension of the likelihood ratio approach. Once again, we will need double subscripts. Following the nomenclature in Example 6-8, V_{ij}, n_{ij}, and p_{ij} are, respectively, the volume, total replicates, and positive replicates in the ith sample at the jth volume. Our null hypothesis is that a single mean characterizes the entire ensemble, versus a separate mean for each sample ($\overline{\mu}_{ML,i}$). The form of the likelihood ratio test is therefore a slight modification of equation 6-27, with r_i being the number of dilutions in sample i and k being the number of samples:

$$-\ln(\Lambda) = \sum_{i=1}^{k} \sum_{j=1}^{r_i} \left[V_{ij}(\overline{\mu}_{ML} - \overline{\mu}_{ML,i})(n_{ij} - p_{ij}) - p_i \ln \frac{1 - \exp(-\overline{\mu}_{ML} V_{ij})}{1 - \exp(-\overline{\mu}_{ML,i} V_{ij})} \right]$$

$$(6\text{-}32)$$

This should be compared against the chi-squared distribution with $k - 1$

TABLE 6-13 **Results for Example 6-9**

$\overline{\mu}_{ML,i}$	$-\ln(\Lambda_i)$
8.633	0.153
5.504	0.117
5.397	0.104
7.950	0.035
4.466	0.228
7.580	0.007
13.799	0.379
	1.023

degrees of freedom (since the alternative hypothesis employs k individual sample means).

Solution of this type of problem requires multiple optimizations to be solved. First, for each sample ($i = 1, \ldots, k$), the sample-wise MPN values ($\bar{\mu}_{ML,i}$) need to be estimated by individual application of equation 6-15. Then, the optimum likelihood ratio (by minimizing equation 6-32) can be used to determine the *ensemble* MPN, and the resultant optimum LR will then be used in a test of significance of the consistency of intersample mean densities.

Example 6-10.   Savage and Halvorson (87) incubated cultures of *Staphylococcus aureus* for 40 hours and recorded dilution counts in repeated samples. They used volumes of 0.1, 0.01, 0.001, 0.0001 and 10^{-5} mL with 10 replicates at each dilution. The following were their results (number of positives at each dilution):

$$A: \ 10–10–5–1–0$$

$$B: \ 10–10–9–3–0$$

$$C: \ 10–10–9–4–0$$

$$D: \ 10–10–8–1–0$$

Is the MPN approach adequate for these data, and (if so), are the four samples characterized by a common mean?

SOLUTION. We start by examining the goodness of fit of the Poisson assumption to each of the four samples individually. For the four samples, we have from application of methods demonstrated in Example 6-9, the results shown in Table 6-14. There is a minor complication in computing Table 6-14. For individual volumes with 100% response (10/10), it should be noted that in equation 6-27 [used to compute the value of $-\ln(L)$], if $p_i = n_i$, the term under summation simplifies to

$$-p_i \ln[1 - \exp(-\bar{\mu}_{ML} V_i)]$$

TABLE 6-14   Computations for Example 6-10

	Row MPN	$-\ln(\Lambda)$ (eq. 6-27)	p Value[a]
A	733.4	0.146	0.99
B	2546.2	0.436	0.929
C	2912.7	0.976	0.745
D	1475.7	0.235	0.976

[a] The fraction of area under a χ^2 distribution with 4 degrees of freedom above $-2 \ln(\Lambda)$.

since the denominator of the second portion of the term in equation 6-27 approaches 1 as the dilution-wise density approaches infinity (i.e., as $p \to n$). Furthermore, if $p_i = 0$, the term in equation 6-31 becomes

$$V_i(\overline{\mu}_{ML} - \overline{\mu}_{ML,i})(n_i - p_i)$$

The results of this analysis show that each sample (A through D) is acceptable, since the p value is in excess of 0.05 (the null hypothesis of acceptable fit cannot be rejected).

We now address the possible consistency of the density between replicate samples. By application of equation 6-32, in conjunction with the row means (values of $\overline{\mu}_{ML,i}$) from Table 6-14, the *ensemble* mean is computed to be 1640.85 mL^{-1}. The individual terms in the likelihood function and the summed likelihood is shown in Table 6-15.

$-2 \ln(\Lambda)$ ($= 9.226$) is tested against a χ^2 distribution with 3 degrees of freedom (equation 6-1). The p value is 0.0264 (this is the area above 9.226). Since this is less than 0.05, the null hypothesis is rejected, and we have evidence that there are, in fact, differences between the means from the four samples. In other words, these data suggest that the variability among the four means is greater than would be expected simply from the variability of the MPN assay itself. Hence, in this case, if the objective is to describe the population from which the four samples were drawn, we need some additional way to describe the between-sample variability.

Confidence Limits: Likelihood. From the results of a sample or a set of samples, we have an estimate for the mean (microbial density) and perhaps other characteristics of the population. However, to do a risk estimate, we seek some measure of the uncertainty with which the mean density is determined. For microorganisms that are distributed according to a Poisson density with constant mean, the process of assessing uncertainty in determination of the population density from plate count experiments is relatively straightforward and can be done exactly. We present this method first, followed by more approximate methods. The approximate methods, although not recommended

TABLE 6-15   Computed Likelihoods for Example 6-10

Sample	$-\ln(\Lambda)$ (eq. 6-28)
A	2.482
B	0.762
C	1.325
D	0.044
	4.613

for data that are truly Poisson, form the basis for determination of uncertainties when more complex distributions must be assumed (e.g., when a test of constant mean is rejected, as in Example 6-5).

For a sum of independent Poisson variates, the resulting distribution can be shown to be Poisson (102). We have shown in equation 6-9 that the ML estimator of the total counts in a series of plate experiments (with total volume $V_t = \Sigma V_i$) is the sum of the observed counts ($N_t = \Sigma N_i$). This is also known to be an unbiased estimator of the population sum of counts from a Poisson distribution (54). Therefore, "exact" confidence limits for the sum of counts can be constructed by the Poisson distribution, where α is the acceptable error rate (i.e., the fraction of time that we are willing to accept a lack of inclusion of the true population value within the interval; this is typically 0.05).

If x_L and x_U are the lower and upper limits to counts (note that these must be integers), they must satisfy the following relations:

$$\sum_{i=0}^{x_L} \frac{\exp(-N_t)}{i!} (N_t)^i = \alpha_1$$

$$\sum_{i=x_U}^{\infty} \frac{\exp(-N_t)}{i!} (N_t)^i = 1 - \sum_{i=0}^{x_U-1} \frac{\exp(-N_t)}{i!} (N_t)^i = \alpha_2 \qquad (6\text{-}33)$$

Since x_U and x_L are restricted to integer values, the value of the lower limit (x_L) is taken as the maximum integer so that the sum does not exceed α_1, and x_U is taken as the smallest integer so that the sum does not exceed α_2. To ensure overall coverage, $\alpha_1 + \alpha_2 = \alpha$.

Note that an additional condition needs to be specified before the formulas of equation 6-33 can be applied. Generally, "symmetric" confidence limits are tabulated such that $\alpha_1 = \alpha/2$—in other words, there is an equal chance of erring on the high side as on the low side. However, it is also possible to compute confidence limits under other circumstances. One-sided intervals are computed where either α_1 or α_2 is allowed to equal zero (we allow all of the error to be on one side or the other)—this might be advisable in some risk assessment applications. Minimum-width intervals are calculated to keep the value of $x_U - x_L$ as small as possible. Numerically, these result in different interval estimates. Table 6-16 provides symmetrical confidence limits to the Poisson distribution for selected total counts ≤ 30. When the total number of counts is large (>15), these intervals may be approximated quite well by the following (54):

$$(N_t + 0.5z_{\alpha/2}^2) \pm z_{\alpha/2}\sqrt{N_t + 0.25z_{\alpha/2}^2} \qquad (6\text{-}34)$$

where $z_{\alpha/2}$ is the normal quantile (e.g., $z_{0.025} = 1.96$).

TABLE 6-16 Confidence Limits to Poisson Counts for Small Total Counts (Exact Values)

Observed Total Counts	Symmetrical Confidence Limits, $\alpha = 0.05$	
	Lower Limit	Upper Limit
1	0	4
2	0	5
3	0	7
4	0	8
5	0	10
6	1	11
7	1	13
8	2	14
9	3	15
10	3	17
15	7	23
30	19	41

Approximate confidence limits can be given by application of the likelihood ratio statistic.[7] In this method, the null hypothesis that an arbitrary density μ provides a fit equal to $\overline{\mu}_{ML}$ (as determined by equation 6-4) is tested against an alternative hypothesis that the fits are significantly different. The likelihood ratio statistic can be constructed as

$$\Lambda(\mu) = \frac{\prod_{i=1}^{k} [(\mu V_i)^{N_i}/N_i!] \exp(-\mu V_i)}{\prod_{i=1}^{k} [(\overline{\mu}_{ML} V_i)^{N_i}/N_i!] \exp(-\overline{\mu}_{ML} V_i)}$$

$$= \prod_{i=1}^{k} \left(\frac{\mu}{\overline{\mu}_{ML}}\right)^{N_i} \exp[-(\mu - \overline{\mu}_{ML})V_i] \qquad (6\text{-}35)$$

All values of μ for which $-2\ln[\Lambda(\mu)] < \chi^2$ at the upper $(1 - \alpha)$th percentile and 1 degree of freedom (since we are varying one parameter—θ) are included within the confidence region. By transformation of equation 6-35, we can write explicitly

$$-2 \ln[\Lambda(\mu)] = 2N_t \ln \frac{\overline{\mu}_{ML}}{\mu} - 2V_t(\overline{\mu}_{ML} - \mu) \qquad (6\text{-}36)$$

[7]These formulations are approximate in that the properties of the likelihood ratio are asymptotic (i.e., becoming more true for larger sample sizes). However, in the applications here, experience has suggested that they are sufficiently close even for the relatively smaller sample sizes used in these analyses.

Since the likelihood ratio formulation still uses a Poisson assumption, if the Poisson distribution is rejected, it cannot be used (nor can the methods embodied in either Table 6-16 or equation 6-34).

Example 6-11. For the two data sets in Example 6-1, assuming the applicability of the Poisson distribution, compute symmetric 95% confidence intervals using the exact Poisson theory and the likelihood ratio approach.

SOLUTION. Each data set had $V_t = 25$ L and $N_t = 27$; with identical values of $\overline{\mu}_{ML}$ of 1.08. Since the confidence limits depend only on these quantities, they will be identical for the two data sets. Solving equation 6-35 for symmetrical conditions gives $x_L = 16$ and $x_U = 38$. Dividing both of these by the volume (25 L) gives a 95% range for density of 0.64 to 1.52 L^{-1}.

Figure 6-3 shows the value of $-2 \ln[\Lambda(\mu)]$ versus μ. The 95th percentile of the chi-squared distribution at 1 degree of freedom is 3.84. The confidence band consists of all values below this critical point; as indicated, this range is 0.72 to 1.54 L^{-1}, identified as the 95% confidence region.

When the count data contains some TNTC information, and can be assumed to be Poisson (or passes a goodness-of-fit-test), the likelihood ratio method can also be used to assess confidence limits for the underlying density of the system from which samples were drawn. However, the likelihood function must now be based on equation 6-10. Modifying this equation to test a hypothesis of inclusion of an arbitrary density yields the following analog to equation 6-35:

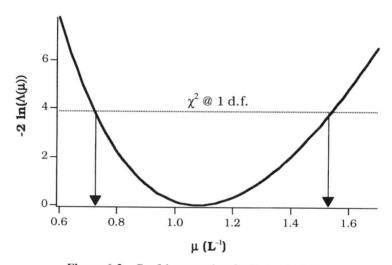

Figure 6-3 Confidence region for Example 6-11.

$$\Lambda(\mu) =$$

$$\frac{\{\Pi_{i=1}^{j}[(\mu V_i)^{N_i}/n_i!] \exp(-\mu V_i)\}\{\Pi_{i=j+1}^{k+j}[1 - \Gamma((N_{L,i} - 1), \mu V_i)]\}}{\{\Pi_{i=1}^{j}[(\overline{\mu}_{ML}V_i)^{N_i}/N_i!] \exp(-\overline{\mu}_{ML}V_i)\}\{\Pi_{i=j+1}^{k+j}[1 - \Gamma((N_{L,i} - 1), \overline{\mu}_{ML}V_i)]\}}$$

$$= \left\{\prod_{i=1}^{j}\left(\frac{\mu}{\overline{\mu}_{ML}}\right)^{N_i} \exp[-V_i)\mu - \overline{\mu}_{ML})]\right\}\left\{\prod_{i=j+1}^{k+j}\frac{1 - \Gamma[(n_{L,i} - 1), \mu V_i]}{1 - \Gamma[(n_{L,i} - 1), \overline{\mu}_{ML}V_i]}\right\}$$

$$(6\text{-}37)$$

This can be simplified to

$$-2 \ln[\Lambda(\mu)] = -2 \sum_{i=1}^{j}\left[N_i \ln\frac{\mu}{\overline{\mu}_{ML}} - V_i(\mu - \overline{\mu}_{ML})\right]$$

$$- 2 \sum_{i=j+1}^{k} \ln\frac{1 - \Gamma((N_{L,i} - 1), \mu V_i)}{1 - \Gamma((N_{L,i} - 1), \overline{\mu}_{ML}V_i)} \quad (6\text{-}38)$$

Application of this equation to the data in Example 6-2 yields a 95% confidence limit of 6.03 to 7.96.

Placing confidence limits around the results of an MPN assay is a topic that has aroused a great deal of interest over many years. Matuszewksi et al. (65) suggested using the method we will describe, employing a likelihood ratio principle.

Woodward (115) devised methods, which remain those that are used in *Standard Methods* (1), in which the probability of occurrence of MPN tube combinations for a given (assumed known) density are computed; these are then ranked by their respective MPN "scores" (density estimates from that tube combination), and the "tails" (upper and lower 2.5% of scores) are discarded. The confidence limits for the particular tube combination are then those assumed densities in which the tube combination remains as a likely (e.g., >95%) occurrence. The Woodward procedure is computationally intensive and has also received criticism for yielding potentially anomalous results (61). Recent proposals to narrow the confidence intervals generated by the Woodward approach have also been made (32).

De Man (20,21) developed MPN confidence limits using Bayesian analysis. However, Loyer and Hamilton (61) critiqued these results on theoretical grounds, and as noted at the end of this chapter, there are deep philosophical objections to the use of Bayesian analysis in simple estimation. Loyer and Hamilton (61) present a procedure conceptually similar to the Woodward method, except that the MPN tube combinations at a given assumed density are ranked by their probability of occurrence rather than the MPN score. These give narrower intervals than Woodward's (37), and in some cases disjoint intervals. The intervals of Loyer and Hamilton have been criticized as giving overly narrow ranges for tube combinations that barely exceed a threshold of being unlikely (6).

Only one study appears to have compared the confidence interval procedures (6). In this study it was found that all methods (likelihood, Woodward, Loyer and Hamilton, and de Man) except for de Man's gave adequate coverage probabilities.[8] On the basis of Best's study as well as on the simplicity of the likelihood method when employed for nontabulated combinations (e.g., the data in Example 6-9), the likelihood method is our method of choice. Furthermore, this is consistent with the other statistical tests that have been presented [and with methods used in dose–response assessment (71)].

In a manner similar to analysis of plate count confidence limits, we will use a likelihood ratio to test the hypothesis that an arbitrary μ is included within a confidence interval. The likelihood ratio (analogous to equations 6-35 and 6-36) is based on equation 6-14 and written as

$$
\Lambda(\mu) = \frac{\Pi_{i=1}^{r} [n_i!/p_i!(n_i - p_i)!][1 - \exp(-\mu V_i)]^{p_i}[\exp(-\mu V_i)]^{n_i - p_i}}{\Pi_{i=1}^{r} [n_i!/p_i!(n_i - p_i)!][1 - \exp(-\bar{\mu}_{ML} V_i)]^{p_i}[\exp(-\bar{\mu}_{ML} V_i)]^{n_i - p_i}}
$$

$$
= \prod_{i=1}^{r} \left[\frac{1 - \exp(-\mu V_i)}{1 - \exp(\bar{\mu}_{ML} V_i)} \right]^{p_i} \exp[-V_i(n_i - p_i)(\mu - \bar{\mu}_{ML})] \qquad (6\text{-}39)
$$

Therefore, the test statistic can be written as

$$
-2 \ln[\Lambda(\mu)] = -2 \sum_{i=1}^{r} \left[p_i \ln \frac{1 - \exp(-\mu V_i)}{1 - \exp(-\bar{\mu}_{ML} V_i)} - V_i(n_i - p_i)(\mu - \bar{\mu}_{ML}) \right]
$$

$$
(6\text{-}40)
$$

All values of μ that yield values of $-2 \ln[\Lambda(\mu)]$ less than the critical value of χ^2 at 1 degree of freedom are within the confidence limit. If the process is repeated at differing values of α, the full confidence distribution can be obtained.[9] For example, using $\alpha = 0.05$ we get the 0.025 and 0.975 percentiles (using a critical chi-squared value of 3.84), using $\alpha = 0.1$ we get the 0.05 and 0.95 percentiles (using a critical chi-squared value of 2.70), and using $\alpha = 0.4$, we get the 0.2 and 0.8 percentiles (using a critical chi-squared value of 0.708). This overall distribution can be termed an uncertainty distribution, since it describes the lack of knowledge about the precise mean density in

---

[8]By this it is meant that in 1000 experiments, we expect that the 95% confidence limits will include the true (assumed) density on 950 occasions.

[9]This method of obtaining the full uncertainty distribution would also be applicable to the plate count data noted earlier as long as underlying consistency with the Poisson distribution at a constant density can be assumed.

the medium from which the sample(s) is(are) drawn. The process is illustrated by example.

Parenthetically, note that equation 6-40 is written for a single sample (with r dilutions). If multiple samples are taken, and it has been shown that collectively they can be characterized by a common mean, a doubly subscripted version of equation 6-40 can be written by summing up terms for $-2 \ln[\Lambda(\mu)]$ for all samples (the degrees of freedom in this case remain equal to 1 for determining confidence limits on μ).

Example 6-12.   Using the data of Example 6-9 for the cell culture assay of Coxsackievirus, determine the uncertainty distribution.

SOLUTION. Trial μ values of 3 to 13 mL^{-1} have been assumed (since it had been already determined that the value of $\overline{\mu}_{ML}$ is 6.949 mL^{-1} (which would be the median, or 0.5 probability point) in Example 6-9. The second column of Table 6-17 is the computed value of equation 6-40. In the third column, the fraction of area exceeding that value (α) is computed. In the fourth column, the cumulative probability is obtained in the following manner:

· If μ is < the maximum likelihood value, then $F = \alpha/2$.
· if μ is > the maximum likelihood value, then $F = 1 - \alpha/2$.

In Figure 6-4, F is plotted against μ. In addition, the derivative of F ($dF/d\mu$) is plotted against μ; this can be interpreted as a probability density function.

Implications for Risk Assessment. If a set of samples has been obtained and has been shown to have a single underlying Poisson mean density, this result can be used as part of an exposure assessment (along with the uncer-

TABLE 6-17   Results for Example 6-12

μ(no./mL)	$-2 \ln(\Lambda(\mu))$	α	Cumulative Probability ≤ F
3	13.024	0.00031	0.000154
5	2.333	0.127	0.0633
6	0.492	0.483	0.242
7	0.00187	0.971	0.514
8	0.498	0.48	0.76
9	1.749	0.186	0.907
10	3.593	0.058	0.971
11	5.919	0.015	0.993
13	11.692	0.000627	0.9996

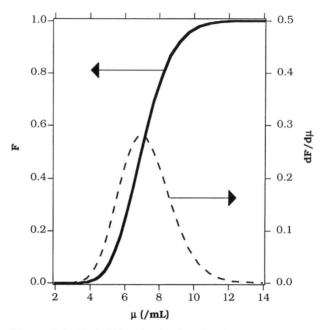

Figure 6-4   Probability density function for Example 6-12.

tainty for the estimate of the mean density, as in Example 6-12). However, if the underlying data do not have a Poisson distribution with a single mean density, we generally would like to describe the variability in the mean density. To do this, we need to consider distributions more complex than the Poisson.

At this point we will state the definitions of uncertainty and variability as used in this book with some precision. Any risk assessment consists not simply of a number (a point estimate), but a distribution of that risk. In computing that distribution, it is necessary to differentiate several characteristics of input distributions that may lead to greater or lesser "spread" in the distribution of the risk estimate (30,72). The terms *variability* and *uncertainty* are used to capture two such characteristics.

The characteristic of variability is an intrinsic property of the risk situation that leads to an irreducible level of "spread" when risk is being assessed for a group of people. Individual persons consume different amounts of water or food. There is not a constant level of contamination (microbial and otherwise) to which all individuals are exposed at all times. Therefore, these inputs must necessarily be captured in the form of a probability distribution. When propagated into the final risk assessment, they necessarily define an irreducible minimum to the precision with which a population risk can be specified.

The characteristic of uncertainty describes elements that with the expenditure of more time, effort, and money (perhaps to develop better analytical methods or to take more samples) can be ever more precisely characterized. For example, if we do microbial analyses with a one-dilution, five-tube MPN assay, we will be a relatively crude measure of microbial densities. However, if we use five binary (factor of 2) dilutions with 50 tubes at each dilution, we will get a much more precise answer. This uncertainty adds the imprecision of the intrinsic variability.

As the example shows, however, frequently it is not easy to separate variability from uncertainty (the results of multiple MPN assays will be used to produce an estimate of the distribution of microbial levels, so the variability will be characterized with some uncertainty). Hence properly attributing the sources of "scatter" in a risk assessment, and presenting such information, is a challenge. This is addressed Chapter 8.

The focus here is to describe distributions that characterize the occurrence of organisms in systems. This will lead to parallel questions of how best to estimate parameters of these alternative (to simple Poisson) distributions, how to determine goodness of fit, and how to determine confidence levels (uncertainties).

Nonpoisson Distributions

We first present a catalog of distributions that have been found potentially useful, or may (in the future) be found potentially useful in characterizing microbial distributions. We then discuss, in general terms, the estimation of distribution parameters from data, examination of goodness of fit, and determination of confidence limits.

First, a general nomenclature is introduced. A probability distribution may be discrete or continuous. A continuous distribution describes the frequency of occurrence of particular random variables, which may assume any value over a particular range (the range may be finite, or it may be infinite). The normal or Gaussian distribution is an example of such a distribution (with a range of $-\infty$ to $+\infty$ The chi-squared distribution is another such example with a range from 0 to $+\infty$. A discrete distribution describes the frequency of occurrence of random variables, which can only assume certain specific values (over a particular, finite, or infinite, range). The Poisson distribution is an example of a discrete distribution, where the random variable can be any integer ≥ 0. The binomial distribution is another example of a discrete distribution, with a random variable that can assume integral values in a finite range.

For a discrete distribution, the function $P(x)$ gives the probability that the random variable assumes the precise value of x. The function $F(x)$ gives the probability that the random variable assumes some value less than or equal to x. Hence these two functions are related by

$$F(x) = \sum_{i=x_L}^{x} P(i) \tag{6-41}$$

where x_L is the lower limit of the range (also termed the *support*) of the distribution. *F* is also termed the *cumulative distribution function* (cdf).

For a continuous distribution, the function $F(x)$ is continuous and is related to a probability density function (pdf) $f(x)$ by the following:

$$F(x) = \int_{x_L}^{x} f(\zeta) \, d\zeta \qquad (6\text{-}42)$$

where, as in equation 6-41, x_L is the lower limit of support for the distribution.

Many of the alternatives to the Poisson distribution can be derived from the Poisson distribution by one of two mechanisms: either as a mixture distribution (29) or as a stopped-sum distribution. In these mechanisms, a set of Poisson distributions is combined with another probability distribution to yield an alternative discrete distribution (that may be used as an alternative to the Poisson in fitting microbial data). The general characteristic of alternative distributions is that they provide for greater variability in the expected count among replicates of the same sample than afforded by a Poisson distribution with constant mean density.

Using the nomenclature above, a discrete mixture distribution [denoted $P_M(x)$] can be derived from the Poisson distribution [denoted as $P_p(x;\mu V)$, where μV is the product of the mean density in a single sample and the volume of that sample] by the following integral:

$$P_M(x;V,\beta) = \int_{0}^{\infty} P_P(x;\mu V)h(\mu;\beta) \, d\mu \qquad (6\text{-}43)$$

where *h*, termed the *mixing distribution*, is the probability density function describing the variability of the sample mean density (μ), and β are parameters of that distribution. For some choices of *h*, the integral in equation 6-43 may be solved analytically; for other choices, it can only be evaluated numerically.

Discrete mixture distributions may be plausible in some circumstances since the mixing distribution, *h*, can be regarded as providing for the variability in the medium in which the organisms are contained. Water, food, and other materials might have a day-to-day (or hour-to-hour) or place-to-place variability. The distribution *h* characterizes this larger heterogeneity, while the Poisson distribution characterizes the within-sample heterogeneity on which the larger variability is superimposed.

A stopped-sum distribution (more particularly a stopped sum of Poisson distributions) is obtained (54,102) by summing a random number (= *k* terms) of Poisson variates (random "drawings" from a Poisson distribution). The distribution of the random number *k* is given by a second discrete distribution.

Since stopped-sum distributions of Poisson form can be interpreted (in most cases, certainly all of interest here) as mixture distributions by theorems of Levy and Maceda (54), we concentrate on the following as Poisson mixture distributions. However, the fact that Poisson mixture distributions can also be derived from alternative mechanisms means that the fact that a particular data set fits a particular mixture distribution, in and of itself does not demonstrate the adherence to the mixture mechanism from which that distribution was derived.

Negative Binomial. The negative binomial distribution has been used to describe microbial count data since at least the 1907 observation by W. Gossett that in certain circumstances, cell counts under a microscope could better be described by negative binomial than by Poisson distributions (104). Greenwood and Yule (34) derived the negative binomial as a gamma mixture of Poisson distributions. Quenouille (80) derived the Poisson distribution as a sum of logarithmic distributions with random length distributed as a Poisson variate.

The negative binomial can be derived by assuming that the mean density of microorganisms in a sample can be written as a gamma distribution[10] with parameters α and β, as follows:

$$ f(\mu) = \frac{1}{\alpha\Gamma(\beta)} \left(\frac{\mu}{\alpha}\right)^{\beta-1} \exp\left(-\frac{\mu}{\alpha}\right) \tag{6-44} $$

The mean of this distribution $(= \bar{\mu})$ is $\alpha\beta$ and the variance is $\beta\alpha^2$. The relative standard deviation in density (square root of variance divided by mean) is equal to $1/\sqrt{\beta}$. Using this as the mixing distribution (h) in equation 6-43, we can write

$$ P_{NB}(x) = \int_0^\infty \frac{(\mu V)^x \exp(-\mu V)}{x!} \left[\frac{1}{\alpha\Lambda(\beta)} \left(\frac{\mu}{\alpha}\right)^{\beta-1} \exp\left(-\frac{\mu}{\alpha}\right)\right] d\mu \tag{6-45} $$

This integral can be evaluated analytically as

$$ P_{NB}(x) = \frac{\Gamma(x + \beta)}{\Gamma(\beta)x!} \left(\frac{\alpha V}{1 + \alpha V}\right)^x (1 + \alpha V)^{-\beta} \tag{6-46} $$

This can also be written as (defining the actual mean density $\bar{\mu} = \alpha\beta$ and $k = \beta$)

$$ P_{NB}(x) = \frac{\Gamma(x + k)}{\Gamma(k)x!} \left(\frac{\bar{\mu} V}{k + \bar{\mu} V}\right)^x \left(\frac{k + \bar{\mu} V}{k}\right)^{-k} \tag{6-47} $$

---

[10]This can be evaluated in EXCEL using the function GAMMADIST, which can also evaluate the cumulative distribution function.

The negative binomial distribution has been widely used in ecological work (54). As early as 1952, it was used (although without recognition that it was a negative binomial) to fit microbial plate counts in water supplies (107). In more recent years it has been used to fit other count data in water and wastewater samples (26,77)

The negative binomial distribution has a number of interesting properties. At the limit of $k \rightarrow \infty$, the negative binomial reduces to the Poisson distribution (equation 6-47 reduces to equation 6-3). Lower values of k (and thus a finite value of the relative standard deviation of the distribution of mean densities) result in a greater variability. This is manifest in a greater number of zero counts and counts at high value. Figure 6-5 compares the Poisson and negative binomial probability distributions at a fixed mean (μV). Note that as k decreases, the decrease from the Poisson distribution increases.

Poisson Lognormal. For many years, microbial density data have been analyzed using lognormal probability plots. Therefore, it is logical to consider that the mixing distribution, h, in equation 6-43 would be a lognormal distribution, which can be written as

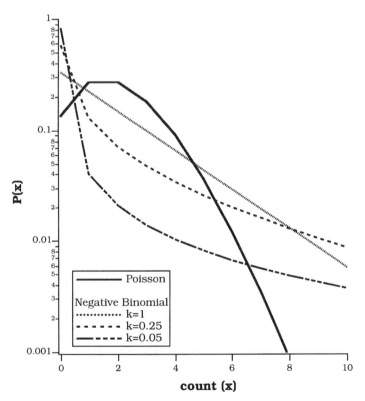

Figure 6-5 Distribution functions for Poisson and negative binomial distributions at fixed $\mu V = 2$.

$$f(\mu) = \frac{1}{\mu s\sqrt{2\pi}} \exp\left\{-\frac{[\ln(\mu) - \zeta]^2}{2s^2}\right\} \qquad (6\text{-}48)$$

where ζ is the mean of the logarithm of density and s is the standard deviation of the logarithm of density. The arithmetic mean density $(\overline{\mu})$ and the variance of the density are (83)

$$\overline{\mu} = \exp\left(\zeta + \frac{s^2}{2}\right)$$

$$\text{var} = \overline{\mu}^2[\exp(s^2) - 1] \qquad (6\text{-}49)$$

Furthermore, the cumulative distribution function of the lognormal distribution can be expressed in terms of the cumulative normal integral, Φ (which is a tabulated function and also built into standard spreadsheets[11] 1) as

$$F(\mu) = \Phi\left(\frac{\ln(\mu) - \zeta}{s}\right) \qquad (6\text{-}50)$$

Substitution of equation 6-48 in equation 6-43 results in the following:

$$P_{\text{PLN}}(x) = \int_0^\infty \frac{(\mu V)x \exp(-\mu V)}{x!} \frac{1}{\mu s\sqrt{2\pi}} \exp\left\{-\frac{[\ln(\mu) - \zeta]^2}{2s^2}\right\} d\mu \qquad (6\text{-}51)$$

This equation has no analytical solution, although it can be evaluated by numerical integration in the foregoing form. We have found the following equivalent form useful, which can be integrated using Gauss–Hermite quadrature (79):

$$P_{\text{PLN}}(x) = \frac{1}{x!\sqrt{\pi}} \int_{-\infty}^\infty \exp(-q^2) \exp[-\varphi(q)V][\varphi(q)V]^x \, dq$$

$$\varphi(q) = \exp(\mu + sq\sqrt{2}) \qquad (6\text{-}52)$$

The Poisson–lognormal distribution has been used to fit species-abundance data (11,83) as well as bibliometric data (101). As s increases (from zero) the distribution deviates from the Poisson in a qualitatively similar fashion to the negative binomial (Figure 6-6); namely, the counts at zero and at high values increase, and counts in the midrange diminish.

Poisson–Inverse Gaussian. The Poisson–lognormal distribution is conceptually interesting, although the inability to express the integral in equation 6-52 in analytical form makes the problem of fitting data somewhat more

---

[11]In EXCEL, the function NRMSDIST(Z) gives the value of $\Phi(Z)$.

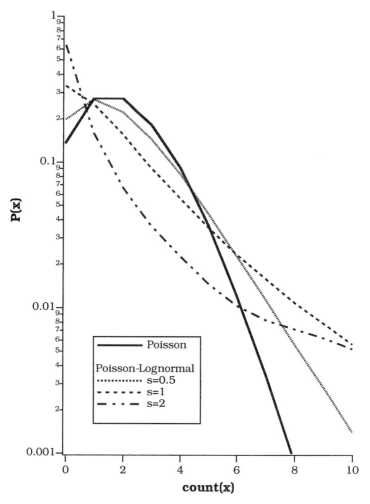

Figure 6-6 Distribution functions for Poisson and Poisson–lognormal distributions at fixed $\mu V = 2$.

difficult; since a minimization process must be coupled to a numerical integration process.[12] A potentially attractive distribution which has some properties similar to the lognormal is the inverse Gaussian distribution (52). This distribution, which is also positively skewed, can be written as

$$f(\mu) = \sqrt{\frac{\overline{\mu}\phi}{2\pi\mu^3}} \, \exp\left\{\phi\left[1 - \frac{1}{2}\left(\frac{\mu}{\overline{\mu}} + \frac{\overline{\mu}}{\mu}\right)\right]\right\} \qquad (6\text{-}53)$$

---

[12]Of course, given the increasing speed and capability of computer processors, the issue of speed becomes less and less of a limitation to the fitting process.

where $\overline{\mu}$ is the mean density, and the variance of the density is equal to $\overline{\mu}^2/\phi$. The cumulative distribution function can be expressed in terms of the normal integral as follows (52):

$$F(\mu) = \phi\left[\left(\frac{\mu}{\overline{\mu}} - 1\right)\sqrt{\frac{\overline{\mu}\phi}{\mu}}\right] + \exp(2\phi)\Phi\left[-\left(\frac{\mu}{\overline{\mu}} + 1\right)\sqrt{\frac{\overline{\mu}\phi}{\mu}}\right] \quad (6\text{-}54)$$

When the inverse Gaussian distribution is used as the mixing distribution for a Poisson, the following results:

$$P_{\mathrm{PIG}}(x) = \int_0^\infty \frac{(\mu V)^x \exp(-\mu V)}{x!} \sqrt{\frac{\overline{\mu}\phi}{2\pi\mu^3}} \exp\left[\phi\left(1 - \frac{1}{2}\left(\frac{\mu}{\overline{\mu}} + \frac{\overline{\mu}}{\mu}\right)\right)\right] d\mu \quad (6\text{-}55)$$

This can be integrated to yield the following form (54):

$$P_{\mathrm{PIG}}(x) = \frac{e^\phi}{x!}(\overline{\mu}V)^x[\phi(\phi + 2\overline{\mu}V)]^{1/4}\left(\frac{\phi}{\phi + 2\overline{\mu}V}\right)^{x/2} K_{x-1/2}[\sqrt{\phi(\phi + 2\overline{\mu}V)}] \quad (6\text{-}56)$$

where $K_v(y)$ is the modified Bessel function of the third kind[13] of order v and argument y. Therefore, if access is available to compute this function with fractional order, the Poisson–inverse Gaussian distribution can be evaluated.

Johnson et al. (54) attribute the introduction of this distribution to Holla, in studying accidents and repeated illness frequencies (46). It has also received extensive attention from Sichel (90–94) in connection with bibliometric studies, marketing, and economic geology. However, it does not yet appear to have been used to study microbial distributions.

The characteristics of the PIG distribution are qualitatively similar to the PLN and NB distributions. As the parameter ϕ decreases, the distribution deviates more markedly from the ordinary Poisson distribution (as the variance of the density increases). This is shown in Figure 6-7. Comparison of Figures 6-6 and 6-7 show a close similarity, indicating that the underlying lognormal and inverse Gaussian distributions bear a close similarity.

Poisson–GIG. The negative binomial, Poisson–lognormal, and Poisson–inverse Gaussian distributions all have only two parameters that completely determine their values. In other words, if the mean and variance of the density are fixed, then (at a fixed volume) the distribution of counts is completely

---

[13]Johnson et al. note that some references call this the modified Bessel function of the second kind.

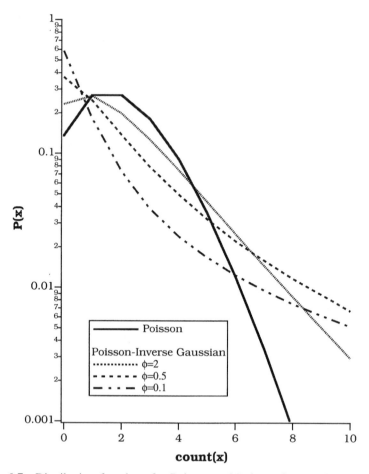

Figure 6-7   Distribution functions for Poisson and Poisson–inverse Gaussian distributions at fixed $\mu V = 2$.

determined. For some data sets, particularly with large numbers of observations and substantial heterogeneity, a greater degree of flexibility in distributional form may be desired. This can be achieved by introducing distributions with greater number of parameters. It is also advantageous (for reasons to be noted below) for these extended distributions to simplify to the two-parameter (or ultimately to the one-parameter Poisson) distribution as special cases.

This flexibility can be achieved by the use of the generalized inverse Gaussian (GIG) distribution, which was studied extensively by Jørgensen (55). This distribution can be written as

$$f(\mu) = \frac{\eta^{-\lambda}\mu^{\lambda-1}}{2\,K_\lambda(\omega)} \exp\left[-\frac{1}{2}\,\omega\left[\left(\frac{\eta}{\mu} + \frac{\mu}{\eta}\right)\right]\right] \qquad (6\text{-}57)$$

The GIG distribution has three parameters: λ can take any real value, while ω and η are nonnegative real numbers. For this distribution, the mean density and the variance are given by (55)

$$\bar{\mu} = \eta \, \frac{K_{\lambda+1}(\omega)}{K_{\lambda}(\omega)}$$

$$\text{var} = \eta^2 \left\{ \frac{K_{\lambda+2}(\omega)}{K_{\lambda}(\omega)} - \left[\frac{K_{\lambda+1}(\omega)}{K_{\lambda}(\omega)} \right]^2 \right\} \tag{6-58}$$

The GIG distribution includes as special cases the inverse Gaussian distribution (for $\lambda = -\frac{1}{2}$), and the gamma distribution (for positive λ with $\omega \to 0$ while keeping ω/η constant).

With the distribution in equation 6-51 substituted (as h) in equation 6-39, the Poisson–GIG distribution is obtained. Sichel (92,93) obtained an analytical integral in terms of Bessel functions as follows:

$$P_{\text{PGIG}}(x) = \frac{(\eta V)^x}{x!} \left(\frac{\omega}{\omega + 2\eta V} \right)^{(x+\lambda)/2} \frac{K_{\lambda+x}[\sqrt{\omega(\omega + 2\eta V)}]}{K_{\lambda}(\omega)} \tag{6-59}$$

Although this distribution has not received a great deal of use in biological applications, it has been used in bibliometric, geological, and marketing applications (90–94,97).

The flexibility of the PGIG distribution is illustrated in Figure 6-8. In this figure, the mean ($\bar{\mu}V$) of the underlying GIG distribution is fixed at 2 and the variance is fixed at 10. By varying the parameter λ and using equations 6-58, the values of ω and φ that give the desired mean and variance can be computed. Table 6-18 provides the parameter combinations. As noted in the figure, even though the mean and variance are fixed, the third parameter of the GIG distribution introduces additional flexibility in terms of the proportion of zero counts as well as the proportions at mid and high values.

Fitting Alternative Distributions. Given a set of data, a first problem is to ascertain whether an alternative distribution more precisely characterizes the data than a Poisson distribution. The first step in examining the significance of possible alternative fits is to determine the best fit of alternative distributions to a data set. This problem can also be cast into a likelihood framework.

We consider a null hypothesis: that the count data are better fit by a particular alternative complex (e.g., negative binomial, Poisson–lognormal, etc.) distribution (with a common set of parameters among all samples). Following the nomenclature in equation 6-43, with the parameter *vector* to be estimated designated as β, the alternative likelihood function can be written as

$$L^0 = \prod_{i=1}^{k} P(N_i; V_i, \beta) \tag{6-60}$$

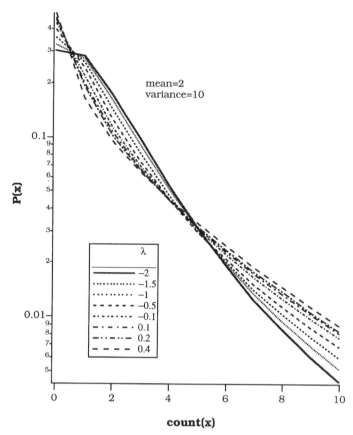

Figure 6-8 Poisson GIG distributions with constant mean ($\mu V = 2$) and variance.

TABLE 6-18   Parameter Combinations of the GIG Distribution Yielding a Mean of 2 and a Variance of 10

λ	ω	ϕ
−2	0.23975	17.409
−1.5	0.41136	6.9585
−1	0.45423	3.75
−0.5	0.39986	1.9996
−0.1	0.28926	1.0366
0.1	0.19516	0.60474
0.2	0.13911	0.40108
0.3	0.073558	0.19749
0.4	0.00069491	0.0017351

The alternative hypothesis is that a set of Poisson distributions with a single mean density for each sample or set of samples (as when individual samples are replicated, such as in Example 6-8) is required to fit the data. The alternative likelihood is then given by

$$L^A = \prod_{i=1}^{k} \frac{(\bar{\mu}_i V_i)^{N_i}}{N_i!} \exp(-\bar{\mu}_i V_i) \tag{6-61}$$

where $\bar{\mu}_i$ is the Poisson mean for sample i. The likelihood ratio can then be taken as L^0 divided by L^A, and the maximum likelihood estimates of β are given as the values that minimize the function $-2 \ln(\Lambda) = -2 \ln (L^0/L^A)$. A goodness-of-fit test compares the optimized value of $-2 \ln(\Lambda)$ versus the chi-squared distribution at $(k - j)$ degrees of freedom, where j is the number of parameters in the vector β (e.g., 2 for the PLN, NB, and PIG distributions and 3 for the PGIG distribution).

Example 6-13.   Consider the *Cryptosporidium* data in Example 6-6. It was shown that these data could be fit to a Poisson distribution with $\bar{\mu} = 0.00516$ L^{-1}, and that the optimum -2 $\ln(\Lambda)$ (for the Poisson fit) was 61.85. Although this was found to pass a goodness-of-fit test to the Poisson distribution, we want to ascertain whether other distributions can also fit this same data set, in particular the NB, PLN, and PIG distributions. Find the best fit of these three distributions to the data set.

SOLUTION. Using numerical optimization, the results shown in Table 6-19 are obtained. We note that the optimized value of $-2 \ln(\Lambda)$ for the alternative distributions are not very different numerically than the optimum value of the likelihood ratio for the Poisson fit. Any of the alternative distributions pass an overall goodness-of-fit test (compare to the chi-squared distribution at $k-j$ degrees of freedom); however, another question that must be asked before

TABLE 6-19   Results for Example 6-13

Distribution	Optimum Parameter Values	$-2 \ln(\Lambda)$
NB	$\bar{\mu} = 5.2077 \times 10^{-3} \; L^{-1}$	60.2429
	$k = 1.7564$	
PLN	$\zeta = 5.4913$	60.3169
	$s = 6.9111 \times 10^{-1}$	
PIG	$\bar{\mu} = 5.229 \times 10^{-3}$	60.296
	$\phi = 1.664$	
PGIG	$\lambda = 1.75628$	60.2428
	$\omega = 0.005017$	
	$\eta = 0.00000748$	

an alternative distribution is accepted is whether the improvement in fit is great enough to justify the inclusion of additional parameter(s) estimated from the data. This question is addressed in the next section.

For data structured as in Table 6-6, in which "tallies" at a particular count, or in a particular range of counts, are given and the sample volume is constant, the approach described above undergoes minor modification. The best-fit parameter values for the alternative (e.g., PLN, NB, etc.) distribution are computed by minimizing $-\ln(\Lambda)$ as defined by equation 6-27, where the values of $f_{ML,i}$ are obtained from the alternative distribution. This type of fitting is demonstrated in the following example.

Example 6-14. For the coliform data in Table 6-6, determine the best fit of the NB, PLN, PIG, and PGIG distributions.

SOLUTION. Table 6-20 summarizes the best-fit parameters for the data using the procedure of minimizing equation 6-27 with the various candidate distributions. Of the two-parameter distributions (PLN, NB, PIG), the negative binomial performs the best [has the lowest value of $-2 \ln(\Lambda)$], and this is substantially lower than the residual likelihood ratio statistic from the Poisson fit. The fit to the PGIG distribution is essentially identical to the NB fit (note the equality of the NB k and the PGIG λ, as well as the likelihood ratios), and as shown in Figure 6-9 and Table 6-21, the predictions from these two distributions are identical (i.e., the optimum PGIG distribution is, in fact, the NB distribution).

As shown in the table and the figure, the PLN distribution predicts too many samples at high counts than is observed. As noted previously, the Poisson distribution underpredicts the low and high count values and puts too much weight in the intermediate range.

Observe, additionally, that the negative binomial distribution, which passes the goodness-of-fit test, gives an estimate for $\bar{\mu}$ of 5.29. The estimate from

TABLE 6-20   Results for Example 6-14

Distribution	Best-Fit Parameters		$-2 \ln(\Lambda)$
Poisson	$\bar{\mu}$	3.93	352.542
PLN	ζ	1.38	15.3751
	s	3.24	
NB	$\bar{\mu}$	5.29	12.8687
	k	0.209	
PIG	$\bar{\mu}$	11.5	20.446
	ϕ	0.0272	
PGIG	λ	0.209	12.8687
	ω	2.16×10^{-12}	
	η	2.73×10^{-11}	

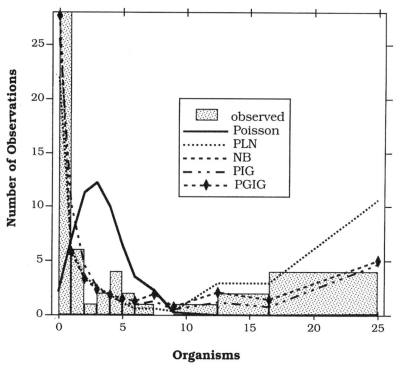

Figure 6-9   Distributions for Example 6-14.

TABLE 6-21   Comparison of Distributions for Example 6-14

Counts	Number of Observations	Poisson	PLN	NB	PIG	PGIG
0	28	1.08	22.04	27.77	25.67	27.77
1	6	4.25	6.06	5.58	10.12	5.58
2	1	8.35	3.45	3.25	4.52	3.25
3	2	10.94	2.58	2.30	2.52	2.30
4	4	10.74	1.84	1.77	1.63	1.77
5	2	8.43	1.12	1.44	1.16	1.44
6	1	5.52	0.60	1.20	0.88	1.20
7–8	0	4.62	0.57	1.91	1.26	1.91
9	1	0.66	0.29	0.78	0.47	0.78
10	0	0.26	0.40	0.69	0.40	0.69
11–14	2	0.14	2.82	2.12	1.16	2.12
15–18	4	8.88×10^{-4}	2.95	1.46	0.75	1.46
>19	4	2.15×10^{-6}	10.30	4.73	4.47	4.73

the Poisson distribution (whose fit was rejected) was 3.93 (Example 6-6). This demonstrates that improper selection of a distribution (i.e., selection, without verifying adequacy of fit) may result in a biased estimate of properties of the system, such as the mean.

Statistical Significance of Alternative Fits. By examining the value of the optimized $-2 \ln(\Lambda)$ against the chi-squared distribution at degrees of freedom equal to the number of observations minus the number of fitted parameters (or, for binned data such as in Example 6-14, the number 'of bins minus the number of fitted parameters), the overall goodness of fit can be ascertained. No distribution should be accepted as adequately descriptive of the data unless it passes this, or analogous, goodness-of-fit tests. For distributions with the same number of parameters (e.g., NB, PLN, PIG), the best distribution should be taken as the one that yields the lowest value of $-2 \ln(\Lambda)$ (56).

However, generally a distribution with a greater number of parameters will provide a value of $-2 \ln(\Lambda)$ no greater than a distribution with fewer parameters—in Example 6-13, the NB, PLN, and PIG distributions all provided a lower value of $-2 \ln(\Lambda)$ than the Poisson distribution. Before these more complex distributions can be accepted, a statistical test must be made to ascertain whether the improvement in fit is sufficiently great to justify the additional parameter.

The formal test to be performed is set up as a likelihood ratio test. There are two distributions to be tested: one is a special (simple case) of the second (and has fewer parameters). For example, the Poisson is a simpler case of the negative binomial (if the NB k approaches infinity, the Poisson is obtained): the PLN (if s approaches 0, the Poisson limit is obtained) and the PIG (if ϕ approaches infinity, the Poisson is obtained). Similarly, the NB and PIG distributions are simplified cases of the PGIG distribution.

The null hypothesis is that the simpler distribution is adequate to fit the data, and the alternative hypothesis is that the more complex distribution is necessary. Let Y_0 be the value of $-2 \ln(\Lambda)$ computed for the best fit of the simpler distribution, and let Y_a be the value of $-2 \ln(\Lambda)$ computed for the more complex distribution. Clearly, Y_a must be less than or equal to Y_0 (more parameters cannot make the fit worse if the null hypothesis is a special case). The statistical significance of the improvement is assessed by comparing difference, $\Delta = Y_0 - Y_a$, against the chi-squared distribution with degrees of freedom equal to the *difference in the number of parameters* between the two distributions; if the tabulated critical chi-squared value is less than Δ, the null hypothesis is rejected, and the use of the more complex distribution produces statistically significant improvement in fit.

As an illustration, consider the analysis in Example 6-13. Considering a null hypothesis of adequacy of the Poisson distribution, the value of Y_0 is 61.85. For any two-parameter distribution, the value of Y_a must be less than $61.85 - 3.84 = 58.01$, since the upper 5% of the chi-squared distribution is 3.84. None of the two-parameter distributions considered (NB, PLN, PIG) reduces the value of Δ below this critical value, and hence none of the alternative two-parameter distributions provide a statistically significant improvement in fit vis-à -is the Poisson distribution. Furthermore, since the Poisson distribution fit to these data passes a goodness-of-fit test (Example 6-6), the Poisson distribution is accepted as the best descriptor for this data set.

For the data in Example 6-14, a similar analysis leads to the finding that all of the two-parameter alternative distributions (NB, PLN, PIG) provide a statistically significant improvement in fit vis-à-vis the Poisson distribution. However, the PGIG does not provide a statistically significant improvement in fit with respect to its two-parameter subset (NB). Also note that the residual likelihood from the NB fit is 12.8687. For 11 degrees of freedom (13 rows minus 2 parameters), the critical chi-squared value (upper 5%) is 19.68; since this is less than the critical value, the null hypothesis of adequacy of the NB distribution is accepted. Hence the NB distribution is adopted as the best descriptor of this data set.[14]

A closely related question to the comparison of one distribution to another is the setting of confidence limits for the parameters of the selected distribution. The likelihood maximization process (which we can also describe as a process of minimizing deviance) results in a point estimate for the parameter(s) of a particular distribution. We would like to know how certain we are of these parameters.

This problem can be addressed in a likelihood framework by considering the deviance with the best parameter(s) as a null hypothesis and computing the deviance at other parameter values.[15] Therefore, $Y_a(\beta)$ can be defined as the value of $-2 \ln(\Lambda)$ at a set of parameter combinations, β. All such parameter combinations which cause the following inequality to be obeyed are included within the confidence region at a confidence coefficient of $1 - \alpha$:

$$Y_a(\beta) - Y_0 \leq \chi^2 \tag{6-62}$$

where χ^2 is evaluated at the upper α percentile at degrees of freedom equal to the number of parameters (i.e., the dimensionality of β). For one parameter, such as the Poisson distribution, the confidence region will be a region in one-dimensional space (Example 6-11), for a two-parameter distribution, it will be a region in two-dimensional space, and so on. Finding the confidence region, particularly for multiparameter cases, involves evaluating the deviance at a range of parameter values to explore the region of space in which inequality 6-62 is satisfied.

In Figure 6-10, the confidence regions for the parameters $(\overline{\mu}, k)$ of the

---

[14]We note in passing that it is possible that some other distribution of which the NB is a special case might prove to offer statistically significant improvement in fit versus the NB distribution. Unless the residual deviance were less than the critical value for one parameter (e.g., 3.84), this cannot be ruled out. In principle, there are an infinite number of ways of generalizing any one distribution by introduction of further parameters.

[15]We have already made use of this principle in the development leading to equation 6-36. The discussion here generalizes that application to multiple parameters.

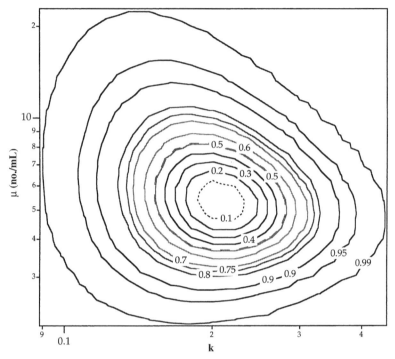

Figure 6-10 Confidence regions for the negative binomial parameters of the fit to data in Example 6-14. Numbers on contours indicate the probability that true parameters are included in the region.

negative binomial fit to the coliform data in Example 6-14 are plotted. To construct this plot, the value of $Y_a - Y_0$ was evaluated over a grid of parameter values. From the inverse chi-squared distribution, the cumulative fractions (p values) were computed. Contours over selected fractions were constructed. For example, the outer closed curve labeled 0.99 is the 99% confidence contour—it is expected to enclose the true values of the parameters with a certainty of 99%.

Note that this graph is plotted on log-log coordinates. On arithmetic coordinates, the curves would look more elongated. Also note that all the contours are closed. For some data sets, particularly where only few observations have been taken, it may be that some (particularly the large confidence value) contours may be open along one or more sides; that is, it is impossible to differentiate that parameter from zero (or infinity) at a certain level of confidence. This would also be true for data where a simplified distribution fits the data as well as a more complex data set. For example, the contours for the *Cryptosporidium* data set of equation 6-15 for the negative binomial distribution are not closed with respect to $k = \infty$ (which would simplify to the Poisson distribution) (42).

The underlying purpose for fitting alternative distributions to such data is to determine the variability of the mean density of microorganisms in repeated samples. Therefore, since the negative binomial distribution can be developed as a gamma mixture of Poisson distributions, it is of interest to examine the cumulative distribution of mean values from the fitted gamma distribution implied by the negative binomial fit. To do this, the values of the gamma distribution are obtained as follows (see equation 6-44):

$$\alpha = \frac{\overline{\mu}}{k}$$

$$\beta = k$$

(6-63)

From the cumulative gamma distribution (which must be obtained by numerical integration[16]), at a given fraction (e.g., 0.2, 0.5, 0.9) the mean density, μ, which is greater than that fraction of samples, may be obtained.

Furthermore, since the parameters (α and β) of the gamma distribution are uncertain (because, by Figure 6-10, the parameters $\overline{\mu}$ and k are uncertain), the cumulative distribution is more properly depicted as a band, as in Figure 6-11.

To construct Figure 6-11, at each proportion, the ML estimates of mean density and k were used to compute the gamma distribution parameters from equation (6-63). Then from the cumulative gamma distribution, this yielded a point on the solid central curve. Then the values of $\overline{\mu}$ and k that minimized or maximized the cumulative gamma quantile at that proportion (providing that they were inside the 95% contour) were used to compute the respective points on the dashed lines. By connecting the series of points, the curves of Figure 6-8 were found.

This process can actually be summarized as a constrained optimization problem (36) as follows. Let f be a particular proportion. Then the point on the lower curve is obtained by solving, where $G(f;\alpha,\beta)$ is the inverse cumulative gamma distribution evaluated at proportion f:

$$\min G\left(p; \frac{\overline{\mu}}{k}, k\right)$$

(6-64a)

subject to

$$Y_a\left(\frac{\overline{\mu}}{k}, k\right) - Y_0 \le \chi^2$$

(6.64b)

---

[16]EXCEL has a built-in function, GAMMADIST, which can compute the cumulative gamma distribution.

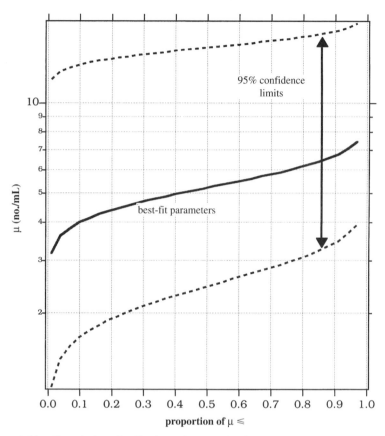

Figure 6-11 Cumulative distribution of mean densities estimated from data of Example 6-16, with 95% confidence limits based on uncertainties in distribution.

From Figure 6-11, the best estimate of the upper 90th percentile of the density is 6.5 mL^{-1} (from the solid curve). However, the 95% confidence limits *of this percentile* are 3.3 to 18 mL^{-1}. Note that the best estimate of the median (50th percentile) density is 5.2. We therefore conclude that the uncertainty of distribution parameters is substantially greater than the variability that those parameters describe.[17]

Fitting Underlying Density Distributions. In some cases, data are available only in the form of density measurements on repeated samples. Although this

---

[17]Which would also lead to the conclusion (as discussed in a later chapter), that if a particular risk assessment were highly sensitive to the microbial input distribution (as opposed to other inputs) it might be important to reduce the uncertainty in knowledge of that distribution (e.g., by taking additional samples).

is less desirable that actual data on organism numbers in tabulated volumes (or masses) of sample, density information can be used to ascertain whether a distribution is a more appropriate descriptor of density than a single value (as would be the case in a Poisson distribution).

This question must be framed using the underlying distribution for the NB, PLN, PIG, and PGIG distributions: namely, the gamma, lognormal, inverse Gaussian, and generalized inverse Gaussian distributions. If the data are available as individual observations (individual density measurements on single samples), the analysis can proceed by writing the likelihood based on the probability density function for the mixing distribution (equation 6-43). If x_i are the observed densities of organisms, the negative likelihood can be written as

$$-2\ln(L) = -2\sum_{i=1}^{n}\ln[f(x_i);\beta] \tag{6-65}$$

where $f(x)$ is the probability density function (e.g., lognormal, gamma, inverse Gaussian, generalized inverse Gaussian) and β is (are) the parameters to be determined from the fit. The value of the β's that minimize $-2\ln(L)$ (which we will term Y) are the ML estimates.

Numerically, if multiple distributions are fitted, then among the distributions with the same number of parameters, the distribution producing the lowest value of $-2\ln(L)$ is to be accepted as preferable. The difference in fit between distributions with different parameters can be tested using the difference between the respective Y values against a chi-squared distribution with degrees of freedom equal to the difference in the number of parameters. The goodness of fit of these data sets can be tested by "binning" data into groups and comparing the predicted versus observed tallies using a statistic of the form of equation 6-27 against a chi-squared distribution of the number of bins minus the number of parameters. In addition, for particular distributions, such as the normal (or lognormal), special-purpose tests of fit that may have greater power have been devised (60,89,98,99). An example of analysis of this type of data is shown in Example 6-15.

Example 6-15. Enterococci—a group of indicator bacteria (12,22) — were measured at the 60-m depth in the vicinity of an ocean outfall discharging chlorinated primary-treated wastewater. The following are 70 density measurements (in no./100 mL) taken during the course of the recreational season (sorted by density). Characterize the density distribution that best describes this data set.[18]

---

[18]For this data set, the volumes of water analyzed for each sample are not known and may vary between samples—hence this must be analyzed as a problem of estimating the distribution of densities rather than the distribution of counts.

```
2 2   6 20
2 2   6 20
2 2   6 22
2 2   6 22
2 2   6 26
2 2   8 28
2 2   8 28
2 2   8 36
2 2   8 38
2 2   8 40
2 2  10 40
2 2  10 42
2 2  10 46
2 4  10 56
2 4  14 58
2 4  16 66
2 4  16 76
4        92
```

SOLUTION. For the lognormal distribution with this type of data (all data observed precisely), the parameters ζ and s are obtained as the mean and standard deviations of the logarithms of the observations. Therefore, we obtain $\zeta = 1.855$ and $s = 1.237$. Using equation 6-48 for the density function of the lognormal distribution along with equation 6-65, we determine that the optimal $-2 \ln(\Lambda)$ for the lognormal distribution is 487.14.

For the inverse Gaussian distribution, using numerical optimization, the best-fit parameters are found to be $\bar{\mu} = 14.17$ and $\phi = 0.356$, with an optimized $-2 \ln(\Lambda)$ of 475.035. For the gamma distribution, using numerical optimization, the best-fit parameters are found to be $\alpha = 0.752$ and $\beta = 18.850$. The optimized $-2 \ln(\Lambda)$ is found to be 506.977. Among the two-parameter distributions, the IG gives the lowest $-2 \ln(\Lambda)$ and is therefore accepted as the best fitting of the two-parameter distributions that are examined.

The generalized inverse Gaussian fit to these data (using numerical optimization in MATLAB) gives an optimized deviance of 473.939, with the following parameters:

$$\lambda = -0.802$$

$$\omega = 0.2762$$

$$\eta = 23.80$$

If we compare the GIG distribution with the IG or the gamma distribution, we get a Δ value of 1.096. Since this is not statistically significant at 1 degree of freedom (using the chi-squared distribution), it is concluded that the GIG cannot be accepted as providing a significant improvement in fit with respect to the IG distribution.

We must now examine the goodness of fit of the IG distribution to the data. The range of support is divided into a number of intervals such that each interval contains an equal frequency of *predicted* observations. Then the number of observations in each of the "bins" is tallied (Table 6-22). Treating the predicted number of observations as f_{ML}, using equation 6-24, the value of $-2 \ln(\Lambda)$ is computed and found to be equal to 15.296. For 2 degrees of freedom (four bins minus two fitted distribution parameters), this is the upper 0.05% of the chi-squared distribution, and hence the fit must be rejected as inadequate. The IG distribution appears to underpredict both the extreme low densities and the high densities.

Therefore, analysis of these data suggests that none of the distributions at hand is capable of providing an adequate fit. This may be symptomatic of different subsets of the population, and perhaps the necessity for using some more complex distribution (e.g., in this data set, there might be a difference in counts between days resulting from meteorological conditions or conditions of ocean current which may alter microbial dispersion patterns).

A major complication with most microbial data sets using the fitting-of-density approach is that many observations occur not as measured densities but as values below a detection limit. A microbial density is a quotient of a count and a volume, or perhaps (as in the MPN method) an assessment based on a quantal assay. If no organisms are observed, or if no positive replicates are found in an MPN assay, it would be inappropriate to record that density as a distinct number. For example, if 1000 mL of sample is analyzed by a plate assay, and if no organisms are found, the density can only be stated as <0.001 mL^{-1}.

Additionally, although it has been common practice, if the density is recorded at precisely the detection limit (in the example above, 0.001 mL^{-1}), substantial estimation biases may result (25,43,45). The most appropriate procedure is to analyze the particular data points in terms of the interval of which their value may best be described. So if the detection limit for sample i is x_i, then for such below-detection-limit (BDL) samples, the term $f(x_i)$ in equation

TABLE 6-22   Results for Example 6-15

Density Range			F_{IG}		
Lower Limit	Upper Limit	Observed	Lower Limit	Upper Limit	Predicted
0	2.79	30	0	0.25	17.5
2.79	6.11	10	0.25	0.5	17.5
6.11	14.96	10	0.5	0.75	17.5
14.96	∞	20	0.75	1	17.5

6-59 should be replaced by the cumulative distribution $F(x_i)$ (see equation 6-38). If the indicator variable δ_i takes a value of 1 if the sample is above a detection limit (i.e., is observed to be a particular precise numerical value), and if it takes a value of 0 if the sample is BDL, the likelihood equation can be written in more general terms as (note the analogy with equation 6-8)

$$-2 \ln(L) = -2 \sum_{i=1}^{n} \{\delta_i [\ln(f(x_i);\beta)] + (1 - \delta_i)[\ln(F(x_i);\beta)]\} \quad (6\text{-}66)$$

Table 6-23 is an example of this type of data—measurements of *Giardia* cysts in a raw water supply. The results of fitting these data to candidate distributions are given in Example 6-16.

Example 6-16. For the data in Table 6-23, compare the fits to the LN, IG, and gamma distributions.

SOLUTION. By finding the optimum deviance (equation 6-66) for the three two-parameter distributions (lognormal, gamma, inverse Gaussian), the results shown in Table 6-24 are obtained. Of the two-parameter distributions, the gamma distribution yields the lowest deviance (Y). The GIG distribution provides no better fit to these data than the gamma distribution. Therefore, the gamma distribution is accepted as the best-fitting distribution to this data set.

For this type of data, in which some counts are BDL, a goodness-of-fit test as in Example 6-15 cannot be done. In the case of the data set in Table 6-23, for example, there is no definitive statement that can easily be made

TABLE 6-23   Presumptive *Giardia* Counts in a Raw Water Supply

x_i (no./100 L)	Quantified, δ_i	x_i (no./100 L)	Quantified, δ_i
0.68	0	25.06	1
3.39	0	30.45	1
4.38	0	30.94	1
5.1	0	38.15	1
5.36	1	40.91	0
11.24	1	50	1
11.51	0	68.83	1
11.55	0	80.65	1
15.2	0	95.63	1
15.71	1	97.36	1
15.95	0	116.38	1
16.46	0	138.46	0
22.47	0	411.53	1

TABLE 6-24   Results for Example 6-16

Lognormal	ζ	2.37
	s	1.99
	Y	181.504
Gamma	α	0.326
	β	134.
	Y	178.25
IG	μ	43.8
	ϕ	0.0834
	Y	187.09

about the number of observations less than 10 per 100 L. Certainly the quantified observation at 5.36 is below this value, as are the nonquantified observations with the four lowest detection limits. However, samples with higher detection limits (including for example the observation that is $<$ a detection limit of 138.46 L^{-1}) may in fact be less than 10 per 100 L. The only firm statement that can be made for these data are for points above the highest nonquantified value (i.e., there is one point that is at 411.38 L^{-1}, and the remaining observations are below this value). This (with only two bins) does not allow the type of goodness-of-fit test as was performed in Example 6-15 to be made. If there were more observations above the highest nonquantified point available, use of the chi-squared goodness-of-fit test as in Example 6-15 would have been possible. To determine confidence limits to the density distribution, equation 6-22 can be used as in the development of Figures 6-10 and 6-11.

For this type of data (density data, or in general data sampled from a continuous distribution) for which some of the observations are "censored" to the left (it is known only that they are below a detection limit), a special-purpose goodness-of-fit test must be performed. We describe a test developed by Hollander and Proschan (47,59). First, the observations, consisting of the values or their detection limits (x) and an indicator variable ($\delta = 1$ for observed precisely, $= 0$ for known only below a detection limit), are ranked from highest ($i = 1$) to lowest ($i = n$, the number of observations). Second, the empirical cumulative distribution function F_i is computed from

$$F = \prod_{j=1}^{i} \left(\frac{n-i}{n-i+1} \right)^{\delta_i} \tag{6-67}$$

This function is frequently termed the *Kaplan–Meir estimator*. The "jump", f_i, is computed by subtracting F_i from the next-highest F_i corresponding to a quantified observation (or, for the highest quantified observation by subtracting from 1.0).

Using the best-fit parameters from the distribution whose fit is to be tested, the value of the cumulative distribution function is computed at each point as $F_i^0 (x_i)$. The following statistic is computed:

$$C = \sum_{i=1}^{n} \delta_i F_i^0 f_i \qquad (6\text{-}68)$$

The following is also computed:

$$\sigma = \frac{1}{4} \sqrt{\sum_{i=1}^{n} \left\{ \frac{n}{n - i + 1} [(F_{i-1}^0)^4 - (F_i^0)^4] \right\}} \qquad (6\text{-}69)$$

where it is understood that $F_0^0 = 1$. Finally, the values of C^* is computed from:

$$C^* = \left(C - \frac{1}{2} \right) \frac{\sqrt{n}}{\sigma} \qquad (6\text{-}70)$$

If the absolute value of C^* is greater than the upper $\alpha/2$th percentile of the cumulative normal distribution, the null hypothesis of goodness-of-fit is rejected (e.g., for a 5% significance level, the critical value is 1.96). The following example illustrates the application of this test.

Example 6-17. Is the best-fitting gamma distribution to the data in Table 6-23 (computed in Example 6-16) adequate to fit the data?

SOLUTION. The layout for the computation is shown in Table 6-25. The data are sorted in decreasing order of x_i (the observed density or the detection limit); the first three columns are the ranks (i), the observations, and the indicator variable for observation (δ_i). Columns 4 and 5 contain the computations for the Kaplan–Meir estimator, F_i. Note that for all quantified observations ($\delta_i = 1$), the estimator F_i is obtained by multiplying the value of F from the next-highest *quantified* observation by $(n - i)/(n - i + 1)$ (column 4). In column 6, the jumps are computed from subtraction of F_i from the value of F for the next-highest *quantified* observation (or from 1 in the case of the first row of the table).

In column 7, the best-fit parameters of the gamma distribution obtained in the previous problem are used to compute the fitted (null hypothesis) distribution. Columns 8, 9, and 10 are the indicated functions of the preceding columns.

The sum of values in column 8 gives the C statistic (equation 6-68), which is equal to 0.4514. To obtain the value of σ, the square root of the sum of

TABLE 6-25 Data for Example 6-17

(1)	(2)	(3)	(4)	(5)	(6)	(7)	(8)	(9)	(10)	(11)
i	x_i	δ_i	$\dfrac{n-i}{n-i+1}$	F_i	f_i	F_i^0	$\delta_i F_i^0 f_{ii}$	$A = (F_{i-1}^0)^4 - (F_i^0)^4$	$B = \dfrac{n}{n-i+1}$	AB
1	411.53	1	0.9615	0.9615	0.0385	0.9932	0.0382	0.0268	1.0000	0.0268
2	138.46	0				0.9111		0.2841	1.0400	0.2955
3	116.38	1	0.9583	0.9215	0.0401	0.8870	0.0355	0.0701	1.0833	0.0760
4	97.36	1	0.9565	0.8814	0.0401	0.8597	0.0344	0.0726	1.1304	0.0821
5	95.63	1	0.9545	0.8413	0.0401	0.8569	0.0343	0.0072	1.1818	0.0085
6	80.65	1	0.9524	0.8013	0.0401	0.8288	0.0332	0.0673	1.2381	0.0833
7	68.83	1	0.9500	0.7612	0.0401	0.8015	0.0321	0.0593	1.3000	0.0770
8	50	1	0.9474	0.7212	0.0401	0.7441	0.0298	0.1060	1.3684	0.1450
9	40.91	0				0.7075		0.0560	1.4444	0.0809
10	38.15	1	0.9412	0.6787	0.0424	0.6948	0.0295	0.0176	1.5294	0.0268
11	30.94	1	0.9375	0.6363	0.0424	0.6569	0.0279	0.0468	1.6250	0.0761
12	30.45	1	0.9333	0.5939	0.0424	0.6540	0.0277	0.0032	1.7333	0.0056
13	25.06	1	0.9286	0.5515	0.0424	0.6195	0.0263	0.0357	1.8571	0.0663
14	22.47	0				0.6006		0.0172	2.0000	0.0344
15	16.46	0				0.5484		0.0396	2.1667	0.0859
16	15.95	0				0.5433		0.0033	2.3636	0.0079
17	15.71	1	0.9000	0.4963	0.0551	0.5409	0.0298	0.0016	2.6000	0.0041
18	15.2	0				0.5356		0.0033	2.8889	0.0095
19	11.55	0				0.4929		0.0232	3.2500	0.0755
20	11.51	0				0.4924		0.0002	3.7143	0.0009
21	11.24	1	0.8333	0.4136	0.0827	0.4888	0.0404	0.0017	4.3333	0.0073
22	5.36	1	0.8000	0.3309	0.0827	0.3882	0.0321	0.0344	5.2000	0.1789
23	5.1	0				0.3821		0.0014	6.5000	0.0090
24	4.38	0				0.3641		0.0037	8.6667	0.0324
25	3.39	0				0.3356		0.0049	13.0000	0.0637
26	0.68	0				0.1998		0.0111	26.0000	0.2882

column 11 is taken and divided by 4 (equation 6-69). This results in $\sigma = 0.3398$.

Substitution of the values for C and σ in equation 6-70 (along with $n = 26$) gives the statistic for fit of $C^* = -0.7296$. Since the absolute value of this is less than the critical value of 1.96, the null hypothesis of adequacy of the gamma distribution fit to the data cannot be rejected.

Microbial assays based on quantal responses, such as the MPN coliform test, can also be used to test whether alternative (non-Poisson) distributions fit the data more adequately. In turn, the interpretation of MPN results may show a substantial bias (even greater than as in Figure 6-2) if a non-Poisson distribution characterizes the system (113).

There are two ways in which non-Poisson behavior may be manifested in the context of an MPN determination. First, if the response of various dilutions within a dilution assay is not governed by Poisson behavior, a test of randomness within a sample, as in Examples 6-9 and 6-10 will indicate this deviation. For Poisson-distributed samples, the probability of a single positive replicate at a particular volume is given by the complement of the zero term of the Poisson distribution, equation 6-13. In the case of the negative binomial distribution, the probability of a positive response is given by the complement of equation 6-47, evaluated at $x = 0$, or

$$\pi = 1 - \left(1 + \frac{\overline{\mu}V}{k}\right)^{-k} \tag{6-71}$$

Similarly, for the PLN, PIG, and PGIG distributions, the single replicate positive probability would differ from the Poisson distribution.

Figure 6-12 compares the frequency of positive samples under the Poisson distribution (equation 6-13) versus a negative binomial distribution (equation 6-71). As noted by Wadley (113), the mean density corresponding to a given fraction of positive responses is less for the Poisson distribution than for the negative binomial, and the discrepancy becomes progressively greater for lower values of the negative binomial k parameter.

A non-Poisson intrasample distribution might result if a facet of the enumeration process introduced a form of variability into the enumeration of counts. If the initial sample had a constant Poisson mean of μ^*, and if the enumeration process resulted in an "effective" density of $\gamma\mu^*$, with γ being an independent random variable distributed according to a lognormal, gamma, IG, or GIG distribution, the distribution of organisms in samples of volume V would be anticipated to be described by the PLN, NB, PIG, or PGIG distributions, respectively.

In general, it would be impossible solely by a dilution or MPN assay to estimate μ^* independently from the parameters describing the distribution of γ; in other words, we estimate the distribution of the effective density ($\gamma\mu^*$).

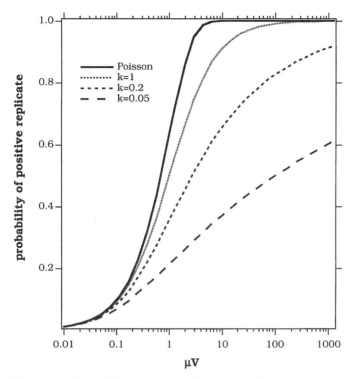

Figure 6-12 Comparison of fraction of positive samples for Poisson versus negative binomial distribution.

Since there would be no way to assess by this assay alone the mean value of γ, the presence of non-Poisson intrasample variability may occur with a concomitant bias in the estimate of mean density (as well as the quantiles) due to the bias of the method.

Hence, although the maximum likelihood estimates of statistics characterizing microbial distributions in an MPN assay that deviates from the Poisson assumption can be derived, we shall not do so, since this can be misleading. Rather, the presence of non-Poisson intrasample results should be used as a quality assessment metric to initiate a change in enumeration methodology.

We would also have the same reservation for count data, such as that presented in Example 6-8. If intrasample replicates showed deviation from Poisson behavior, the assessment of intersample distributional properties is problematic due to possible methodological deficiencies.

Quite a different scenario is presented when the intrasample testing shows Poisson behavior of an MPN assay, but the intersample tests show that there is not an overall uniform Poisson mean. This was exactly the situation depicted in the data of Example 6-10 (for *Staphylococcus aureus*).

The alternative scenario envisioned is that a single sample is drawn and has a mean density, μ, which is distributed (among samples) by a gamma, IG, GIG, or LN distribution. We seek to characterize the parameters of these density distributions in a similar manner to analysis of count data, as in Examples 6-13 through 6-16.

The starting points for this analysis are equation 6-14 (which gives the probability of observing a set of positive tube scores given a density) and an analog of equation 6-43 (which gives the impact of a mixing distribution on the probability of an observed result). If non-Poisson intersample variability is present, the likelihood function for a quantal assay is written as equation 6-14 integrated with a mixing distribution as follows:

$$L = \int_0^{\infty} \left\{ \prod_{i=1}^{k} p_i! \, \frac{n_i!}{(n_i - p_i)!} \, [1 - \exp(-\overline{\mu}V_i)]^{p_i} [\exp(-\overline{\mu}V_i)]^{n_i-p_i} \right\} h(\overline{\mu};\beta) \, d\overline{\mu}$$

(6-72)

In equation 6-72, the mixing distribution, $h(\overline{\mu};\beta)$, could be one of the density distributions used previously, such as the lognormal, inverse Gaussian, gamma, or GIG. Thomas (107) suggested analysis of MPN results using a gamma mixing distribution, however, he noted the mathematical intractability of the problem and approximated the results with an analysis by method of moments. In principle, the integral defined by equation (6-72 can be used in an optimization scheme to fit the parameters of the mixing distribution; however, this is currently a bit beyond reasonable computing power. None of the mixing distributions discussed here is known to give an analytical solution to the type of integral noted in equation 6-72.

Estimation of Mean Density from Negative Binomial and Poisson–Lognormal Distribution.

Once the density distribution has been estimated, quantities derived from that density distribution can also be estimated. In the case of the Poisson, negative binomial, and Poisson–inverse Gaussian distributions, the mean density appears as a parameter of the distribution. For the sake of completeness, Table 6-26 gives the formulas for the mean density and the standard deviation of the density based on for data whose underlying densities fit the LN, gamma, IG, and GIG distributions based on parameters used to fit the analogous Poisson mixture distributions. These should be regarded as the maximum likelihood estimators of the mean and standard deviation.[19] The mean density, in particular, may be used as an input into a risk

[19]Note that particularly in small samples, the maximum likelihood estimators may have substantial bias. There is a considerable body of work on *unbiased minimum variance estimators* (UMVEs), which depending on the application, may be preferable. However, typically, these are more complicated to compute and may be more sensitive to small deviations from fit to the presumed distribution (52,103).

TABLE 6-26 Mean and Standard Deviation of Density from Various Mixing Distributions

Distribution of Density	Mean Density	Standard Deviation of Density
Gamma	$\overline{\mu}$	$\dfrac{\overline{\mu}}{\sqrt{k}}$
Lognormal	$\exp\left(\zeta + \dfrac{s^2}{2}\right)$	$\sqrt{\exp(s^2) - 1}\ \exp\left(\zeta + \dfrac{s^2}{2}\right)$
Inverse Gaussian	$\overline{\mu}$	$\dfrac{\overline{\mu}}{\sqrt{\phi}}$
Generalized inverse Gaussian	$\eta\,\dfrac{K_{\lambda+1}(\omega)}{K_\lambda(\omega)}$	$\eta\,\sqrt{\dfrac{K_{\lambda+2}(\omega)}{K_\lambda(\omega)} - \left[\dfrac{K_{\lambda+1}(\omega)}{K_\lambda(\omega)}\right]^2}$

estimate (forming, along with an estimate for consumption, a portion of the exposure estimate). Based on the fit-to-density distributions, as in Figure 6-11, the variability of density, along with its' uncertainty, can be computed. If a risk assessment is to be performed incorporating variability and/or uncertainty, information such as in Figures 6-10 and 6-11 would form one input.

Application to Direct Analysis of the Finished Product

The framework presented to this point in the chapter is applicable to assessing the concentration distribution in material at the point of exposure. In some cases, information to assess this finished concentration distribution is readily available. However, in other cases, the finished concentration distribution must be inferred from a distribution in a raw material and information on the transformation processes that occur in passage from raw to finished product.

For example, in going from cattle to a finished, cooked hamburger, a variety of processes (cattle slaughter, primal cut preparation, ground beef preparation, holding, cooking, etc.) occur. Each of these processes can alter the microbial levels. Therefore, if information is available only on the raw material distribution and the attenuation processes, the final distribution can be inferred. In the case of drinking water, as another example, the relatively low levels of pathogens in finished drinking water (82) necessitate such indirect inferences, rather than a direct measurement of concentration distributions in the material to which humans are directly exposed. We discuss methods to characterize such transformation processes.

Experimental Difficulties in Pathogen Assessment

As another aside, it should be stressed again that for pathogens, in particular, it must not necessarily be assumed that measurement techniques are perfect.

For example, in a performance evaluation study conducted under the sponsorship of the U.S. Environmental Protection Agency, spiked samples of *Cryptosporidium* oocysts were sent to laboratories for execution of standard oocyst enumeration methods. Figure 6-13 presents a histogram of the fractional recovery efficiency. Also shown on this figure is a best fit of these data to a beta distribution. The beta distribution is a continuous probability distribution for a variable x, defined between 0 and 1, with parameters α and β, and has a density function as follows:

$$f(x) = \frac{\Gamma(\alpha + \beta)}{\Gamma(\alpha)\Gamma(\beta)} \, x^{\alpha-1}(1 - x)^{\beta-1} \tag{6-73}$$

These data show that the average recovery is quite poor, and in addition that there is considerable variability in recovery between samples (and although it is not shown, there is a systematic laboratory-to-laboratory difference in performance). These findings have also been reported by other investigators

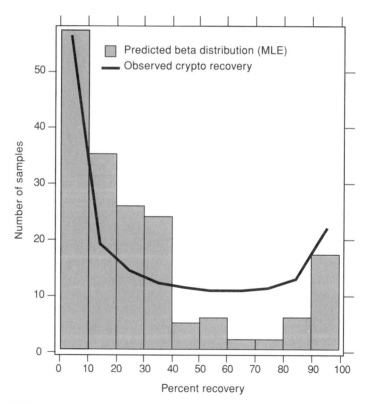

Figure 6-13   Recovery of *Cryptosporidium* oocysts in laboratory performance evalution studies.

(15). Unfortunately, such recovery studies do not exist for most pathogenic microorganisms (nor for many indicators), and therefore the assumptions of adequate experimental technique have not been tested sufficiently.

Estimation by Analysis of Process Performance

We seek to characterize the various processes from raw material to finished, consumed material that can result in the transformation (either growth or decline) of infectious pathogens. Figure 6-14 presents a conceptual process sequence for the production of hamburger consumed at home (top) and the production of drinking water delivered to households (bottom). As suggested, there are some important parallels in the process sequences. Both contain segments of bulk processing, that is, some industrial manipulations at a large-scale facility such as the slaughterhouse/packinghouse, and the drinking water treatment plant. Both contain elements of bulk transport: frozen transport of ground beef to the user and piped transmission of drinking water. Both contain elements unique to the particular point of consumption. For each of the boxes in this sequence, in principle, one could define the change of microorganism levels such that given the input levels, the output (at the point of consumption) could be obtained (and its distribution).

As another example of this approach, Haas et al. (40) investigated the risk from infectious viruses to people who consume groundwater (without treatment) that has been contaminated with microorganisms emanating from sanitary landfill leachate. As shown in Figure 6-15, in considering this scenario, a number of intervening processes beginning with the initial virus contamination in solid waste being emplaced in the landfill must be considered. These include inactivation in the landfill, potential for desorption from solid waste

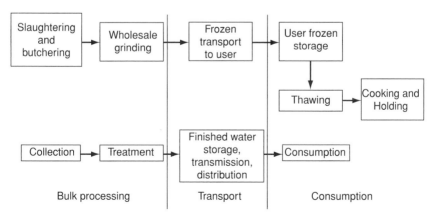

Figure 6-14   Conceptual process sequences in the hamburger production consumption chain (top) and the drinking water treatment consumption chain (bottom).

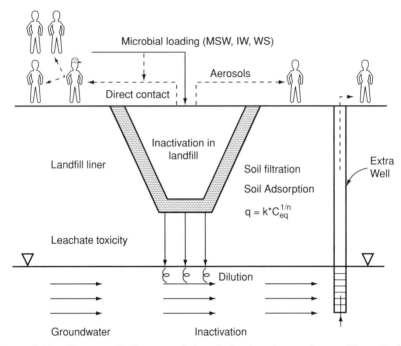

Figure 6-15 Conceptual diagram of the viral migration pathway. (From Ref. 40; reproduced by permission of Water Environment Federation.)

into mobile leachate, adsorption into landfill liners, adsorption onto aquifer solids, and inactivation in groundwater.

In general terms, if there are k processes in series, and if μ_0 is the initial microbial concentration entering the system, and μ_i is the concentration exiting process i, then in rather general terms, the following can be written:

$$\mu_1 = f^{(1)}(\mu_0, y_1^{(1)}, y_2^{(1)}, \ldots, y_{c_1}^{(1)})$$

$$\mu_2 = f^{(2)}(\mu_1, y_1^{(2)}, y_2^{(2)}, \ldots, y_{c_2}^{(2)})$$

$$\cdot$$
$$\cdot$$
$$\cdot$$

$$\mu_k = f^{(k)}(\mu_{k-1}, y_1^{(k)}, y_2^{(k)}, \ldots, y_{c_2}^{(k)})$$

(6-74)

In essence, these equations describe a situation in which the concentration of organisms exiting process i (μ_i) is a function of the organisms entering the process (μ_{i-1}) and other variables (the y's) that characterize process performance (these may be constant or may be variable; for example, the temperature,

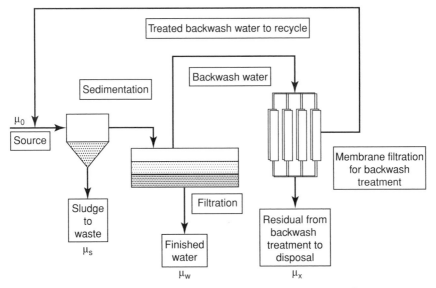

Figure 6-16 Schematic diagram of a water treatment plant.

pH, or level of intensity—chemical dosage). Note that the particular functional forms (the f's) may differ between stages.

In fact, this formulation may not be sufficiently general to describe all situations, particularly when a potential for recycle between stages exists. In this situation, a detailed process model must be constructed based on appropriate mass balances (on microorganisms) and reaction rates or partition factors. As one example of where this may be necessary, we cite the situation of a water treatment plant itself.

Figure 6-16 shows a schematic of a water treatment plant that contains a membrane treatment system for backwash water reclamation, The recycle of membrane filter effluent from the backwash water treatment system to mix with the source water entering the sedimentation tank means that the entire system mass balance must be considered simultaneously (the membrane filter effluent is recycled and in part influences the influent to the membrane filter.

Frequently, in simple situations, the processes are all simple first-order systems. In other words, the effluent from one process is a multiple (perhaps dependent on the ancillary process variables—the y's) of the influent. In this situation, in the absence of variability and uncertainty, if we know the initial microbial concentration (μ_0) and the transformation ratios[20] for each of these

---

[20]The term *transformation ratio* is used in a general sense, and it may have a value >1 if there is growth or increase in a particular process, as in regrowth in a water distribution system or multiplication in a food product subject to lack of proper refrigeration.

processes (ε_1, ε_2, . . . , ε_k, for k processes), a simple multiplication would yield the final density (μ) as follows:

$$\mu = \prod_{i=1}^{k} \varepsilon_i \qquad (6\text{-}75)$$

However, the individual ε values may have some variability and uncertainty, and hence we need to estimate these distributions (which we discuss in this chapter) and then use the set of distributions to estimate the overall exposure distribution (which we discuss in a subsequent chapter).

When the ε values have variability and uncertainty, we can in general approach their description in two ways. In the first approach we seek to describe the information on ε by a probability distribution. This raises issues of determination of the best probability distribution, the best parameters, and the uncertainty of the parameters.

In the second approach, we presume that ε is a number determined absolutely as a function of ancillary information (temperature, time, etc.). The deterministic model may have some parameters that are uncertain, and the intrinsic variability of the ancillary information needs to be taken into account in assessing the effective overall span of ε.

Describing Process Transformation Distributions. There are a number of situations in which the dynamics of organisms through a process can be described by simple first-order relationships. Figure 6-17 summarizes measurements of animal virus on Buffalo Green Monkey (BGM) cells in the influent to lime treatment, and the lime treatment effluent, at the Water Factory 21 water reclamation facility (51). The top panel shows a scatter plot of the paired observations. In the bottom panel, the transformation ratios are plotted on a lognormal probability plot. This suggests that for this process, the transformation ratio may be adequately modeled by a random variable that is lognormally distributed.

However, there are a number of other possible distributions that may adequately characterize the data, and therefore we need to formally address the issue of how to select the best distribution, and how to estimate the best set of parameters of the distribution. For the purpose of this section it is assumed that all data are observed precisely (there are no binned data, or data above or below detection limits; for these types of modifications, approaches similar to those used for microbial monitoring data analysis should be used).

For transformation ratios that are bounded by 0 and 1, the beta distribution (equation 6-73) can be used to describe data. This distribution can be generalized by additional parameters to cover arbitrary intervals as discussed by Vose (112), who should be consulted for a broad catalog of distributions useful in risk analysis. For transformation ratios that may take on a broader

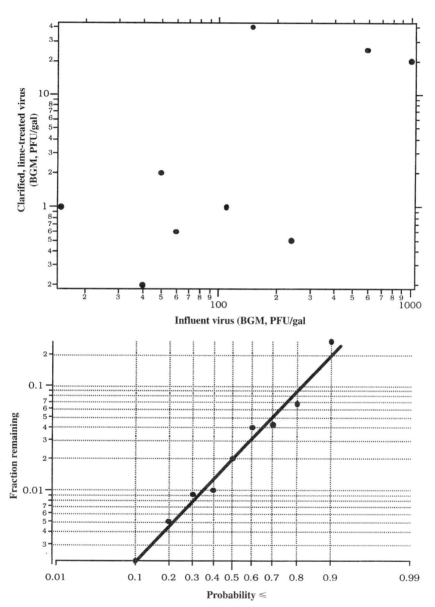

Figure 6-17 Virus removal by lime treatment at Water Factory 21 reclamation facility. (From Ref. 51.)

range over all positive real numbers,[21] other possible distributions include the lognormal (equation 6-48), gamma (equation 6-44), inverse Gaussian (equation 6-53), and generalized inverse Gaussian (equation 6-57). One other distribution that may be useful in some situations is the Weibull distribution. The density function $f(x)$ and cumulative distribution function $F(x)$ are given by

$$f(x) = \frac{c}{\alpha} \left(\frac{x}{\alpha} \right)^{c-1} \exp\left[-\left(\frac{x}{\alpha} \right)^c \right] \tag{6-76}$$

$$F(x) = 1 - \exp\left[-\left(\frac{x}{\alpha} \right)^c \right] \tag{6-77}$$

The mean and variance of the Weibull distribution are given by

$$\overline{\mu} = \alpha\Gamma\left(\frac{1}{c} + 1 \right)$$

$$\text{var} = \alpha^2 \left\{ \Gamma\left(\frac{2}{c} + 1 \right) - \left[\Gamma\left(\frac{1}{c} + 1 \right) \right]^2 \right\} \tag{6-78}$$

Given a distribution density function $f(x;\beta)$, which gives the probability density at an observed value x given parameter values of β, the optimum values of β can be obtained by minimizing a negative log-likelihood function of the following form (where the observations are $x_1, \ldots, x_k$):

$$Y = -2 \sum_{i=0}^{k} \ln[f(x_i;\beta)] \tag{6-79}$$

This has the same form as has been used earlier. Distributions with the same number of parameters can be ranked by the optimal Y value, and those with the lowest Y should be retained. Distributions with additional parameters can be compared for statistically significant improvement by use of the difference in Y values compared to a χ^2 distribution with degrees of freedom equal to the difference in number of parameters.

Parenthetically, it should be noted that the identification of distributional form may be subject to error if there are only a small number of data points used and if the relative standard deviation is only moderate to small (38). In

---

[21]These data, even if not physically restricted between 0 and 1, may still be adequately described by the beta distribution provided that the upper tails of the observed data are not broad.

this case, the risk analysis should be careful not to place too much weight on the far tails of the input distribution for this parameter.[22]

Once having selected the distributional form, the optimum parameter values for the selected distribution are those that had resulted in the minimum value of Y for that distribution. If there are sufficient data points to do a classical "binned" chi-squared goodness-of-fit test, this can be used to assess overall goodness-of-fit to that distribution. Otherwise, special-purpose tests must be used.[23] Confidence limits to parameters of the distribution can be obtained by application of equation 6-62. The overall fitting process is illustrated by the following example.

Example 6-18. The data points in the lower panel of Figure 6-17 show virus removal ratios, $k_{10}(1/\text{day}^{-1})$, as follows:

0.00208333	0.04
0.005	0.0423729
0.00909091	0.0666667
0.01	0.266667
0.02	

Given this information, select among the beta, lognormal, gamma, Weibull, and inverse Gaussian distribution and report the best-fit parameters. Discuss the goodness of fit to the distribution selected.

SOLUTION. From the data, we obtain the fits shown in Table 6-27 (ranked by value of Y). The inverse Gaussian yields the lowest value of Y among the distributions considered, and hence this distribution is accepted.

To perform a goodness-of-fit test, we need to construct a tabular histogram of the observed frequencies and the predicted frequencies. To do a goodness-of-fit test using this approach, it is best to select class intervals such that there are is an equal predicted frequency in each bin of the table. Five bins are selected, and therefore the class intervals are computed so as to have 1.8 predicted observation per row ($= 9/5$). From the class ranges, the number of observations falling into that range are tallied, and the contribution to the

---

[22]Later we discuss the importance of doing a sensitivity analysis for inputs, and if the particular parameter has only a low impact on the final results, the specification of distributional form *for that parameter* may not be as critical as when the parameter has a major impact on the final result.

[23]In any event, it is inappropriate to use a test such as the Kolmogorov–Smirnov test described earlier unless specific corrections have been made for the significance levels to take into account the estimation of parameters. Table 6-8 would be *inappropriate* in this application. For the normal distribution (and lognormal), Lilliefors (60) has provided a modified version of Table 6-8. Stephens (99) has reviewed applications of these types of test.

TABLE 6-27 Results for Example 6-18

Distribution	Best-Fit Parameters		Y
Inverse Gaussian	$\bar{\mu} = 0.0513$	$\phi = 0.211$	-39.019
Lognormal	$\zeta = 3.899$	$s = 1.388$	-38.755
Weibull	$c = 0.734$	$\alpha = 0.0409$	-37.328
Gamma	$\alpha = 0.656$	$\beta = 0.0783$	-36.715
Beta	$\alpha = 0.615$	$\beta = 11.045$	-36.424

goodness-of-fit statistic is calculated. Table 6-28 results. The sum of the last column is 0.510, which is the goodness-of-fit statistic. Since this is less than the critical chi-squared distribution at 3 degrees of freedom (five rows minus two parameters for the fitted distribution), the null hypothesis (of the acceptability of the fit) is accepted. If the computed goodness-of-fit statistic was in excess of 7.84 (the upper 5th percentile), the fit would be rejected. However, the analyst should be warned that with the small number of samples, no goodness-of-fit test will have substantial power against alternatives, and therefore some caution must be warranted in use of this data set for extrapolation to very high or low quantiles (if sensitivity analysis shows that this is an influential input).

Transformation Ratio as a Function of Other Parameters. As an example of a first-order transformation ratio which may depend on ancillary factors (as well as other random variables), we cite the inactivation of virus during passage in saturated groundwater. Based on a review of published data (primarily on poliovirus), it was concluded (40) that the transformation ratio for passage in groundwater could be described by

$$\varepsilon = 10^{-k_{10}t} \tag{6-80}$$

where t is the travel time in the groundwater. Furthermore, based on several studies, the inactivation rate constant (at ambient temperature and other con-

TABLE 6.28 Goodness-of-Fit Calculations for Example 6-18

Class Range		Number of Observations f_{obs}	Predicted Frequency f_{pred}	$-2f_{obs} \ln \dfrac{f_{pred}}{f_{obs}}$
Lower Limit	Upper Limit			
0	0.00556	2	1.800	0.421
0.00556	0.0113	2	1.800	0.421
0.0113	0.02328	1	1.800	-1.176
0.02328	0.05975	2	1.800	0.422
0.05975	∞	2	1.800	0.422

ditions) in base 10 logarithms (k_{10}) was found to be lognormally distributed (Figure 6-18). From this analysis, the parameters of the best-fitting lognormal distribution of k ($\log_{10}/d$) are found to be $\zeta = -2.07$ and $s = 0.754$.

Now, the transit time in groundwater prior to consumption (t) is itself a probability distribution. For example, in a survey performed for the American Water Works Association, a distribution of distances from water supply wells (of community systems) to the points of nearest contamination was determined (Figure 6-19). With this information, and also with a distribution of velocities of groundwater, t (or more correctly, its distribution) can be determined.

Obviously, the information in Figure 6-19 shows a complex distributional; shape. Although it is possible, in principle, to analytically compute the probability distribution of a function of random variables (as in equation 6-70) (102,103), this is generally mathematically quite intractable. Therefore, the entire process of characterizing inputs which are in turn functions of random variables is handled using the process of Monte Carlo analysis. This method is discussed in Chapter 8.

What can be done at this stage, however, with this type of information, is to calculate the transformation ratio using particular values of the inputs. For example, assuming that the groundwater flow velocity is 1 ft/day, with the modal distance to contamination of 60 ft (Figure 6-19), the time would be computed as 60 days. Using the median k value of 0.02 day^{-1}, ε would be computed as $10^{-(0.02)(60)} = 0.063$. However, this calculation gives only a single value for k, and we have no way of knowing whether this is a mean, median,

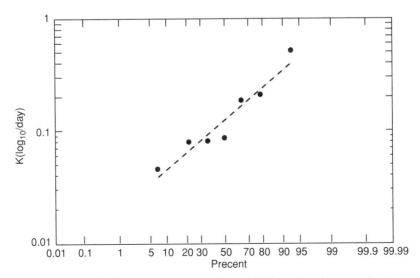

Figure 6-18 Inactivation rate constants (base 10) for virus in groundwater under ambient conditions.

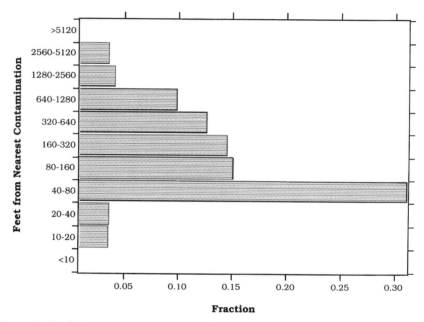

Figure 6-19   Distance to nearest source of contamination of community water supply wells. (From an unpublished survey conducted by C. Dery and C. N. Haas.)

or some other particular predetermined percentile of the transformation ratio. Therefore, although this method is relatively easy to carry out, it provides little information and may not provide a controlled degree of conservatism in computations.

For some processes, extensive databases on microbial removal are available. For example, under the U.S. Surface Water Treatment Rule, requirements for achieving defined reduction of *Giardia* and viruses have led to development of a database on the disinfection of these agents in water by chlorine, ozone, and chlorine dioxide (63). In the food industry, there is an extensive history of describing inactivation of microorganisms by high-temperature treatment in terms of decimal reduction times (the time required for reduction by a factor of 10) (114). In recent years, the development of the field of predictive microbiology has focused on describing the growth of pathogenic microorganisms in food during conditions in which there is a mild temperature rise (called "temperature abuse" situations, when, for example, there is a breakdown in refrigeration). In these studies, basic equations are used to describe the dynamic behavior of microorganism counts in food. A number of authors have successfully applied the Gompertz equation (2,8,85,95,117). This equation predicts the mean microbial concentration $\mu(t)$ at a given time by the following function:

$$\mu = \exp\{A + C \exp[-\exp(-B(t - M))]\} \qquad (6\text{-}81)$$

where A, B, C, and M are characteristics of the growth environment.

As part of an extensive data collection effort, the parameters in this equation has been correlated against physical–chemical conditions, such as temperature, acidity, preservative concentrations, and so on. As a partial example of the existing data, Table 6-29 summarizes some of the studies that have been conducted. Unfortunately, a comparable database does not exist for microbial growth in the water environment.

Figure 6-20 plots the Gompertz equation for constant A and C values but varying M and B values. A and C depict the initial microbial levels and asymptotic levels, respectively, while B and M control the lag time until growth begins and the slope of the most rapid portion of the growth curve.

Use of the Gompertz model in the form of equation (6-81), even when the effect of temperature (and other conditions) on the underlying parameters has been incorporated, has limitations when temperature and the other conditions that may affect these parameters vary. These limitations include the following (110): (1) the Gompertz model accounts only for growth, not for inactivation; (2) there is no consideration of the prior history of growth; and (3) there is no capability for simulation of the transition between growth and inactivation (e.g., with high temperatures).

After consideration of these difficulties, van Impe et al. (110,111) developed a generalization of the Gompertz model suitable for estimation of microbial population dynamics under variable temperature conditions. This model, which simplifies to the Gompertz model under constant temperature, is defined by two differential equations (where n is the natural logarithm of the population density, and the variable n_0 can be viewed as a fictive popu-

TABLE 6-29   Examples of Some Available Data on Pathogen Growth Kinetics Using the Gompertz Model

Organism(s)	Variables Considered	Reference
Yersinia enterocolitica	Temperature, pH, sodium chloride, sodium nitrite	7
E. coli O157:H7	Temperature, pH, sodium chloride, oxygen tension	9, 105
Listeria monocytogenes, Aeromonas hydrophila, Yersinia enterocolitica	Temperature, oxygen tension	48, 49
Clostridium botulinum	Temperature, salt added/water activity	33
Staphylococcus aureus	Temperature, initial pH, salt, sodium nitrite, aerobic versus anaerobic conditions	10

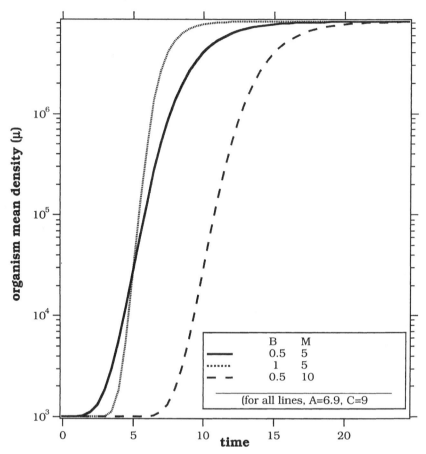

Figure 6-20 Plot of the Gompertz growth equation for constants A and C and different values of B and M.

lation density, which would equal the actual initial log density if no intervening periods of inactivation occurred) as follows:

$$\frac{dn}{dt} = [1 - f_{trans}(T)]c(n - n_0)\ln\frac{A_0 - n_0}{n - n_0} - f_{trans}(T)k(T) \qquad (6\text{-}82)$$

$$\frac{dn_0}{dt} = \gamma f_{zero}(T)\{n - n_0 - A\exp[-\exp(b)]\} \qquad (6\text{-}83)$$

The second term in equation 6-82 depicts thermal death, where k is the thermal death rate constant and is a function of temperature. Initial conditions ($t = 0$) for this problem are

$$n = \ln(N_0) + A \, \exp[-\exp(b)]$$

$$n_0 = \ln(N_0)$$

The temperature dependence of the basic parameter set may be given by an empirical model of the Ratkowsky square root type or from response surface modeling determinations. The parameter k, which may be temperature (and environment) dependent, is the first-order microorganism decay rate. The functions f_{trans} and f_{zero} account for the transition from growth to zero inactivation, and are defined by

$$f_{trans}(T) = \exp\{-\exp[\alpha(T_{trans} - T)]\}$$

$$f_{zero}(T) = \exp\{-\exp[\beta(T_{zero} - T)]\}$$

(6-84)

where T_{zero} is the root of the following equation (the right-hand side of equation 6-82):

$$[1 - f_{trans}(T_{zero})]c\,(n - n_0) \ln \frac{A_0 - n_0}{n - n_0} - f_{trans}(T_{zero})k(T_{zero}) = 0 \quad (6\text{-}85)$$

The parameters α, β, γ, and T_{trans} characterize the width and position of the transition zones between positive and negative net growth rate. Given a particular temperature versus time history for a product (or more broadly speaking, environmental condition versus time history), these equations can be integrated given parameter values for the particular organism of interest. As an example of this, Figure 6-21 shows the result of integrating the van Impe model (the parameters used were those in the original paper) for a particular time versus temperature profile, also shown in the figure.

It can be seen that the van Impe modification of the Gompertz model is capable of modeling changes in microbial concentration during abrupt temperature changes. Therefore, given estimates for the conditions that characterize each stage in a food-handling pathway, the mean microbial composition of the food at the point of ultimate consumption can be estimated.

It should be noted that there are other approaches to describing the effect of varying environmental conditions, particularly temperature, on microbial dynamics in food. In particular, Baranyi et al. have been developing an alternative quasi-mechanistic approach (2–4); the relative adequacy of the two approaches has not yet been evaluated. Furthermore, neither of these approaches has yet addressed microbial dynamics under conditions where chemical changes (e.g., pH, oxygen concentration, water activity) in addition to, or instead of, temperature changes occur.

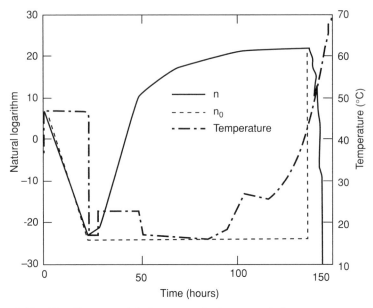

Figure 6-21 Van Impe model for a particular set of time versus temperature information.

CONSUMPTION DISTRIBUTIONS

The assessment of exposure to microbial agents requires, in addition to knowledge of the concentration of organisms in various items to which a person is exposed, the amount of the medium containing the organisms that a person may ingest or otherwise contact (equation 6-1). In this section we review some available data on consumption or contact distributions. Fortunately, since exposure to food, water, air, and so on does not simply pose a microbial risk but may also pose risks from other (e.g., chemical) agents, there is a fairly large database on ingestion/contact distribution. Furthermore, many of these distributions are achieving the status of "consensus" values, so for a particular risk assessment, the need to develop new information for these inputs is lessening.

There are a variety of exposure routes. Each route of exposure can be characterized by a contact rate and an exposure frequency. The *contact rate* is the amount of that particular material that is ingested, inhaled, or results in other direct exposure for each time the activity that results in exposure occurs. The *exposure frequency* is the rate at which the activity resulting in the exposure occurs. Both the contact rate and the exposure frequency may be characterized by point estimates and also by distributions reflecting uncertainty and variability. Table 6-30, summarizes point estimates for a number of exposures that may carry microbial agents.

TABLE 6-30 Point Estimates for Consumption/Contact Factors

Pathway	Contact Rate	Exposure Frequency
Ingestion of drinking water	1.4 L/day	365 days/yr
Ingestion of surface water while swimming	50 mL/h 2.6 h/swim	7 swims/year
Soil ingestion by children	200 mg/day (under 6 years in age) 100 mg/day (over 6 years)	Depends on context
Inhalation	20 m³/day (adult) 15 m³/day (child)	365 days/year
Inhalation during showering	0.07 m³/shower	365 showers/year
Ingestion of fish	0.113 kg/meal	48 meals/year

Source: Adapted from Ref. 18.

For microbiological risk it is important to characterize some routes of exposure in greater detail than in chemical risk assessment, and there are some routes of exposure that are virtually unique to microbial risks. For foodborne exposures, these exposure frequencies and contact rates are fairly well characterized. For example, the U.S. Department of Agriculture (USDA) Agriculture Research Service sponsors an ongoing *Continuing Survey of Food Intake by Individuals.* For several raw foods whose consumption may pose an excess microbial risk, estimates of consumption frequency reported as a fraction of person-days is reported as in Table 6-31. Using these multiple exposures, it is possible to estimate overall daily "loadings" of particular pathogens to a population (at least on an average or aggregate basis). This is shown by the following example.

Example 6-19. Consider a population subject to microbial exposure from drinking water, swimming in a surface water, and consuming raw shellfish.

TABLE 6-31 Consumption Frequency (U.S. Average) for Raw Foods, 1989–1990

Food	Fraction of Person-Days of Consumption
Raw beef	$<10^{-4}$
Raw fish	4×10^{-4}
Raw shellfish	9×10^{-4}
Raw eggs[a]	9×10^{-4}

Source: Ref. 81.

[a]Excluding consumption of raw eggs as an ingredient on sauces, dressings, or mayonnaise.

Use the information in Tables 6-30 and 6-31, and the following assumptions, to compute the average daily loading of a particular viral pathogen. The average viral loading in drinking water is 10^{-3} L^{-1}, in surface recreational water it is 0.1 L^{-1}, and in shellfish it is 1 g^{-1}. It is assumed that 150 g of shellfish is consumed per occurrence.

SOLUTION. For drinking water, with 1.4 L/day consumption, the viral load is estimated as 1.4×0.001 (1.4×10^{-3}) virus/day.

For swimming, on a per day basis we expect 7/365 "swim," with each swim being 2.6 h, and resulting in ingestion of 50 mL/h. Therefore, the daily average ingestion rate is

$$\frac{7}{365} \times 2.6 \times 0.05 = 0.0025 \text{ L/day}$$

With the viral concentration of 0.1 L^{-1}, the viral load from this route is 2.5×10^{-4} day^{-1}. For shellfish, there is a daily consumption of $(9 \times 10^{-4})(150 \text{ g}) = 0.135$ g/day. Given the virus concentration, the loading from this route is 0.135 virus/day. Therefore, overall, given these assumptions, the viral exposure is vastly dominated by the shellfish exposure. The total exposure is

$$1.4 \times 10^{-3} + 2.5 \times 10^{-4} + 0.135 = 0.136 \text{ virus/day}$$

The distribution of contact rates and exposure frequencies has been studied by a number of workers. It should be noted that the issue of specification of distributional form for these factors is as much (if not even more) of an issue than for densities of microorganisms. Therefore, when these distributions are used to develop probability distributions for risk, the particular most sensitive inputs should be carefully assessed to determine whether a different specification of distributional form would result in substantially altered results.

Table 6-32 summarizes a number of reported distributions for exposures that are of particular relevance to microbial risk. These distributions are for average U.S. populations (primarily adult). However, as we note below, there is considerable heterogeneity within these distributions, and so in some circumstances precise distributions for the particular subpopulations of relevance may be more appropriate.

For some types of exposure that may be of concern in a particular risk assessment, insufficient data may be available to precisely determine parameters of a best-fitting distribution. Nevertheless, it is possible to incorporate some indication of possible uncertainty and variability into the risk assessment by using expert knowledge, and intuition, to assign relatively crude distributions to the input parameters. For this purpose, two distributions have been widely employed: the uniform and triangular distributions.

TABLE 6-32 Reported Distributions for Exposures

Exposure	Distribution	Best-Fit Parameters	Reference
Consumption of finished drinking water (mL/day)	Lognormal	$\zeta = 7.49$ $s = 0.407$	84
Consumption of fish from all sources[a] (g/day)	Lognormal[b]	$\zeta = 3.682$ $s = 0.463$	73
Consumption of self-caught fish[a] (g/day)	Lognormal[c]	$\zeta = 3.862$ $s = 0.531$	73
Adherence of soil to skin[d] (mg/cm² per exposure)	Lognormal	$\zeta = -1.059$ $s = 1.448$	31

[a] For people in Michigan.
[b] 45% ate fish with this distribution; the remaining 55% did not eat fish.
[c] 8% ate self-caught fish; 92% did not.
[d] Based on hand exposure; units are mass of soil per unit area of skin in contact.

A *uniform distribution* (53) is one that has the following density function:

$$f(x) = \begin{cases} \dfrac{1}{b-a}, & a \le x \le b \\ 0, & x < a \text{ or } x > b \end{cases} \tag{6-86}$$

This has a corresponding cumulative distribution function of

$$F(x) = \begin{cases} 0, & x < a \\ \dfrac{x-a}{b-a} & a \le x \le b \\ 1, & x > b \end{cases} \tag{6-87}$$

This distribution has a mean of $(a + b)/2$ and a standard deviation $(b - a)$ $\sqrt{\frac{1}{12}}$. To a risk assessor, this distribution has the advantage that only an upper and a lower bound need be specified, and the distribution assigns uniform risk over that range.

The *triangular distribution* derives its name from the shape of its density function (Figure 6-22), which (112) can be written in terms of a minimum (*a*), mode (*b*) and maximum (*c*) as follows:

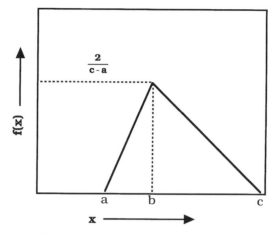

Figure 6-22 Density function for a triangular distribution.

$$f(x) = \begin{cases} \dfrac{2(x - a)}{(b - a)(c - a)}, & a \leq x \leq b \\[3mm] \dfrac{2(c - x)}{(c - a)(c - b)}, & b \leq x \leq c \end{cases}$$

(6-88)

The cumulative distribution function for this distribution is

$$F(x) = \begin{cases} \dfrac{(x - a)^2}{(c - a)(b - a)}, & a \leq x \leq b \\[3mm] \dfrac{x(2c - x) - c(a + b) + ab}{(c - a)(c - b)}, & b \leq x \leq c \end{cases}$$

(6-89)

For the triangular distribution, the mean and variance are given by

$$\text{mean} = \frac{a + b + c}{3}$$

(6-90)

$$\text{var} = \frac{a^2 + b^2 + c^2 - ab - ac - bc}{18}$$

If $(c - b) \neq (b - a)$, the distribution is asymmetric. The triangular distribution affords flexibility to the risk analyst in that only modal and upper and lower values need to be specified. For example, consider the information in Table 6-31. If we were doing a risk analysis on the consumption of raw egg and we wanted to investigate the uncertainty and variability in raw egg consumption, it might be plausible (in the absence of other information) to use a triangular distribution for the fraction of person-days in which eggs were consumed[24] in which the modal value (b) was 9×10^{-4}, with a range given perhaps by a factor of 1.25 on either side (i.e., $a = 7.2 \times 10^{-4}$, $c = 1.13 \times 10^{-3}$).

There are a number of routes of exposure for microorganisms to humans that are rather different from chemical routes of exposure. For these particular routes, much more limited knowledge of the quantitative transfer efficiencies are available, nor in fact is it clear that there is a comprehensive catalog of such exposure routes available. We enumerate some of these routes of exposure and quantitative estimates of extensiveness below. There is a full expectation that many of these microbial transfer processes will be more adequately quantified in the coming years.

In food preparation, many raw foods, especially raw animal foods (meat, poultry, fish), have potential microbial hazards. The inadequate sanitation of surfaces in contact with raw foods may result in an ultimate transfer to a person, and subsequent hand-to-mouth transfer. As an example of the potential for this, a case control of *E. coli* O157:H7 cases in New Jersey showed two statistically significant risk factors being failure to wash hands with soap and water after handling raw ground beef, and not washing work surfaces with soap and water after contact with raw ground beef (66). To characterize exposure properly, in addition to knowing the microbial loading in ground beef we would like to know:

- For a given surface area of exposure (e.g., of a cutting board) to raw beef what is the amount of beef mass transferred per contact? (If we knew this and we knew the microbial density—organisms per gram—in the beef, we could estimate the transferred microorganism level.)
- If a second food product (e.g., a vegetable) or a hand contacts the now-contaminated surface, what fraction of the mass on the surface would be transferred to the hand or second food product?

In a child day-care center, the prevalence of diarrhea has been found to be associated with microbial contamination of hands and of washroom sinks and faucets (57). To properly characterize the risk process in this environment, we would like to know the propensity for hand-to-surface and surface-to-hand transfer of contaminated fecal material.

In both of the situations above, we would also like to know the propensity to inadvertently ingest contaminated material from hands that have been in

---

[24]Although we reiterate that if a subsequent sensitivity analysis shows this assumption to be of major import, then some reexamination of its plausibility (and perhaps the acquisition of additional data to support a distributional fit) must be performed.

contact with a contaminated surface (i.e., what fraction of material on the hand would be inadvertently ingested per day?). Unfortunately, these transfer phenomena are not yet well characterized.

Marples and Towers (64) developed a controlled protocol to test transfer of microorganisms to and from hands. From a moist cloth containing microorganisms, 10% of the available cells (based on contact area) were transferred to hands by a single contact. From wet hands to moist fabric, there was an 85% transfer of microorganisms. Handwashing by water reduced transfer by 95%, and washing by 70% ethanol reduced transfer by 99.99%. Pether and Gilbert (76) determined that organisms on fingertips could be transferred to cooked foods by contact; although their data are difficult to interpret as a contact probability.

Mackintosh and Hoffman (62) extended the work of Marples to several different bacteria. Results of their experiments are summarized in Table 6-33. In general, transfer of organisms from a contaminated cloth to hands is less efficient than from the hands to a cloth. In addition, there are clearly organism-to-organism fairly complex differences. It is clear that *Staphylococcus saprophyticus* has a greater relative affinity for hands than does *Klebsiella aerogenes,* for example, since the former organism is transferred to hands more efficiently and to cloth less efficiently than the latter organism. It should be noted that intervening time between the cloth-to-hand and hand-to-cloth transfers could alter the overall transfer process by providing time for microbial growth and/or decay to occur, and also time for organism migration into less (or more?) accessible skin pores. Furthermore, the efficiency of transfer between hands and other moist surfaces needs to be studied. These are additional areas ripe for investigation.

Scott and Bloomfield (88) investigated transfer to hands and stainless steel bowls from a work surface contaminated with *E. coli, Salmonella* spp., or *Staphylococcus aureus.* While these experiments were not designed in a way to compute transfer probabilities as in the manner of Table 6-33, they nonetheless support the existence of a pathway for exposure along the following lines:

TABLE 6-33 Transfer of Bacteria Between Skin and Hands

Organism	Fraction Transferred	
	Cloth → Hand	Hand → Cloth
Staphylococcus saprophyticus	0.0167	0.17
E. coli	0.0047	0.88[a]
Pseudomonas aeruginosa	0.0036	0.76
Klebsiella aerogenes	0.0029	0.86
Streptococcus pyogenes	0.00007–0.00021	Not reported
Serratia marcescens	0.0046	Not reported

Source: Ref. 62.

[a] Suspect due to loss of viability.

- Contamination of a washcloth in the kitchen environment, perhaps by contact with raw animal products
- Survival and/or growth of microorganisms on the cloth
- When the cloth is used to wipe an additional surface, spread of microbial contamination to other surfaces
- Secondary contamination of other foodstuffs (e.g., vegetables or cheeses to be eaten uncooked), or contamination of hands followed by hand to mouth contact

Direct contact of surfaces with ground beef may represent a source of transfer. Miller et al. (68) found that contact of wooden and plastic cutting boards with ground beef containing aerobic heterotrophic microorganisms at a level of $10^{-6.7}$ g^{-1} resulted in a surface loading of about 1000 organisms/cm^2 on the cutting boards immediately after transfer. This implies that the contact of ground beef with a cutting board surface (and perhaps with other surfaces) results in an adherence of about 0.2 mg/cm^2. This is of a similar magnitude to the soil-to-skin transfer rates reported in other studies (see Table 6-32), and is deserving of follow-up work.

Humphrey et al. (50) conducted a study in which eggs were inoculated with known doses of pathogenic *Salmonella enteridis*. Human volunteers were asked to break the eggs, and the presence of *Salmonella* on "finger washes" was determined. Table 6-34 summarizes the results. This can be viewed as an MPN experiment. Let y_i be the average density of cells per egg (at dose level i). If f is the fraction of egg transferred per exposure, then from the, zero term of a Poisson distribution, the probability of at least one organism being transferred to a fingertip from contact can be written as $1 - \exp(-fy_i)$ (this is based on analogy with equation 6-14). The best value of f can be obtained by maximum likelihood fitting. However, a lack-of-fit test in the manner of equation 6-31 leads us to reject this fit.

Rather, we surmise that the nature of the experiment introduces some variability into the process (perhaps from the person-to-person variation in how eggs are broken). We therefore examine fitting the data with a negative bi-

TABLE 6-34   Effect of Level of *Salmonella enteriditis* in Fresh Eggs on the Probability of Acquiring *Salmonella* on Fingertips upon Breaking the Egg

Cells/Egg	Total Subjects	Positive Subjects
1,000	20	1
10,000	50	6
100,000	40	10
1,000,000	40	8

Source: Ref. 50.

nomial equation, based on equation 6-71, which can be developed by assuming that the fraction of egg transferred per event is a gamma-distributed variable. The probability of a positive response then is written as

$$\pi_{\text{NB},i} = 1 - \left(1 + \frac{fy_i}{k}\right)^{-k} \tag{6-91}$$

An empirical response probability based on the response observed at a single dose is defined as follows based on the observed positives (P) and total subjects (N):

$$\pi_i^0 = \frac{P_i}{N_i} \tag{6-92}$$

With these definitions, a deviance statistic (by analogy to the derivation of equation 6-31) for the negative binomial fit can be constructed. This considers a null hypothesis of negative binomial fit versus an alternative hypothesis that the individual empirical response probabilities are better descriptors of the experiment:

$$Y = -2 \sum_{i=1}^{k} \left[P_i \ln \frac{\pi_{\text{NB},i}}{\pi_i^0} + (N_i - P_i) \ln \frac{1 - \pi_{\text{NB},i}}{1 - \pi_i^0} \right] \tag{6-93}$$

The value of f and (the negative binomial parameter) k that minimize the deviance are accepted as the best, and the value of the deviance tested against 2 degrees of freedom (four doses minus two parameters) using the chi-squared distribution provides a test of goodness of fit. It is found that the optimal values of the parameters and the deviance statistic are

$$f = 1.919 \times 10^{-4}$$
$$k = 0.03282$$
$$Y = 1.484$$

Since the Y value is not in excess of the critical chi-squared value (5.99), the fit is accepted. We therefore conclude that the average fraction of an egg transferred per exposure is 1.919×10^{-4}; or using 50 g as a mass of an average egg, this implies a transfer of 9.6 mg per exposure. Given the underlying gamma distribution, confidence limits can also be computed.

In related work, these authors (50) also determined the potential for spatter of egg-containing batter to contaminate adjacent surfaces during hand mixing and electric mixing. However, the data presented are less amenable to a quantitative analysis.

Systematic Subpopulation Differences

As suggested by Table 6-32, there are distributions that have been fit to **a** number of potential "consumptions" (e.g., for water consumption). However, the distribution of activity (such as consumption, or transfer of material from food to skin) (as well as any statistic, such as the average or the median exposure) may be a function of discernible characteristics of the population exposed to a risk. By incorporating particular information about the population into computing the consumption distribution, the overall accuracy of the risk assessment may be enhanced.

Note that specific subpopulations may have different risks in that they are intrinsically more or less susceptible to a microbial agent or have a greater or lesser degree of exposure. The former differences are discussed in Chapter 7. In this section, the nature of some identified factors leading to different exposures are enumerated.

There are first, some aspects of lifestyle choice that may influence exposure. People may choose to alter their exposure to drinking water by using alternative sources of fluid intake, such as bottled water or home water purifiers. They may choose to change the nature of the foods they consume (perhaps from processed foods to fresh foods, for example). There may be cultural, regional, or ethnic norms that alter preferences in consumption. The USDA surveys indicate substantial regional and ethnic variation in consumption (aggregate of raw and cooked) of chicken, ground beef, shellfish, and eggs (Table 6-35). For example, this table demonstrates that consumption of eggs is substantially greater in the western United States, and among Hispano-Americans and "other" Americans than either the overall U.S. average or other categories. Hence, for microbial agents that might be associated with consumption of eggs, the potential for risk may be greater in the identifiable subpopulations.

TABLE 6-35 Variability of Food Consumption Patterns by Region and by Ethnic Group

Region	Ethnic Group	Percent of Person-Days with Consumption			
		Chicken	Ground Beef	Shellfish	Eggs
All U.S.	All U.S.	17.6	15.5	1.8	8.8
Northeast	All U.S.	21.3	12.7	1	6.7
North central	All U.S.	4.3	15.9	1	7.8
South	All U.S.	19.8	16.8	1.6	7.1
West	All U.S.	14.4	15.4	2.3	14.8
All U.S.	Hispanic Americans	22.5	16	1.3	15.7
	European Americans	15.7	15.4	1.7	7.8
	African Americans	26.2	17.3	2.4	9.4
	Others	20.4	6.6	2.3	16.3

Source: Ref. 81.

There are also occupational issues that may alter exposure. For example, in adult males the respiration (breathing) rate increased from 7.5 L/min at rest to 23 L/min when engaged in light activity (17); hence any microbial exposure via an inhalation route will clearly be different depending on the activity level.

Tap water consumption may be a function of age and other factors. A Canadian study found tap water consumption similar to that in U.S. populations (27). However, this consumption is less than found in some European countries, in particular Holland and the United Kingdom. Hence inferences of exposure from U.S. and North American data to other countries may introduce a source of systematic bias.

Roseberry and Burmaster (84) provided water consumption distributions by age. They found the distributions to be lognormal, with parameters as shown in Table 6-36. There tends to be an increase in water consumption with age, and therefore the elderly may tend to have greater exposure, by virtue of water consumption. In addition, in pregnant women, there is an increase in the upper tail of the water consumption distribution: 15% of pregnant women consuming greater than 2 L/day versus 10% of nonpregnant adult women (28).

TABLE 6-36   Age-Specific Probability Distribution Parameters (Lognormal) for Water Consumption (mL/Day) in U.S. Populations

Age Group (years)	ζ	s
0–1	6.979	0.291
1–11	7.182	0.34
11–20	7.49	0.347
20–65	7.563	0.4
>65	7.583	0.36

Source: Ref. 84.

AFTERWORD

As an afterword to this chapter and to Chapter 7, it is noted that a frequentist or likelihood-based approach to parameter estimation has been taken. An alternative approach to estimation can be taken using Bayes' theorem. In this approach the probability distribution of a set of unknown parameters (β) given a set of observations (y), p ($\beta|y$), termed the *posterior distribution,* is given by the following formula:

$$p(\beta/y) = \frac{L(y/\beta)\pi(\beta)}{\int_{\beta} L(y/\beta)\pi(\beta)\, d\beta} \tag{6-94}$$

where $L(y|\beta)$ is the likelihood function (e.g., equation 6-60), and $\pi(\beta)$ is termed the *prior distribution* of the parameters. This procedure raises a number of problems. First, the integral in the denominator of equation 6-94 is often very difficult to evaluate except by numerical methods; and hence to determine the optimum set of parameters, which occurs at the maximum value of $p(\beta|y)$, an optimization of a function with a numerical integral must be performed, which can be computationally very challenging.

Second, the best estimate is a function of the prior distribution as well as the data. Strict Bayesian philosophy dictates that one's prior state of knowledge should be embodied in the prior distribution. However, since different analysts may have different perceptions of their prior state of knowledge, the result is that there is no single "Bayesian estimate." To alleviate some of this subjectivity, uninformative prior distributions can be used, but in this case the ultimate result is frequently no different than the result of a straight likelihood estimate. Due to these difficulties, this book eschews the Bayesian approach to estimation. For those who wish to explore Bayesian approaches to estimate, several entry points are offered (5,23).

APPENDIX

A number of estimation problems presented in this chapter and in Chapter 7, involve the use of numerical optimization procedures. These can readily be programmed in many spreadsheets if an optimization routine is available. We illustrate the solution procedure for two types of problems using Microsoft EXCEL. All screen snapshots are from version 4 of the Apple Macintosh edition. We illustrate the solution of a problem requiring unconstrained optimization. In the appendix to Chapter 7 we illustrate solution of a constrained optimization problem using a spreadsheet.

Solution of the Unconstrained Optimization Problem

The solution of Example 6-9 is illustrated. The objective was to estimate the mean density in an MPN test. The numerical values of the spreadsheet are shown in Figure 6-23, with the problem data in columns A through C. Figure 6-24 gives a screen snapshot of the formulas that are inserted to solve the problem. In cell E14, the user inserts an estimated value for ln $(\overline{\mu})$. Cell E13 contains the computed value for $\overline{\mu}$. In column D the value of the row-wise MPN, $\overline{\mu}_{\text{ML},i}$, is computed. In column E, from rows 2 to 8, the value of $-\ln(\Lambda_i)$ is computed by the term in equation 6-31 using the row-wise MPN and the value in E13 (as the estimated MPN). Cell E10 contains the sum, which is to be minimized.

We use an estimated value for the logarithm of the mean density rather than the mean density itself since the SOLVER add-in within EXCEL does unconstrained optimization over the entire real line (i.e., the range $-\infty$ to

	A	B	C	D	E
					unconstr
1	Volume	total tubes	positive tubes	μ_ml_i	objective function
2	0.032	29	7	8.633	0.153
3	0.026	30	4	5.504	0.117
4	0.021	28	3	5.397	0.104
5	0.018	30	4	7.950	0.035
6	0.016	29	2	4.466	0.228
7	0.011	25	2	7.580	0.007
8	0.005	30	2	13.799	0.379
9					
10				sum	1.023
11					
12				assumed ln(mu)	1.939
13				mu	6.949

Figure 6-23 Screen snapshot showing numerical values for solution of Example 6-9.

$+\infty$). The logarithm can assume values over this entire range, and by exponentiating the logarithm, we assure ourselves of an MPN that is nonnegative.

The SOLVER add-in is used to search for the value of ln(MPN): the value in cell E12 that optimizes a particular cell in the spreadsheet (in our case, we want to minimize the value in E10). The SOLVER macro is invoked (in EXCEL version 4; there may be somewhat different menu commands in later versions) under the Formula menu.[25] In Figure 6-25, the commands to the SOLVER macro are shown. The cell to be minimized is E10 (note that the button labeled Mim has been checked) by varying cell E12.

If there were multiple parameters that we would want to allow to vary, rather than entering simply one cell (El2), a range of cell addresses (using commas and/or colons) could be entered. At this point we can click on the button labeled SOLVE for the solution to be found.

	D	E
1	μ_ml_i	objective function
2	=-(1/A2)*LN((B2-C2)/B2)	=+A2*($E$13-D2)*(B2-C2)-C2*LN((1-EXP(-$E$13*A2))/(1-EXP(-D2*A2)))
3	=-(1/A3)*LN((B3-C3)/B3)	=+A3*($E$13-D3)*(B3-C3)-C3*LN((1-EXP(-$E$13*A3))/(1-EXP(-D3*A3)))
4	=-(1/A4)*LN((B4-C4)/B4)	=+A4*($E$13-D4)*(B4-C4)-C4*LN((1-EXP(-$E$13*A4))/(1-EXP(-D4*A4)))
5	=-(1/A5)*LN((B5-C5)/B5)	=+A5*($E$13-D5)*(B5-C5)-C5*LN((1-EXP(-$E$13*A5))/(1-EXP(-D5*A5)))
6	=-(1/A6)*LN((B6-C6)/B6)	=+A6*($E$13-D6)*(B6-C6)-C6*LN((1-EXP(-$E$13*A6))/(1-EXP(-D6*A6)))
7	=-(1/A7)*LN((B7-C7)/B7)	=+A7*($E$13-D7)*(B7-C7)-C7*LN((1-EXP(-$E$13*A7))/(1-EXP(-D7*A7)))
8	=-(1/A8)*LN((B8-C8)/B8)	=+A8*($E$13-D8)*(B8-C8)-C8*LN((1-EXP(-$E$13*A8))/(1-EXP(-D8*A8)))
9		
10	sum	=SUM(E2:E9)
11		
12	assumed ln(mu)	1.938576980883
13	mu	=+EXP(E12)

Figure 6-24 Screen snapshot for formulas to solve Example 6-9.

[25]If it has not already been installed, it may also be necessary to open the SOLVER macro file prior to invoking the macro itself.

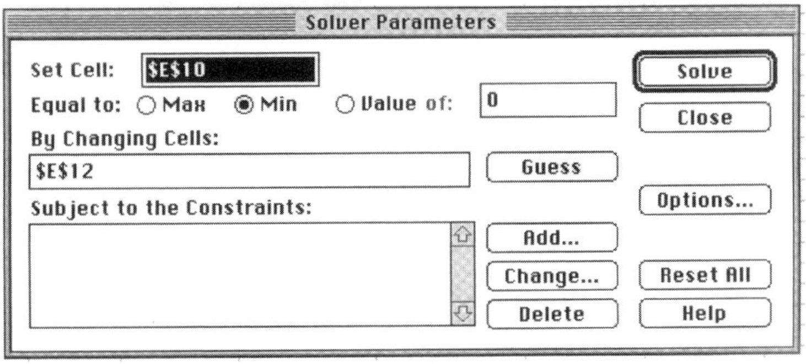

Figure 6-25   Setup of Solver for solution of Example 6-9.

Generally, we would want to run the problem from several diverse starting guesses to assure that we have found the true "global" optimum. Numerical search routines such as those in use in the SOLVER only converge to local optima; and particularly in a problem with multiple parameters, there may be several local optima, only one of which is global.

PROBLEMS

6-1. For each sample in Example 6-1, test whether the Poisson distribution with a constant mean density (for each sample) fits the data.

6-2. Test the improvement in fit (versus the Poisson distribution) for fits of the data in Example 6-1 to the PLN and NB distributions; is the improvement statistically significant?

6-3. For the MPN estimate of *Salmonella* in chicken computed in Example 6-3, compute a bias-corrected estimate using equation 6-17.

6-4. Consider a four-dilution (10, 1, 0.1 and 0.01 mL) MPN experiment, with five tubes at each dilution. Compute the MPN for the following codes:

$$3-0-0-0$$
$$5-4-1-0$$
$$1-2-0-0$$
$$0-3-0-1$$
$$1-1-0-0$$

6-5. Example 6-8 found that the mean densities between sets in the study were not consistent with a simple Poisson distribution. Test the following alternative hypotheses:

   (a) Each sample (A, B, and C in the table) is characterized by a single density, regardless of volume analyzed.

   (b) Each volume analyzed is characterized by a single density, regardless of volume analyzed.

6-6. Can the entire data set in Example 6-8 be characterized by the negative binomial distribution with a single mean and k value for the entire data set?

6-7. For each of the MPN codes in Problem 6-4, do a test of goodness of fit. At the 5% significance level, which codes are considered unlikely?

6-8. Compute the total uncertainty distribution for the mean density using the MPN data for *Salmonella* in chicken presented in Example 6-3.

6-9. Given the *Cryptosporidium* data in Example 6-6 and the best-fit Poisson distribution, compute the following:

   (a) The fraction of 1-L samples that you would expect to see with two or more organisms.

   (b) The 95% confidence limits to that fraction.

6-10. For the data in Example 6-5, which were found not to be fit with a Poisson distribution with a single mean, determine:

   (a) The statistical significance of an improvement in fit using the PLN, NB, PIG and PGIG distributions.

   (b) The overall goodness of fit of any of the alternative distributions for this data set.

6-11. For the Poisson–lognormal, Poisson–inverse Gaussian, and Poisson–generalized inverse Gaussian, write formulas for the expected proportion of positive samples for a given value of μV (analogous to equation 6-71 for the negative binomial). Is it also true for these distributions that the fraction of positives at a given μV value is less than for a Poisson distribution?

6-12. A study (109) of occurrence of *Listeria monocytogenes* on beef carcasses showed the following distribution of density:

Range of Counts (no. / cm^2)	Number of Samples
<0.03	53
0.03–0.30	20
0.3–3.0	4
>3	5

Which, if any, distribution fits data. Estimate the mean density and the density that would be exceeded only by 1% of samples.

6-13. From your analysis of the data in Problem 6-12, compute the 95% confidence limits to the estimate of the upper 1 percentile of densities.

6-14. Using the negative binomial fit to the data on egg breaking in Table 6-34, prepare a graph similar to that of Figure 6-10 for the 95% confidence limits to f and k.

6-15. For the virus ratios presented in Example 6-18, compute the upper 99th percentile for the best fits from the inverse Gaussian and lognormal fits to the data.

6-16. Use the inverse Gaussian fit to the data in Example 6-18, and compute the 95% confidence limits to the upper 99th percentile of the ratio.

REFERENCES

1. American Public Health Association, American Water Works Association, and the Water Pollution Control Federation. 1989. Standard methods for the examination of water and wastewater, 17th ed. APHA, AWWA, and WPCF, Washington, DC.

2. Baranyi, J., and T. Roberts. 1994. A dynamic approach to predicting bacterial growth in food. Int. J. Food Microbiol. 23:277–294.

3. Baranyi, J., T. A. Roberts, and P. McClure. 1993. A non-autonomous differential equation to model bacterial growth. Food Microbiol. 10:43–59.

4. Baranyi, J., T. P. Robinson, A. Kaloti, and B. M. Mackey. 1995. Predicting growth of *Brochothrix thermosphacta* at changing temperature. Int. J. Food Microbiol. 27(1):61–75.

5. Berger, J. O. 1985. Statistical decision theory and Bayesian analysis, 2nd ed. Springer-Verlag, New York.

6. Best, D. J. 1990. Optimal determination of most probable numbers. Int. J. Food Microbiol. 11:159–166.

7. Bhaduri, S., C. O. Turner-Jones, R. L. Buchanan, and J. G. Phillips. 1994. Response surface model of the effect of pH, sodium chloride and sodium nitrite on growth on *Yersinia enterocolitica* at low temperatures. Int. J. Food Microbiol. 23:233–245.

8. Buchanan, R. L. 1992. Predictive microbiology: mathematical modeling of growth in foods, pp. 250–260. *In* J. W. Finley, S. F. Robinson, and D. J. Armstrong, eds., Food safety assessment, Vol. 484. American Chemical Society, Washington, DC.

9. Buchanan, R. L., L. K. Bagi, R. V. Goins, and J. G. Phillips. 1993. Response surface models for the growth kinetics of *Escherichia coli* O157:H7. Food Microbiol. 10(4):303–315.

10. Buchanan, R. L., J. L. Smith, C. McColgan, B. S. Marmer, M. Golden, and B. Dell. 1993. Response surface models for the effects of temperature, pH, sodium chloride and sodium nitrite on the aerobic and anaerobic growth of *Staphylococcus aureus* 196E. J. Food Safety 13(3):159–175.

11. Bulmer, M. G. 1974. On fitting the Poisson lognormal distribution to species abundance data. Biometrics 30:101–110.

12. Cabelli, V. J. 1983. Health effects criteria for marine recreational waters. U.S. Environmental Protection Agency, Washington, DC.

13. Chang, S. L., G. Berg, K. A. Busch, R. E. Stevenson, N. A. Clarke, and P. W. Kabler. 1958. Application of the "most probable number" method for estimating concentrations of animal viruses by the tissue culture technique. Virology 6:27–42.

14. Chase, G. R., and D. G. Hoel. 1975. Serial dilutions: error effects and optimal designs. Biometrika 62:329–334.

15. Clancy, J. L., W. Gollnitz, and Z. Tabib. 1994. Commercial labs: how accurate are they? J. Am. Water Works Assoc. 86(5):89–97.

16. Cochran, W. G. 1950. Estimation of bacterial densities by means of the most probable number. Biometrics 6:105–116.

17. Cohrssen, J. J., and V. T. Covello. 1989. Risk analysis: a guide to principles and methods for analyzing health and environmental risks. National Technical Information Service, U.S. Department of Commerce, Springfield, VA.

18. Covello, V. T., and M. W. Merkhofer. 1993. Risk assessment methods. Plenum Press, New York.

19. Crohn, D. M., and M. V. Yates. 1997. Interpreting negative virus results from highly treated water. J. Environ. Eng. 123(5):423–430.

20. de Man, J. C. 1977. MPN tables for more than one test. Eur. J. Appl. Microbiol. 4:307–316.

21. de Man, J. C. 1975. The probability of most probable numbers. Eur. J. Appl. Microbiol. 1:67–78.

22. Dufour, A. P. 1984. Health effects criteria for fresh recreational waters. U.S. Environmental Protection Agency, Washington, DC.

23. Efron, B. 1986. Why isn't everyone a Bayesian? Am. Stat. 40:6–7.

24. Eisenhart, C., and P. W. Wilson. 1943. Statistical methods and control in bacteriology. Bacteriol. Rev. 7:57–137.

25. El-Shaarawi, A. H. 1989. Inferences about the mean from censored water quality data. Water Resour. Res. 25(4):685–690.

26. El-Shaarawi, A. H., S. R. Esterby, and B. J. Dutka. 1981. Bacterial density in water determined by Poisson or negative binomial distributions. Appl. Environ. Microbiol. 41(1):107–116.

27. Environmental Health Directorate. 1981. Tapwater consumption in Canada. 82-EHD-80. Minister of National Health and Welfare, Ottawa, Ontario, Canada.

28. Ershow, A. G., L. M. Brown, and K. P. Cantor. 1991. Intake of tapwater and total water by pregnant and lactating women. Am. J. Public Health 81:328–334.

29. Everitt, B. S. 1985. Mixture distributions, pp. 559–569. *In* S. Kotz and N. L. Johnson, eds., Encyclopedia of statistical sciences, Vol. 5. Wiley, New York.

30. Finkel, A. M. 1990. Confronting uncertainty in risk management. resources for the future. Center for Risk Management, Washington, DC.

31. Finley, B. L., P. K. Scott, and D. A. Mayhall. 1994. Development of a standard soil-to-skin adherence probability density function for use in Monte Carlo analyses of dermal exposure. Risk Anal. 14(4):555–569.

32. Garthright, W. E., and R. J. Blodgett. 1996. Confidence intervals for microbial density using serial dilutions with MPN estimates. Biom. J. 38(4):489–505.

33. Gibson, A. M., N. Bratchell, and T. A. Roberts. 1987. The effect of sodium chloride and temperature on the rate and extent of growth of *Clostridium botulinum* type A in pasteurized pork slurry. J. Appl. Bacteriol. 62:479–490.

34. Greenwood, M., and G. U. Yule. 1920. An inquiry into the nature of frequency distributions representative of multiple happenings with particular reference to the occurrence of multiple attacks of disease or of repeated accidents. J. R. Stat. Soc. Ser. A 83:255–279.

35. Greenwood, M., and G. U. Yule. 1917. On the statistical interpretation of some bacteriological methods employed in water analysis. J. Hyg. 16:36–56.

36. Haas, C. N. 1994. Dose–response analysis using spreadsheets. Risk Anal. 14(6) 1097–1100.

37. Haas, C. N. 1989. Estimation of microbial densities from dilution count experiments. Appl. Environ. Microbiol. 55(8):1934–1942.

38. Haas, C. N. 1997. Importance of distributional form in characterizing inputs to Monte Carlo risk assessment. Risk Anal. 17(1): 107–113.

39. Haas, C. N., and B. A. Heller. 1986. Statistics of enumerating total coliforms in water samples by membrane filter procedures. Water Res. 20:525–530.

40. Haas, C. N., J. Anotai, and R. S. Engelbrecht. 1996. Monte Carlo assessment of microbial risk associated with landfilling of fecal material. Water Environ. Res. 68(7):1123–1131.

41. Haas, C. N., and B. A. Heller. 1988. Test of the validity of the Poisson assumption for analysis of MPN results. Appl. Environ. Microbiol. 54(12):2996–3002.

42. Haas, C. N., and J. B. Rose. 1996. Distribution of *Cryptosporidium* oocysts in water supplies. Water Res. 30(10):2251–2254.

43. Haas, C. N., and P. A. Scheff. 1990. Estimation of averages in truncated samples. Environ. Sci. Technol. 24:912–919.

44. Haldane, J. B. S. 1939. Sampling errors in the determination of bacterial or virus density by the dilution method. J. Hyg. 39:289–293.

45. Hartley, H. O., and R. R. Hocking. 1971. The analysis of incomplete data. Biometrics 27:783–823.

46. Holla, M. S. 1966. On a Poisson-inverse Gaussian distribution. Metrika 11:115–121.

47. Hollander, M., and M. A. Proschan. 1979. Testing to determine the underlying distribution using randomly censored data. Biometrics 35:393–401.

48. Hudson, J. A., and S. J. Mott. 1993. Growth of *Listeria monocytogenes, Aeromonas hydrophila* and *Yersinia enterocolitica* on cooked beef under refrigeration and mild temperature abuse. Food Microbiol. 10(5):429–437.

49. Hudson, J. A., and S. J. Mott. 1993. Growth of *Listeria monocytogenes, Aeromonas hydrophila* and *Yersinia enterocolitica* in pate and a comparison with predictive models. Int. J. Food Microbiol. 20(1):1–11.

50. Humphrey, T. J., K. W. Martin, and A. Whitehead. 1994. Contamination of hands and work surfaces with *Salmonella enteritidis* PT4 during the preparation of egg dishes. Epidemiol. Infect. 113:403–409.

51. James M. Montgomery Consulting Engineers Inc. 1979. Water Factory 21 virus study. Orange County Water District, Orange County, CA.

52. Johnson, N. L., S. Kotz, and N. Balakrishnan. 1994. Continuous univariate distributions, Vol. 1, 2nd ed. Wiley-Interscience, New York.

53. Johnson, N. L., S. Kotz, and N. Balakrishnan. 1995. Continuous univariate distributions, Vol. 2, 2nd ed. Wiley, New York.

54. Johnson, N. L., S. Kotz, and N. Balakrishnan. 1994. Discrete univariate distributions, Vol. 2, 2nd ed. Wiley-Interscience, New York.

55. Jørgensen, B. 1982. Statistical properties of the generalized inverse Gaussian distribution. Lecture notes in statistics, Vol. 9. Springer-Verlag, New York.

56. Kappenman, R. F. 1982. On a method for selecting a distributional model. Commun. Stat. Theor. Methods 11(6):663–672.

57. Laborde, D. J., K. A. Weigle, D. J. Weber, and J. B. Kotch. 1993. Effect of fecal contamination on diarrheal illness rates in day care centers. Am. J. Epidemiol. 138(4):243–255.

58. Lawless, J. F. 1982. Statistical models and methods for lifetime data. Wiley, New York.

59. Lee, E. T. 1992. Statistical methods for survival data analysis, 2nd ed. Wiley-Interscience, New York.

60. Lilliefors, H. W. 1967. On the Kolmogorov–Smirnov test for normality with mean and variance unknown. J. Am. Stat. Assoc. 62:399–402.

61. Loyer, M., and M. Hamilton. 1984. Interval estimation of the density of organisms using a serial dilution experiment. Biometrics 40:907–916.

62. Mackintosh, C. A., and P. N. Hoffman. 1984. An extended model for transfer of microorganisms via the hands: differences between organisms and the effect of alcohol disinfection. J. Hyg. 92:345–355.

63. Malcolm Pirnie, and HDR Engineering. 1991. Guidance manual for compliance with the filtration and disinfection requirements for public water systems using surface water sources. American Water Works Association, Denver, CO.

64. Marples, R. R., and A. G. Towers. 1979. A laboratory model for the investigation of contact transfer of microorganisms. J. Hyg. 82:237–248.

65. Matuszewski, T., J. Neyman, and J. Supinska. 1935. Statistical studies in questions of bacteriology: the accuracy of the "dilution method." J. R. Stat. Soc. Suppl. 2:63–82.

66. Mead, P. S., L. Finelli, M. A. Lambert-Fair, D. Champ, J. Townes, L. Hutwagner, T. Barrett, K. Spitalny, and E. Minz. 1997. Risk factors for sporadic infection with *Escherichia coli* O157:H7. Arch. Intern. Med. 157(2):204–208.

67. Mehrabi, Y., and J. N. S. Matthews. 1995. Likelihood-based methods for bias reduction in limiting dilution assays. Biometrics 51:1543–1549.

68. Miller, A. J., T. Brown, and J. E. Call. 1996. Comparison of wooden and polyethylene cutting boards: potential for the attachment and removal of bacteria from ground beef. J. Food Prot. 59(8):854–858.

69. Moran, P. A. P. 1954. The dilution assay of viruses. J. Hyg. 52:189–193.

70. Moran, P. A. P. 1954. The dilution assay of viruses, II. J. Hyg. 52:444–446.

71. Morgan, B. J. T. 1992. Analysis of quantal response data. Chapman & Hall, London.

72. Morgan, M. G., and M. Henrion. 1990. Uncertainty: a guide to dealing with uncertainty in quantitative risk and policy analysis. Cambridge University Press, Cambridge.

73. Murray, D. M., and D. E. Burmaster. 1994. Estimated distributions for average daily consumption of total and self-caught fish for adult in Michigan angler households. Risk Anal. 14(4):513–519.

74. Norden, R. H. 1972. A survey of maximum likelihood estimation. Int. Stat. Rev. 40(3):329.

75. Norden, R. H. 1973. A survey of maximum likelihood estimation: part II. Int. Stat. Rev. 41(1):39.

76. Pether, J. V. S., and R. J. Gilbert. 1971. The survival of *Salmonellas* on fingertips and transfer of the organisms to foods. J. Hyg. 69:673–681.

77. Pipes, W. O. 1977. Frequency distributions for coliform bacteria in water. J. Am. Water Works Assoc. 69:664.

78. Plummer, R. A. S., S. J. Blissett, and C. E. R. Dodd. 1995. *Salmonella* contamination of retail chicken products sold in the UK. J. Food Prot. 58(8):843–846.

79. Press, W. H., B. P. Flannery, S. A. Teukolsky, and W. T. Vetterling. 1989. Numerical recipes in Pascal: the art of scientific programming. Cambridge University Press, New York.

80. Quenouille, M. H. 1949. A relation between the logarithmic, Poisson and negative binomial series. Biometrics 5:162–164.

81. Ralston, K. 1995. Identifying frequent consumers of foods associated with foodborne pathogens, pp. 41–50. *In* T. Roberts, H. Jensen, and L. Unnevehr, eds., Tracking foodborne pathogens from farm to table: data needs to evaluate control options. Miscellaneous Publication 1532. Economic Research Service, U.S. Department of Agriculture, Washington, DC.

82. Regli, S., J. B. Rose, C. N. Haas, and C. P. Gerba. 1991. Modeling risk for pathogens in drinking water. J. Am. Water Works Assoc. 83(11):76–84.

83. Reid, D. D. 1981. The Poisson lognormal distribution and its use as a model of plankton aggregation. Stat. Distrib. Sci. Work 6:303–316.

84. Roseberry, A. M., and D. E. Burmaster. 1992. Lognormal distributions for water intake by children and adults. Risk Anal. 12(1):99–104.

85. Ross, T., and T. McMeekin. 1994. Predictive microbiology. Int. J. Food Microbiol. 23:241–264.

86. Salama, I., G. Koch, and H. Tolley. 1978. On the estimation of the most probable number in a serial dilution experiment. Commun. Stat. Theory Methods A7(13): 1267–1281.

87. Savage, G. M., and H. O. Halvorson. 1941. The effect of culture environment on results obtained with the dilution method of determining bacterial population. J. Bacteriol. 41:355–362.

88. Scott, E., and S. F. Bloomfield. 1990. The survival and transfer of microbial contamination via cloths, hands and utensils. J. Appl. Bacteriol. 68:271–278.

89. Shapiro, S. S. 1990. How to test normality and other distributional assumptions. American Society for Quality Control, Milwaukee, WI.

90. Sichel, H. S. 1992. Anatomy of the generalized inverse Gaussian–Poisson distribution with special applications to bibliometric studies. Inf. Process. Manag. 28(1):5–17.

91. Sichel, H. S. 1985. A bibliometric distribution which really works. J. Am. Soc. Inf. Sci. 30(5):314–321.

92. Sichel, H. S. 1975. On a distribution law for word frequencies. J. Am. Stat. Assoc. 70(351):542–547.

93. Sichel, H. S. 1974. On a distribution representing sentence length in written prose. J. R. Stat. Soc. A 137(1):25–34.

94. Sichel, H. S. 1982. Repeat-buying and the generalized inverse Gaussian–Poisson distribution. Appl. Stat. 31(3):193–204.

95. Skinner, G. E., J. W. Larkin, and E. J. Rhodehamel. 1994. Mathematical modeling of microbial growth: a review. J. Food Saf. 14(3):175–217.

96. Spika, J. S., J. E. Parsons, and D. Nordenberg. 1986. Hemolytic uremic syndrome and diarrhea associated with *Escherichia coli* O157:H7 in a day care center. J. Pediatr. 109:287–291.

97. Stein, G., W. Zucchini, and J. Juritz. 1987. Parameter estimation for the Sichel distribution and its multivariate extension. J. Am. Stat. Assoc. 82(399):938–944.

98. Stephens, M. A. 1983. Anderson–Darling test for goodness of fit, pp. 81–85. *In* S. Kotz and N. Johnson, eds., Encyclopedia of statistical sciences, Vol. 1. Wiley, New York.

99. Stephens, M. A. 1986. Tests based on EDF statistics, pp. 97–193. *In* R. B. D'Agostino and M. A. Stephens, eds., Goodness of fit techniques. Marcel Dekker, New York.

100. Stevens, W. L. 1958. Dilution series: a statistical test of technique. J. R. Stat. Soc. B Methodol. 20:205–214.

101. Stewart, J. 1994. The Poisson-lognormal model for bibliometric/scientometric distributions. Inf. Process. Manag. 30(2):239–251.

102. Stuart, A., and J. K. Ord. 1987. Kendall's advanced theory of statistics, 5th ed., Vol. 1: Distribution theory. Oxford University Press, New York.

103. Stuart, A., and J. K. Ord. 1987. Kendall's advanced theory of statistics, 5th ed., Vol. 2: Classical inference and relationship. Oxford University Press, New York.

104. Student. 1907. On the error of counting with a hemocytometer. Biometrika. 5:351–360.

105. Sutherland, J. P., A. J. Bayliss, and D. S. Braxton. 1995. Predictive modeling of growth of *Escherichia coli* O157:H7: The effects of temperature, pH and sodium chloride. Int. J. Food Microbiol. 25:29–49.

106. The MathWorks Inc. 1994. MATLAB, 4.1 ed. MathWorks, Natick, MA.

107. Thomas, H. A., Jr. 1952. On averaging results of coliform tests. Boston Soc. Civil Eng. J. 39:253–270.

108. Thomas, H. A. 1955. Statistical analysis of coliform data. Sewage Ind. Wastes 27:212–222.

109. U.S. Department of Agriculture. 1994. Nationwide beef microbiological baseline data collection program: steers and heifers. Food Safety and Inspection Service, USDA, Washington, DC.

110. van Impe, J. F., B. M. Nicolai, T. Martens, J. de Baerdemaeker, and J. Vandewalle. 1992. Dynamic mathematical model to predict microbial growth and inactivation during food processing. Appl. Environ. Microbiol. 58(9):2901–2909.

111. van Impe, J. F., B. M. Nicolai, M. Schellekens, T. Martens, and J. de Baerde-maeker. 1995. Predictive microbiology in a dynamic environment: a system theory approach. Int. J. Food Microbiol. 25:227–249.

112. Vose, D. 1996. Quantitative risk analysis: a guide to Monte Carlo simulation modeling. Wiley, New York.

113. Wadley, F. M. 1954. Limitations of the zero method of population counts. Science 119:689–690.

114. Whiting, R. C., and R. L. Buchanan. 1994. Scientific status summary: microbial modeling. Food Technol. 48(6):113–120.

115. Woodward, R. L. 1957. How probable is the most probable number? J. Am. Water Works Assoc. 49:1060–1068.

116. Wyshak, G., and K. Detre. 1972. Estimating the number of organisms in quantal assays. Appl. Microbiol. 23:784–790.

117. Zwietering, M. H., I. Jongenburger, F.M. Rombouts, and K. Van't Riet. 1990. Modeling of the bacterial growth curve. Appl. Environ. Microbiol. 56(6):1875–1881.

CHAPTER 7

CONDUCTING THE DOSE–RESPONSE ASSESSMENT

The objective of dose–response assessment is to develop a relationship between the level of microbial exposure and the likelihood of occurrence of an adverse consequence. In general, dose–response analysis would not be necessary if the level of microbial risk that was acceptable was sufficiently high so that actual experimentation (on humans, or perhaps on animals) could be conducted so as to permit the direct assessment of risk in the observable range. However, since frequently the level of risk from a single exposure is much lower than $1/1000$, the use of direct experimentation to assess this risk becomes impractical (as well as ethically questionable) since >1000 subjects would be needed to ascertain the "acceptable" dose. Therefore, the use of a parametric dose–response curve to facilitate low-dose (and low-risk) extrapolation becomes necessary.

A dose–response model is, in a most general sense, a mathematical function that takes as an argument a measure of dose—which can be any nonnegative number—and yields the probability of the particular adverse effect—which is bounded by zero (no effect) and one (complete conversion to adverse state). There are an infinite number of such *possible* functions. Even restricting the universe to those functions that are monotonic (as dose increases, response probability is nondecreasing) and bounded by zero and 1 (at dose = 0, no response; as dose $\rightarrow \infty$, complete response), there remain an infinite number of *possible* functions. In particular, the last set of conditions is identical to those required of cumulative distribution functions, so any cumulative distribution function with support over $\langle 0, \infty \rangle$ can be a candidate dose–response function. Conversely, any dose–response function that is monotonic and bounded by zero and 1, with support over $\langle 0, \infty \rangle$ is a cumulative distribution function. This is convenient, since for many distribution functions, the mathematical properties have been well studied.

It is most desirable that the dose–response model(s) chosen for use are biologically plausible. Therefore, we first review some attributes of a biologically plausible microbial dose–response model for infectious agents. Then some classes of these models are derived and analyzed. Subsequently, some empirical (albeit lacking in plausibility) models are presented. The remainder of this chapter then deals with issues of fitting data to models and assessing validity using independent data.

PLAUSIBLE DOSE RESPONSE MODELS

There are two key distinguishing features of infectious microorganisms from other human health risks (e.g., chemicals, ionizing radiation). Any dose–response model failing to consider these factors lacks biological plausibility. First, particularly at low levels, the statistics of microbial distribution (as discussed in Chapter 6) dictate that a population of humans exposed to infectious agents will of necessity receive a distribution of actual doses. For example, if a group of people each consume exactly 1 L of water containing an average concentration of 0.1 organism/L, we expect (assuming random, Poisson, distributions between doses) that about 90% of persons [$\exp(-0.1)$] would actually consume zero organisms, about 9% of persons [$0.1 \times \exp(-0.1)$] would consume one organism, 0.45% of persons would consume two organisms, and 0.015% would consume three or more organisms. If the distribution of organisms between doses were nonrandom (e.g., negative binomial), the percentages would obviously be different. Any biologically plausible dose–response framework should consider this phenomenon. This is very different than with many chemicals, which are present at high enough concentrations (when viewed in terms of molecules), such that the quantal nature of the substance delivered does not affect the actual delivered dose between subjects exposed.

The second distinguishing aspect of infectious microbial agents is their ability to propagate within a susceptible host at an appropriate location within the body. There are, in fact, suggestions that the very trait of microbial pathogenicity may result from coevolutionary processes between infectious agents and humans (46). Although there are quite specific mechanisms of pathogenicity and methods of circumventing the numerous antimicrobial defense mechanisms of the human body, the infection and disease process represents an overcoming of these barriers by the infectious agent (12). The time course of microbial infection can be described by competing processes of birth and death within the host; infection resulting when birth is sufficient to produce a body burden above some critical level to induce the effect (2,72).

The exposure to microbial agents can result in a series of endpoints, as noted in Figure 7-1. A fraction of individuals, P_1 will become infected. Infection is manifest as the multiplication of organisms within the host, followed by excretion. In addition, measurable rises in serum antibodies or an increase in body temperature can indicate infection. Of those infected, a fraction of

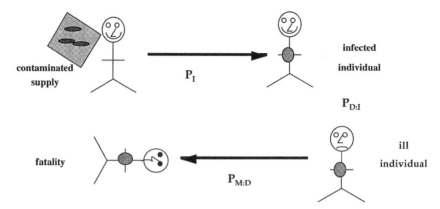

Figure 7-1   Schematic of endpoints of microbial exposure.

these, $P_{D:I}$ (termed the *morbidity ratio*), will become ill. Of those becoming ill, a fraction of these, $P_{M:D}$ (termed the *mortality ratio*), will die.

For some infectious agents, the development of consequences can be quite complex, involving multiple disease symptomatologies and endpoints. This is illustrated in Figure 7-2 for the case of *E. coli* O157:H7, based on overall U.S. prevalence statistics (10). This figure illustrates another aspect of many infectious agents; often, the frequency of more severe consequences is better characterized than the frequency of less severe endpoints, including infection, since the latter endpoints represent individuals who may not be recognized by the health care system (and they may not seek medical attention).

In dose–response assessment we focus heavily on the early processes, particularly infection. The relationship of P_I to dose will be characterized. Focus

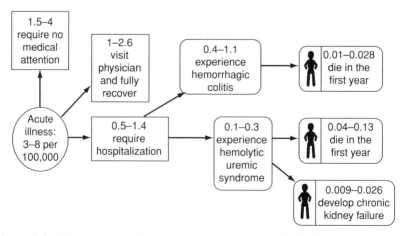

Figure 7-2   Development of consequences from *E. coli* O157:H7. (Adapted from Ref. 10.)

on infection as an initial object of study is justified since from a public health point of view, the minimization of infections may introduce an element of conservatism into the analysis.

FRAMEWORK FOR MECHANISTIC DOSE–RESPONSE RELATIONSHIPS

The process of infection may be considered as requiring two sequential sub-processes to occur (33):

1. The human host must ingest one or more organisms that are capable of causing disease.
2. The organisms undergo decay or are impaired from multiplying to cause infection/disease by host responses, and only a fraction of the ingested organisms reach a site where infection can begin.

The probability of ingesting precisely j organisms from an exposure in which the mean dose (perhaps the product of volume and density, both estimated via the methods' previously described) is d (i.e., the first of the sequential processes) is written as $P_1(j|d)$, the probability of k organisms ($\leq j$) surviving to initiate an infectious process (the second step) is written as $P_2(k|j)$. If these two processes are regarded as independent, the overall probability of k organisms surviving to initiate infectious foci is given by (the law of independent events)

$$P(k) = \sum_{j=1}^{\infty} P_1(j|d)P_2(k|j) \tag{7-1}$$

The function P_1 incorporates the individual-to-individual variation in actual numbers of organisms ingested or otherwise exposed, and the function P_2 expresses the factors of organism–host interaction in vivo that allow some organisms to survive to initiate infectious foci.

Infection occurs when at least some critical number of organisms survive to initiate infection. If this minimum number is denoted as k_{min}, the probability of infection (i.e., the fraction of subjects who are exposed to an average dose d who become infected) may be written as

$$P_I(d) = \sum_{k=k_{min}}^{\infty} \sum_{j=k}^{\infty} P_1(j|d)P_2(k|j) \tag{7-2}$$

We emphasize that the use of k_{min} in the precise sense of equation 7-2 does not correspond to the often used term *minimal infectious dose* (12,16). The latter term refers to the average dose administered, and most frequently really relates the average dose required to cause half of the subjects to experience a response; we would prefer the term *median infectious dose* for this. As we will show, one cannot ipso facto infer from the magnitude of the median infectious dose whether or not k_{min} is large or small (as small as 1). The following observation of Meynell and Stocker (50) is illuminating:

> The relationship between inoculated bacteria might take one of two extreme forms. Either they could be assumed to be acting cooperatively, death being a consequence of their joint action, or they could be regarded as acting independently, more than one bacterium usually being needed because the probability of a given bacterium being lethal is less than unity. The situation when the LD_{50} dose contains many organisms is analogous to that of a poor marksman firing at a bottle. Since his aim is poor, the bottle is unlikely to have been broken after a small number of shots has [*sic*] been fired but if he persists he will probably hit the bottle eventually. A local observer might be aware that the bottle was broken by the action of one bullet. On the other hand, a distant observer, informed only of the total number of shots fired before the bottle broke, would not be able to exclude the hypothesis that the breakage was due to the accumulated stresses produced by all of the bullets fired.

The two conceptual alternatives have been termed the *hypothesis of independent action,* in which in principle k_{min} equals 1, and the *hypothesis of cooperative interaction,* in which k_{min} is some number greater than 1.

If it is understood that k_{min} may not be a single number but may in fact be a probability distribution (we generalize equation 7-2 to admit this possibility subsequently), equation 7-2 or its generalization is expected to be sufficiently broad to encompass all plausible dose–response models. By specifying functional forms for P_1 and P_2, as well as numerical values of k_{min}, we can derive a number of specific useful dose–response relationships.

Exponential Dose–Response Model

The simplest dose–response model that can be formulated assumes that the distribution of organisms between doses is random (i.e., Poisson), that each organism has an independent and identical survival probability,[1] r, and that k_{min} equals 1. From the Poisson assumption we have

$$P_1(j|d) = \frac{d^j}{j!} \exp(-d) \qquad (7\text{-}3)$$

The assumption with respect to survival means that the binomial distribution can be used as

---

[1]Strictly, this is the probability that the organism survives to initiate an infectious focus.

$$P_2(k|j) = \frac{j!}{k! \, (j|k)!} \, (1 - r)^{j-k} r^k \qquad (7\text{-}4)$$

Substituting equations 7-3 and 7-4 into equation 7-2 yields

$$P_1(d) = \sum_{k=k_{min}}^{\infty} \sum_{j=k}^{\infty} \left[\frac{d^j}{j!} \exp(-d) \right] \left[\frac{j!}{k! \, (j - k)!} \, (1 - r)^{j-k} r^k \right]$$

$$= \sum_{k=k_{min}}^{\infty} \frac{(dr)^k \exp(-dr)}{k!} \sum_{j=k}^{\infty} \frac{[d(1 - r)]^{j-k}}{(j - k)!} \exp[-d(1 - r)] \qquad (7\text{-}5)$$

The second summation equals unity (it is the sum of a Poisson series), and therefore we have

$$P_1(d) = \sum_{k=k_{min}}^{\infty} \frac{(dr)^k}{k!} \exp(-dr) = 1 - \left[\sum_{k=0}^{k_{min}-1} \frac{(dr)^k}{k!} \exp(-dr) \right] \qquad (7\text{-}6)$$

Finally, with the assumption of $k_{min} = 1$, this yields[2]

$$P_1(d) = 1 - \exp(-rd) \qquad (7\text{-}7)$$

This is the exponential dose–response relationship. It has one parameter, r, which characterizes the process. The median infectious dose (N_{50}) can be given by

$$N_{50} = \frac{\ln(0.5)}{-r} \qquad (7\text{-}8)$$

Note that the mean dose is multiplied by a constant. This in effect means than any measurement of microbial dose can be used which is strictly proportional to microbial dose. Given the fact that as noted in Chapter 4, some measurement methods may be imperfect, this means that we can nonetheless use these methods for dose–response assessment with the exponential model, recognizing that the parameter, r, is specific for that method. This same observation will hold true for the other dose–response models to be considered below.

The exponential dose–response relationship has the property of low dose linearity. If $rd \ll 1$, then $\exp(-rd) \approx 1 - rd$, and equation 7-7 can be approximated as

[2] We will write this in an alternative form as well, $P_1(d) = 1 - \exp(-d/k)$, where obviously $k = 1/r$.

$$P_1(d) \approx rd \qquad \text{for } rd \ll 1 \tag{7.9}$$

Another property of this and other dose–response curves that we examine is the slope of the curve at the median point ($P_1 = 0.5$). Differentiation of equation 7-7 produces

$$\frac{dP_1}{dd} = r \exp(-rd) \tag{7.10}$$

Since at the median point, $\exp(-rd) = 0.5$ (see equation 7-7), this can also be written as

$$\left.\frac{dP_1}{d(rd)}\right|_{P_1=0.5} = 0.5 \tag{7-11}$$

By similar analysis, the slope of a log-log plot at the median point for the exponential dose–response equation can be determined to be

$$\left.\frac{d\,\ln(P_1)}{d\,\ln(d)}\right|_{P_1=0.5} = -\ln(0.5) = 0.69 \tag{7-12}$$

Beta-Poisson Dose–Response Model

The exponential model assumes constancy of the pathogen–host survival probability (r). For some agents and populations of human hosts, there may be variation in this success rate. Such variation may be due to diversity in human responses, diversity of pathogen competence, or both. This variation can be captured by allowing r to be governed by a probability distribution. This phenomenon of host variability was perhaps first invoked by Moran (51). Armitage (3) appears to have been the first to characterize this variability by a beta distribution; however, computational limitations precluded his use of this model: beta-Poisson and other tolerance distributions. Furumoto and Mickey (23,24) appear to be the first to have used this model in the context of microbial dose–response relationships.

To introduce variability in the interaction probability, equation 7-4 is replaced by a mixture distribution in a fashion similar to the mixture distributions introduced in Chapter 6. However, here (equation 7-13) a mixture distribution of the binomial with respect to the parameter r is used:

$$P_2(k|j) = \int_0^1 \left[\frac{j!}{k!\,(j-k)!}\,(1-r)^{j-k}r^k\right]f(r)\,dr \tag{7-13}$$

The mixing distribution $f(r)$ should have (its only) support over the interval $\langle 0,1 \rangle$, corresponding to the allowable range of variability of r itself. Use of equation 7-13 is identical to applying the mixture operation directly to equation 7-7 if the Poisson distribution for dose-to-dose variation is assumed, thus yielding

$$P_I(d) = \int_0^1 [1 - \exp(-rd)]f(r)\ dr$$

$$= \int_0^1 f(r)\ dr - \int_0^1 \exp(-rd)f(r)\ dr$$

$$= 1 - \int_0^1 \exp(-rd)f(r)\ dr \tag{7-14}$$

A logical distribution, which offers a great deal of flexibility, is the beta distribution (equation 6-73). Incorporating this into equation 7-13 yields

$$P_I(d) = 1 - \int_0^1 \left[\frac{\Gamma(\alpha + \beta)}{\Gamma(\alpha)\Gamma(\beta)} r^{\alpha-1}(1 - r)^{\beta-1} \right] \exp(-rd)\ dr \tag{7-15}$$

The integral in equation 7-15 can be expressed as a confluent hypergeometric function:

$$P_I(d) = 1 - {}_1F_1(\alpha, \alpha + \beta, -d) \tag{7-16}$$

Properties of this function are given in standard references (1,43). In particular, it can be written as a series expansion:

$${}_1F_1(\alpha, \alpha + \beta, -d) = 1 + \frac{\Gamma(\alpha + \beta)}{\Gamma(\alpha)} \sum_{j=1}^{\infty} \left[\frac{\Gamma(\alpha + j)}{\Gamma(\alpha + \beta + j)} \frac{(-d)^j}{j!} \right] \tag{7-17}$$

Therefore, the exact solution to the beta-Poisson model can be written as

$$P_I(d) = \frac{\Gamma(\alpha + \beta)}{\Gamma(\alpha)} \sum_{j=1}^{\infty} \left[\frac{\Gamma(\alpha + j)}{\Gamma(\alpha + \beta + j)} \frac{(-1)^{j-1}(d)^j}{j!} \right] \tag{7-18}$$

Furumoto and Mickey (23,24) derived the following approximation to equation 7-18:

$$P_I(d) = 1 - \left(1 + \frac{d}{\beta}\right)^{-\alpha} \tag{7-19}$$

The closeness of equation 7-18 to 7-19 can be seen by a comparing a Taylor series expansion of 7-19 with the corresponding terms in equation 7-18. The approximation becomes poorer at small values of β or large values of d (and risk). However, both forms (exact and approximate) have the property of low-dose linearity. Furthermore, numerical studies show that the magnitude of discrepancy between the exact and approximate solutions in fitting dose–response data are not substantial (65). Hence we shall use equation 7-19 as the beta-Poisson model. Parenthetically, equation 7-19 describes the cumulative distribution function for a Pareto type II or Lomax distribution (41).

It is convenient to rewrite equation 7-19 by redefining the parameters in terms of the median infectious dose. By solving, it can be determined that

$$N_{50} = \frac{\beta}{2^{1/\alpha} - 1} \tag{7-20}$$

By rearranging equation 7-20 to solve for β, and substituting the result in equation 7-19, a reparameterized beta-Poisson model can be written in the form

$$P_{\mathrm{I}}(d) = 1 - \left(1 + \frac{d}{N_{50}} (2^{1/\alpha} - 1) \right)^{-\alpha} \tag{7-21}$$

By differentiating equation 7-21, it is found that the slope at median dose is

$$\left. \frac{dP_{\mathrm{I}}}{d(d/N_{50})} \right|_{P_{\mathrm{I}}=0.5} = \frac{\alpha}{2} (1 - 2^{-1/\alpha}) \tag{7-22}$$

On log-log coordinates, the slope is

$$\left. \frac{d \ln(P_{\mathrm{I}})}{d \ln(d)} \right|_{P_{\mathrm{I}}=0.5} = \alpha(1 - 2^{-1/\alpha}) \tag{7-23}$$

Since α is nonnegative, equations 7-22 and 7-23 always yield slopes less than the respective exponential, equations 7-11 and 7-12. In other words, the beta-Poisson model is shallower than the exponential model. This is shown in Figure 7-3, in which it is also shown that as $\alpha \rightarrow \infty$, the beta-Poisson model approaches the exponential model. Figure 7-3 (top) also shows that all models at sufficiently low doses yield a slope of 1, indicating linearity, on a log-log plot.

Simple Threshold Models

The beta-Poisson dose–response model modifies the exponential model by allowing for a distribution of microorganism–host interaction probabilities. Another way of modifying the exponential model is by stipulating that some minimum number of surviving organisms other than 1 (i.e., $k_{\min} > 1$) is re-

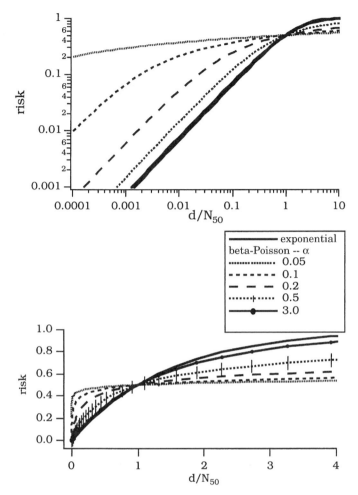

Figure 7-3   Comparison of exponential and beta-Poisson models on log-log (top) and arithmetic (bottom) scales.

quired for the infection process to occur. This model has already been given in equation 7-6. However, the sum of Poisson terms in equation 7-6 can be replaced by an incomplete gamma function, in the manner of equation 6-6c to yield the more compact form

$$P_1(d) = \Gamma(k_{min}, dr) \tag{7-24}$$

This functional form is identical to that for a gamma probability distribution (41), and therefore properties of the gamma distribution can be used to make inferences about threshold models. This threshold model was invoked in the context of microbial dose response assessment (albeit in cell culture assays) as early as 1945 (45). Unlike the beta-Poisson and all mixture models that can be written in the form of equation 7-14, multihit models with $k_{min} > 1$ do not have low-dose linearity (or slopes of 1:1 on log-log plots), and their slope at the median infectious dose is in excess of the simple exponential. On a log-log scale, the slope at low doses is in fact equal to k_{min}. On an arithmetic scale, the slope of risk at low doses is proportional to the k_{min} power of dose. The behavior of these curves is shown in Figure 7-4; the curves are computed such that the median infectious dose is 10 organisms.

The term *threshold* is used to indicate that at the level of the interaction between microbial pathogens and the human host, more than one organism

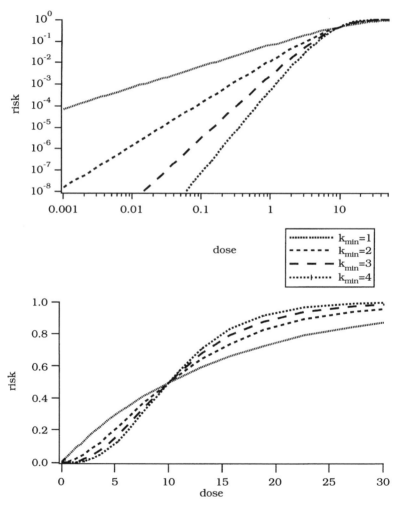

Figure 7-4 Behavior of threshold models as a function of k_{min}. All models computed to have a median infective dose of 10 organisms.

is required to survive to initiate infection. This is the hypothesis of *cooperativity*, or in the analogy of Meynell and Stocker presented earlier, the multiple-bullet scenario. The term *hit* is also used in connection with these models, since they are identical to the single-target multihit model used in analysis of radiation dose–response (69).

In general, threshold models have not been widely used in microbial risk assessment work, since virtually all experimental data show slopes at the median infectious dose equal to or less than the simple exponential. Therefore, the nature of differences most frequently is in the direction shown by Figure 7-3 rather than that shown by Figure 7-4.

Thresholds with Variable Host Sensitivity. The threshold approach can also be used to modify the models that incorporate variable host sensitivity (the beta mixture of r values). We can start with equation 7-6 for a threshold number k_{min}. This is now integrated with respect to a beta distribution of r values in the manner of equation 7-15 to yield

$$P_I(d) = 1 - \sum_{k=0}^{k_{min}-1} \left\{ \int_0^1 \left[\frac{\Gamma(\alpha + \beta)}{\Gamma(\alpha)\Gamma(\beta)} r^{\alpha-1}(1 - r)^{\beta-1} \right] \left[\frac{(dr)^k}{k!} \exp(-dr) \right] dr \right\}$$

(7-25)

Each of the integrals in braces in equation 7-25 is of the same functional form as in equation 7-15. Hence the result can be written as a series of confluent hypergeometric functions as follows:

$$P_I(d) = 1 - \frac{\Gamma(\alpha + \beta)}{\Gamma(\alpha)} \sum_{k=0}^{k_{min}-1} \frac{\Gamma(\alpha + k)}{\Gamma(\alpha + \beta + k)} d_1^k F_1(\alpha + k, \alpha + \beta + k, - d)$$

(7-26)

On a log-log plot, this equation has a slope of k_{min}, so it also produces a substantially lower low-dose risk estimate than the nonthreshold ($k_{min} = 1$) variants.[3]

---

[3]This is not easy to deduce. However, the equivalent result can be obtained by starting with equation 7-24 and integrating using the beta distribution. By expanding the incomplete gamma function as a power series, and integrating term by term, the following result can be obtained:

$$P_I = \frac{d^{k_{min}}}{B(\alpha,\beta)\Gamma(k_{min})} \sum_{i=0}^{\infty} \left[\frac{(-d)^i}{i! (k_{min} + i)} B(\alpha + k_{min} + i, \beta) \right]$$

where $B(x,y)$ is the beta function defined by $B(x,y) = \Gamma(x)\Gamma(y)/\Gamma(x + y)$. If the series is then truncated at the leading term ($i = 0$), it is now obvious that the leading term, which dominates at small dose, is of the k_{min} power.

Negative Binomial Dose Distributions

Another possible variant on the general dose–response relationship (equation 7-2) is when the actual dose distribution differs from Poisson. This might be due, to systematic variability in the medium in which organisms are administered to exposed individuals. We consider the negative binomial distribution as an alternative to the Poisson; however, any alternative discrete distribution can be used.

Recall that the negative binomial distribution can be obtained as a gamma mixture of Poisson distributions. The dose–response when a mixture of Poisson distributions characterize the dose distribution can be given by averaging the corresponding Poisson dose–response relationship with the mixture distribution itself. In other words, if $P_{I,Poisson}$ is the dose–response relationship corresponding to the simple Poisson dose-to-dose distribution of microorganisms, the dose–response relationship assuming a mixture of Poisson distributions with mixing distribution density function $h(d)$ is given by

$$P_{I,mixture}(\overline{d}) = \int_0^\infty P_{I,Poisson}(q)h(q) \, dq \tag{7-27}$$

With Constant Host Sensitivity. If we assume a negative binomial, with constant host sensitivity (constant r), the effect of non-Poisson dose distributions is obtained by mixing equation 7-7 (simple exponential) with a gamma distribution. For reasons to become apparent, we reverse the usual parameter designations in the gamma and describe the distribution of mean dose as (and recall that the average dose delivered would be $\overline{d} = \alpha\beta$)

$$
\begin{aligned}
h(d) &= \frac{1}{\beta\Gamma(\alpha)} \left(\frac{d}{\beta}\right)^{\alpha-1} \exp\left(-\frac{d}{\beta}\right) \\
&= \frac{1}{(\overline{d}/\alpha)\Gamma(\alpha)} \left(\frac{d\alpha}{\overline{d}}\right)^{\alpha-1} \exp\left(-\frac{d\alpha}{\overline{d}}\right) \\
&= \frac{1}{\Gamma(\alpha)} \left(\frac{d\alpha}{\overline{d}}\right)^{\alpha} \exp\left(-\frac{d\alpha}{\overline{d}}\right)
\end{aligned}
\tag{7-28}
$$

Using equation 7-7 as $P_{I,Poisson}$ along with equation 7-28 in equation 7-27 produces

$$
\begin{aligned}
P_I(\overline{d}) &= \int_0^\infty [1 - \exp(-rd)] \left[\frac{1}{\Gamma(\alpha)} \left(\frac{d\alpha}{\overline{d}}\right)^{\alpha} \exp\left(-\frac{d\alpha}{\overline{d}}\right)\right] dd \\
&= 1 - \int_0^\infty \exp(-rd) \left[\frac{1}{\Gamma(\alpha)} \left(\frac{d\alpha}{\overline{d}}\right)^{\alpha} \exp\left(-\frac{d\alpha}{\overline{d}}\right)\right] dd \\
&= 1 - \left(1 + \frac{r\overline{d}}{\alpha}\right)^{-\alpha}
\end{aligned}
\tag{7-29}
$$

Note that the end result of this model is functionally identical to the beta-Poisson model (equation 7-19). Therefore, if a data set fits equation 7-19 or 7-29, in the absence of additional information it is impossible to differentiate between two hypotheses whereby this functional form may be obtained. Furthermore, there is no assurance that additional derivations with alternative assumptions may produce the same model.

With Variable Host Sensitivity. It is also possible to couple the negative binomial dose-to-dose distribution assumption with an assumption of beta-distributed pathogen survival probabilities between doses. Using the approximate beta-Poisson form as $P_{\text{I,Poisson}}$ and the gamma mixing distribution as parameterized above (except that we use α^* for the gamma parameter so that we can retain α as the beta-Poisson exponent), we can obtain the following:

$$
P_{\text{I}}(\bar{d}) = \int_0^\infty \left[1 - \left(1 + \frac{d}{N_{50}} (2^{1/\alpha} - 1) \right)^{-\alpha} \right] \left[\frac{1}{\Gamma(\alpha^*)} \left(\frac{d\alpha^*}{\bar{d}} \right)^{\alpha^*} \right]
$$
$$
\times \exp\left(-\frac{d\alpha^*}{\bar{d}} \right) \right] dd
\tag{7-30}
$$

With the substitution $q = d/N_{50}$ in the integral, this can be transformed to

$$
P_{\text{I}}(d) = 1 - \frac{1}{\Gamma(\alpha^*)} \left(\frac{N_{50}\alpha^*}{\bar{d}} \right)^{\alpha^*} \int_0^\infty \frac{q^{\alpha^*-1}}{[1 + q(2^{1/\alpha} - 1)]^\alpha} \exp\left(-\frac{q\alpha^* N_{50}}{\bar{d}} \right) dq
$$
$$
\tag{7-31}
$$

Unfortunately, there does not appear to be a closed-form solution to the integral in this equation. It can, however, be evaluated by numerical integration. Parametric study of this equation, which we do not show, suggests that the incorporation of negative binomial variation in dose-to-dose aliquots has little influence on the shape of the dose–response relationship, either near or at the median infectious dose or at low doses. Therefore, we will not pursue this model further.

Variable Threshold Models

The last class of mechanistic models that will be derived considers the possibility of variable thresholds. Suppose that the population of persons who are exposed have a distribution with respect to the minimum number of infectious foci that must be initiated prior to a response. In other words, the parameter k_{min} is characterized by a distribution. Since it is discrete, we must use a discrete distribution. Note that if there are any hosts (persons) for whom k_{min} equals zero, there will be infections in the absence of any microorganisms; hence we need to use a discrete distribution with a range of support over the nonzero positive integers.

The starting point for this model are the individual terms from equation 7-6, giving the fraction of hosts who receive sufficient organisms to initiate k infectious centers (assuming independent identical binomial survival probabilities), $[(dr)^k/k!] \exp(-dr)$. We consider an infection to occur if the host has a threshold of k_{min}. However, k_{min} is given by a discrete probability distribution such that $P^*(k_{min})$ are the fraction of subjects who have a threshold equal to k_{min}. From this, the total proportion of subjects who receive an actual dose in excess of their threshold can be written as

$$P_I(d) = \sum_{i=1}^{\infty} \left[\frac{(dr)^k}{k!} \exp(-dr) \sum_{j=1}^{i} P^*(j) \right] \quad (7\text{-}32)$$

There are a number of possible distributions for $P^*(j)$. One of the simplest may be the zero-truncated Poisson, which is a Poisson distribution with the zero term remaining. For $x = 1, \ldots, \infty$,

$$P^*(x) = \frac{1}{\exp(\Theta) - 1} \exp(-\Theta) \frac{\Theta^x}{x!} \quad (7\text{-}33)$$

When Θ approaches 1, this distribution approaches one with 100% weight at $x = 1$ (i.e., the model in equation 7-32 approaches the exponential model). Figure 7-5 presents graphs of the resulting dose–response relationship. These graphs are drawn such that the median infectious dose in all cases is 200 organisms. Like the simple threshold models, as long as Θ is sufficiently greater than 1, the slope at the median infectious dose is greater than the simple exponential model. However, on a log-log plot there is a clear change in slope (for higher values of Θ). Low doses result in an effect primarily with those persons who are most sensitive (having a low k_{min}). Higher doses must infect the less sensitive individuals (with higher k_{min}) and therefore result in a higher slope. This is quite different from the pattern in Figure 7-4, in which the slopes at low dose are constant on a log-log plot.

Other Mixture Models

There are clearly many other variants to plausible microbial dose–response models that can be developed. Recalling equation 7-2, we have a choice of two probability distributions (P_1 and P_2) as well as of the numerical value or distribution (and associated parameters) used for k_{min}. The previous discussion outlines some options for these aspects of the model.

In Table 7-1, a taxonomy of dose–response models is presented. A number of these have been presented; however, a number of these models (of greater complexity) can be derived as indicated. In addition, by using a mixing distribution other than the gamma to derive the dose-to-dose variability of delivered microorganisms (e.g., an inverse Gaussian, etc.), or by using a distribution other than the beta to describe the variability in r [e.g., Johnson

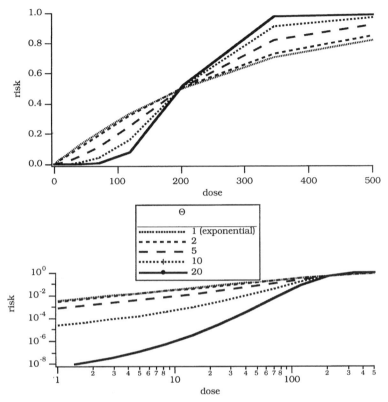

Figure 7-5   Dose–response model for constant host–microorganism interaction ratio (r) and truncated Poisson distribution for k_{min}. All curves are fixed at a median infectious dose of 200.

et al. (42) describe a generalized beta distribution, characterized by three rather than two parameters], or by using a distribution other than the truncated Poisson to describe the variability in k_{min}, a rich spectrum of possibilities emerges. As an indicator of the possible breadth of models, the work of Turner (69) in developing "hit theory" models for radiation effects can serve as a paradigm.

Biological Arguments for One-Hit Models

As seen, for example, in Figures 7-4 and 7-5, there is a considerable difference between one hit (whether or not there is Poisson dose-to-dose variability, or constancy of r) and threshold models. This difference exists particularly with respect to low-dose extrapolation. Comparing the one-hit versus two-hit simple threshold models (Figure 7-4), the dose corresponding to a risk of 10^{-4} differs by over $1\frac{1}{2}$ orders of magnitude between these two models, despite the fact that the models have an identical median infectious dose. For exper-

TABLE 7-1 Taxonomy of Potential Mechanistic Dose–Response Models

k_{min}	Functional Form of P_2	Functional Form of P_1	
		Poisson	Negative
$k_{min} = 1$	Binomial	Exponential equation 7-7	Yields beta-Poisson form equation 7-27
	Binomial mixture (beta)	Beta-Poisson equation 7-19	Equation 7-29
k_{min} constant, > 1	Binomial	Multihit equation 7-24	Gamma mixture of equation 7-24
	Binomial mixture (beta)	Equation 7-26	Gamma mixture of equation 7-26
k_{min} variable	Binomial	Equation 7-32	Gamma mixture of equation 7-32
	Binomial mixture (beta)	Beta mixture of equation 7-32	Beta mixture of equation 7-32 followed by gamma mixture of result

iments in which small numbers of subjects have been used and where the doses are close to the median dose, it becomes hard to differentiate acceptability of fit between these models (note the bottom panel of Figure 7-4). Hence it is desirable to obtain some additional mechanistic information supporting or refuting the possibility that one organism, administered to a host, is sufficient to initiate an infection.[4] Of course, if data are sufficient to support a model that may be derived from a one-organism hypothesis and to reject a model that can be derived from a threshold process, this is supportive of the idea of a single organism sufficing to initiate infection; however, in view of the fact that individual models may be derived from multiple assumptions, the observation that a model is consistent with a set of data does not serve to prove that a particular set of assumptions used to derive the model are true.

Nonetheless, there has been strong evidence for over 50 years that the single-organism hypothesis produces results consistent with observed infection dose–response. This is elaborated on in the case studies in this and subsequent chapters. Some early historical work on this point are the studies of Lauffer and Price on virus (45), Goldberg et al. (29) on respiratory bacterial and viral agents, and Meynell and Stocker (50) on *Salmonella*.

---

[4]Note that we are not saying that one organism will always initiate an infection, merely that it is sufficient at least under some circumstances to initiate an infection. To refute this hypothesis, evidence would be needed to show that two or more organisms are *always* necessary to initiate an infection.

From a basic point of view, an individual viable organism contains all the information necessary to reproduce. In fact, some organisms that may not be recognized as viable, called *viable nonculturable bacteria,* may also initiate a disease process (40,59,62). Perhaps the most persuasive lines of evidence for the single-organism hypothesis other than pure fits to dose–response curves have been reviewed by Rubin (60). A number of prior studies have been conducted in which multiple pathogens at low doses have been used to expose susceptible (animal) hosts. After the infection or disease process has been initiated, the relative prevalence of the individual strains in vivo has been assessed. Under the assumption of independence, one would expect (it is argued) that "the compositions of the samples from all [animals] given [a large dose] should be similar to that of the challenge dose; that samples from [animals] given smaller doses should vary in composition from [animal to animal]; and that most samples from [animals] given 1 or <1 [median lethal dose] should each contain only a single variant" (50). This is in fact what has been found, and therefore the single-organism hypothesis is regarded as being supported.

EMPIRICAL MODELS

In contrast to the models developed from mechanistic assumptions, it is also possible to use empirical models that lack the criteria of plausibility as defined earlier. These are typically models that have been used for analysis of chemical toxicity. We present three of these; however, there are numerous examples, These models stem from ideas initiated by Gaddum (25) in which it is presupposed that the population of susceptible hosts (human or animal) has an intrinsic tolerance distribution for an adverse agent. If the population is exposed to the agent at a certain level, all members of the population who have a tolerance less than or equal to the dosed level will exhibit the adverse effect. Hence the problem of assessing the dose–response curve is simultaneously that of assessing the tolerance distribution of the susceptible population.

We do not consider these models intrinsically plausible (although as will be seen, certain tolerance distributions may be developed from a biologically plausible rationale). Furthermore, there is no assurance that at some point a derivation from certain of the tolerance distribution models will not be found, thereby rendering them biologically plausible, but no such derivation for the three empirical models is currently known to exist.

A tolerance distribution for microbial dose–response is a probability density function with respect to microbial dose (average number of organisms in a dose), $f(d)$. In principle, any density function with support over the positive line can be a tolerance distribution. Then the dose–response curve for infection is defined by the integral

$$P_1(d) = \int_0^d f(y)\, dy \qquad (7\text{-}34)$$

For convenience, to permit easy evaluation of $P_1(d)$, it would be desirable if the integral of the tolerance distribution (i.e., the corresponding cumulative distribution function) were an analytical integral. In fact, since any of the dose–response functions derived in previous sections are differentiable with respect to d, and since they each have the properties of a cumulative distribution-function, they can be considered as implicitly specifying a tolerance distribution. For example, since the cumulative distribution function of the gamma distribution is an incomplete gamma function (41), the simple threshold model (equation 7-24) can also be considered to be a tolerance distribution model with gamma-distributed tolerances.

Table 7-2 gives the dose–response relationships for three empirical dose–response models that have received some attention (although primarily for chemical agents) (8,9,22,54). The log-logistic model uses the log-logistic distribution as a tolerance distribution. The log-probit model uses the lognormal as a tolerance distribution. The Weibull uses the Weibull distribution. Figure 7-6 compares the three distributions at low doses. All distributions are computed to have the same median infectious dose (200 organisms) and the same 10% infectious dose (100 organisms). At high doses, all three models are quite similar. At low doses, however, while the log-logit and Weibull models are linear on a log-log scale, the log-probit model has a substantial curvature and gives a much lower risk estimate.

FITTING AVAILABLE DATA

Given a set of data on infectivity or some other endpoint (such as illness, or even mortality), we would like to ask some basic questions with respect to dose–response estimation:

TABLE 7-2   Empirical Dose–Response Functions

Name	Dose–Response Relationship
Log-logistic	$P_1(d) = \dfrac{1}{1 + \exp[q_1 - q_2\ln(d)]}$
Log-probit	$P_1(d) = \Phi\left(\dfrac{1}{q_2}\ln\dfrac{d}{q_1}\right)$ where $\Phi(y) = \dfrac{1}{\sqrt{2\pi}}\displaystyle\int_{-\infty}^{y}\exp\left(-\dfrac{x^2}{2}\right)dx$
Weibull	$P_1(d) = 1 - \exp(-q_1 d^{q_2})$

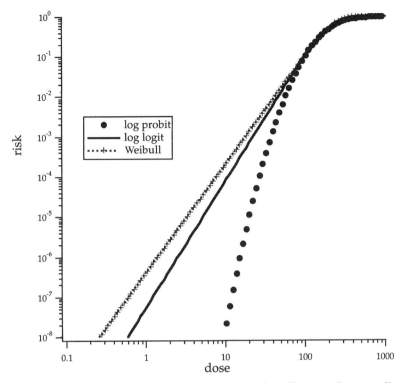

Figure 7-6   Log-probit, log-logit, and Weibull models. All curves have median infectious dose of 200 and 10% infectious dose at 100 organisms in average dose.

1. Assuming a given dose–response model, what are the best parameter estimates using the experimental data at hand?
2. How do we determine which of a set of plausible models provides the best fit to the data?
3. Is the best-fitting model itself adequate or is there still a significant amount of unexplained variance?
4. What is the uncertainty in the parameters estimates of a particular model for the data at hand?
5. Are the results from two or more data sets adequately describable by a common set of dose-response parameters?
6. How can we explain lack of fit?

We address these issues in turn and in the process introduce some real dose–response data (as well as some hypothetical data).

Types of Data Sets

In general, the data set available for dose–response analysis is one of two types. In the first type of data, several sets of subjects (either human or animal) are fed replicates with a known mean dose, and the subsequent response, in terms of infection, illness, or mortality, is determined. In the second type of data set, typically resulting from the investigation of an outbreak study, the average dose administered to individual sets of subjects is estimated, and the attack rate in each group is assessed, Both of these data sets can be described using the same nomenclature. To stress the parallelism to the problem of estimating densities in the dilution assay (studied in Chapter 6), nomenclature similar to that in Table 6-3 is used. Nomenclature for dose–response studies is given in Table 7-3.

It should be noted that while many data are available using human subjects, in some cases the only data available are animal data. Principles for extrapolating the results from animal studies to human risk assessment have not yet been thoroughly studied. When animal data are used, it is particularly important to perform validation by comparison with outbreak investigations. This issue of validation is addressed later in the chapter, and examples of animal-to-human extrapolation are given in subsequent chapters.

Best-Fit Estimation. A particular dose–response model is selected for study. This model may be characterized by a function which predicts the proportion of positive responders given dose and the values of one or several dose–response parameters. In a generic sense, we can write the predicted response as $\pi_i = P_1(d_i;\Theta)$, where Θ is the set of dose–response parameters. We also define the response for each set based strictly on the observations as $\pi_i^0 = p_i/n_i$. If the individual subjects have independent responses, the overall system can be characterized using a similar likelihood framework as was used for analyzing dilution assays.

We will use the symbol Y in this context to denote the value of a -2 log-likelihood ratio; we will now term this the *deviance*. For the dose–response assay in Table 7-3 we can define this deviance statistic as

$$Y = -2 \sum_{i=1}^{k} \left[p_i \ln \frac{\pi_i}{\pi_i^0} + (n_i - p_i) \ln \frac{1 - \pi_i}{1 - \pi_i^0} \right] \qquad (7\text{-}35)$$

The optimum, maximum-likelihood estimates of dose–response parameters

TABLE 7-3   Schematic Layout of Dose–Response Assay

Set	Average Dose of Microorganisms	Number of Subjects in the Set	Positive Subjects[a]
1	d_1	n_1	p_1
2	d_2	n_2	p_2
3	d_3	n_3	p_3
4	d_4	n_4	p_4

[a]With infection, illness, death, or some other positive indicator of response.

are obtained by finding the values of Θ that through their influence on π_i, minimize the deviance, Y (52).

This problem is one of unconstrained minimization, albeit possibly in several parameters (if the dose–response model has several parameters). This problem can be solved in EXCEL using a method similar to that described in the appendix of Chapter 6 (32). Use of this method is shown in the next example. The data set used will be a human dose–response study using rotavirus, with infection as an endpoint (70). The data are shown in Table 7-4 where response is infection.

Example 7-1. Analyze the data in Table 7-4 using the exponential, beta-Poisson, and log-probit models.

SOLUTION. The optimization is performed using the SOLVER in EXCEL. Table 7-5 summarizes the parameters of the best fits. Clearly, as indicated by the deviance, the beta-Poisson model produces the minimum deviance. In the next section we consider whether the fit of any of these models is acceptable and whether the improvement in fit going from the exponential to the beta-Poisson (since the latter is a special case of the former model) is statistically significant.

The data are compared to the models in Figure 7-7. Clearly, the exponential model provides too sharp of a rise in comparison to the data. Qualitatively, within the observable range, the beta-Poisson model and the log-probit model appear similar, and they both describe the more gradual rise between zero and complete response better than does the exponential. However, as shown in the inset graph, there is a considerable difference between the three models at low doses. Interestingly, in this case, the best-fit log-probit model gives greater estimated risk at low dose than the other two models (contrast the low-dose behavior in Figure 7-7 to that in Figure 7-6).

One additional question we may ask is the adequacy of the approximation to the beta-Poisson model (equation 7-19) to the exact solution using the

TABLE 7-4 Human Dose–Response Study of Ward et al. (70)

Dose	Total Subjects	Positive Subjects
90,000	3	3
9,000	7	5
900	8	7
90	9	8
9	11	8
0.9	7	1
0.09	7	0
0.009	7	0

TABLE 7-5   Best-Fit Parameters for Example 7-1

Model	Best-Fit Parameters	Deviance, Y
Exponential	$r = 0.0126$	129.48
Beta-Poisson	$\alpha = 0.265$	6.82
	$N_{50} = 5.597$	
Log-probit	$q_1 = 10.504$	11.87
	$q_2 = 4.137$	

confluent hypergeometric function (equation 7-16). The data in Table 7-4 are fit to the model of equation 7-16, and the following fit is found:

$$N_{50} = 10.08 \qquad \alpha = 0.173 \qquad Y = 5.05$$

Note that the residual deviance (Y) is even lower for the exact solution than

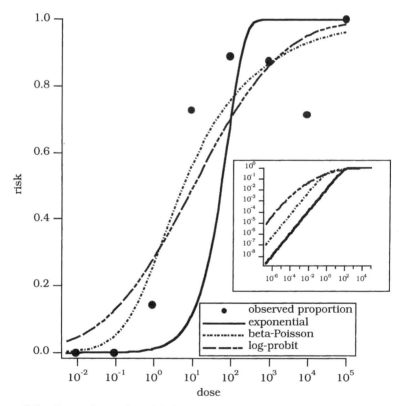

Figure 7-7 Comparison of model fits to rotavirus data. Inset graph shows log-log plot of best-fit models.

for the approximate solution. There is some numerical difference in the parameters. Figure 7-8 compares the approximate and exact fits in the observable range and at low dose. The exact solution has a computed β value of 0.188, which is quite low (in view of the criterion that β must be large for the approximation to the beta-Poisson model to be a good one. Nevertheless, although there is some difference in the functional form within the observable range, there is virtually no effect on the extrapolated low-dose risk. This is reassuring from an application point of view, since equation 7-19 is much simpler to fit and to compute than equation 7-16.[5]

Goodness of Fit Determinations. No model should be accepted for use without examination of goodness of fit. We apply the methods of Chapter 6 to this problem. The value of the optimized deviance, which we shall now denote as Y^*, is compared to a χ^2 distribution with $k - m$ degrees of freedom (where k is the number of doses and m is the number of parameters in the dose–response model). The null hypothesis of fit acceptability is rejected (i.e., the dose–response model is rejected) if Y^* is in excess of an upper (e.g., 5th) percentile of the distribution.

Example 7-2. For the dose–response fits to the rotavirus data in Example 7-1, determine the goodness of fit for the three alternative models.

SOLUTION. Table 7-6 presents the degrees of freedom, the upper fifth percentile of the chi-squared distribution, and the p value (i.e., the probability that a fit as poor or poorer would be found if the null hypothesis was in fact true). Since the optimized deviance for the exponential model is substantially greater than the critical value [i.e., since the p value is extremely small ($<<$ 0.05)], the fit of the exponential model to these data is rejected. Similarly, since the residual deviances from the beta-Poisson and log-probit models are

---

[5]As a technical note, the exact fit was computed in MATLAB. However, since the confluent hypergeometric function is not currently a built-in function, the value of $_1F_1(a,c,x)$ was computed by solving the defining differential equation (41) for the value of y at the desired x:

$$x\frac{d^2y}{dx^2} + (c - x)\frac{dy}{dx} - ay = 0$$

with initial conditions

$$x = 0 \qquad y = 1 \qquad \frac{dy}{dx} = \frac{a}{c}$$

This is an initial value problem, which can be solved by numerical integration. To avoid division by zero, the initial value problem is begun and initial conditions are applied at a very small ($\approx 10^{-16}$) value of x.

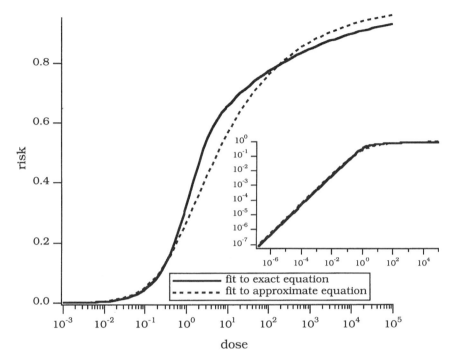

Figure 7-8   Comparison of fit of rotavirus dose–response to exact and approximate beta-Poisson equations.

less than the critical value, or the p value is >0.05, the fits cannot be rejected on statistical grounds.

The goodness-of-fit test based on comparing Y^* to a chi-squared distribution is actually an asymptotic one, and at small values of $k - m$ (as well as, perhaps, number of subjects per group) may have some bias (64). This may be corrected by determining exact significance levels using a Monte Carlo approach, in the manner of Crump et al. (11) for carcinogenic dose–response assessment. Further work on the small-sample properties of this goodness-of-fit test in the context of dose–response analysis is needed.

TABLE 7-6   Results for Example 7-2

Model	Deviance, Y^*	$k - m$	Critical χ^2	p Value
Exponential	129.48	7	14.067	8.07×10^{-25}
Beta-Poisson	6.82	6	12.591	0.338
Log-probit	11.87	6	12.591	0.0649

Comparison of Nested Models. In a manner similar to Chapter 6, if we have two dose–response models, where model 2 is a special case of model 1, with numbers of parameters m_1 and m_2, where $m_2 < m_1$, we can compare the statistical significance of the improvement in fit by examining $Y_2^* - Y_1^*$ against a χ^2 distribution with $m_1 - m_2$ degrees of freedom. The null hypothesis (that the fits are indistinguishable) is rejected if the difference in deviances exceeds the critical value; that is, if the difference is in excess of the critical value, we are justified in accepting the more complex model (with more parameters) rather than to the more parsimonious model.

For example, with the data analyzed in Examples 7-1 and 7-2, the exponential model is a special case of the beta-Poisson. Hence we compare the difference, $Y_2^* - Y_1^* = 129.48 - 6.82 = 122.66$, which is clearly greater than the critical value at 1 degree of freedom $(2 - 1)$ (3.84), so the more parsimonious (exponential) model is rejected in favor of the beta-Poisson model.

As a corollary to this observation, if a data set can be fit by a dose–response model with residual deviance less than 3.84 (i.e., the upper 5 percentile of the χ^2 distribution with 1 degree of freedom), no more complex model, of which the more parsimonious model is a subset, can provide a statistically significant improvement in fit. This is illustrated by use of the data set in Table 7-7, depicting the response of human volunteers to controlled oral doses of *Cryptosporidium parvum* oocysts (13,34).

Example 7-3. Examine the fit of the exponential dose–response relationship to the *C. parvum* data in Table 7-7.

SOLUTION. Fitting the exponential dose–response relationship to the data in Table 7-7 results in a best-fit estimate of $r = 0.00419$. The residual deviance (Y^*) from this fit is 0.503. This is clearly less than a critical chi-squared

TABLE 7-7 *Cryptosporidium parvum* **Infectivity in Human Volunteers by Oral Dosing**

Dose	Positive Subjects	Total Subjects
30	1	5
100	3	8
300	2	3
500	5	6
1,000	2	2
10,000	3	3
100,000	1	1
1,000,000	1	1

Source: Ref. 13.

value for $8 - 1 = 7$ degrees of freedom; therefore, it is acceptable. Since the residual deviance is less than 3.84, no more complex model (e.g., the beta-Poisson) can provide an improvement on this fit.

The adequacy of the fit can be seen qualitatively in Figure 7-9, which compares the model predictions and the observations. Also shown in this figure (as the dashed lines) are the likelihood based 95% confidence limits, developed using the method outlined next.

Confidence Intervals and Regions: Likelihood. To assess the precision with which the dose–response curve has been determined (i.e., the uncertainty), we would like to determine the confidence limits to the parameters of the dose–response curve and also the upper and lower envelopes (at a certain confidence coefficient) around the dose–response curve. A likelihood ratio–based approach can be used as in equation 6-62; this approach appears to yield much more accurate results than simpler linearization methods (52). Let $Y(\Theta)$ be the deviance evaluated at an arbitrary combination of parameters. Then the $1 - \alpha$ confidence limit is set for all Θ satisfying the following inequality:

$$Y(\Theta) - Y^* \leq \chi^2 \qquad (7\text{-}36)$$

where the χ^2 distribution is evaluated at m degrees of freedom (the dimensionality of the parameter vector) and the upper α percentile. For a model with m parameters, this results in an m-dimensional region.

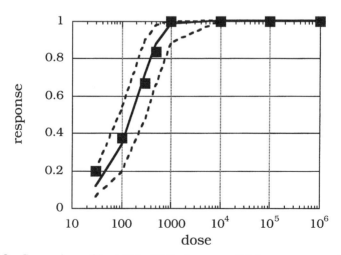

Figure 7-9 Comparison of best-fit (solid line) exponential dose–response relationship to data on oral infectivity of *Cryptosporidium* to humans (Table 7.7). Dashed lines give 95% confidence limits.

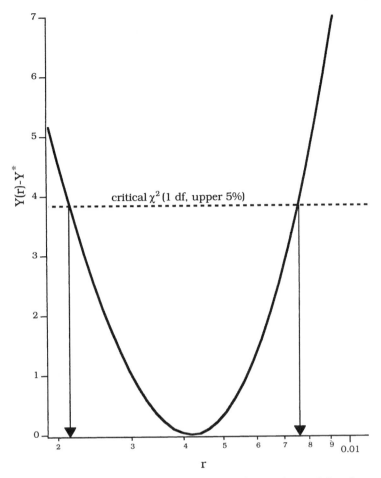

Figure 7-10 Determination of likelihood ratio confidence interval for *Cryptosporidium* data.

In the case of a one-parameter model, a direct plot of the excess deviance, $Y(\Theta) - Y^*$, versus Θ can be made.[6] This is illustrated in Figure 7-10. From this plot, the 95% confidence interval for the r value of the exponential dose response model for this data set is determined to be 0.00215 to 0.00757. Using these values, the upper and lower limits on the dose–response curve are plotted in Figure 7-9. In addition, using a method directly following that of Chapter 6, a density function for r can be obtained.

[6]In EXCEL this can be accomplished using the data table function to determine the value for the excess deviance as a function of different values of the parameter r.

For dose–response models with more than one parameter, determination of the likelihood-based confidence interval becomes somewhat more complex, since the value of $Y(\Theta) - Y^*$ must be evaluated over a region of two dimensions or more. The intersection of this surface (or hypersurface) with a level plane (or hyperplane) at the critical χ^2 defines the likelihood based confidence region.

Figure 7-11 presents the confidence contours for the rotavirus (Table 7-4) data set fit to the beta-Poisson model. To obtain these curves, the excess deviance was computed over a two-dimensional grid of α and N_{50} values. From this, the p values were computed using the cumulative χ^2 distribution. Contours were then drawn at specific p values. Although this computation can, in principle, be performed using spreadsheets, it is tedious. To produce the graph in Figure 7-11, the MATLAB (67) program was used to generate the grid of excess deviances. From that, a contouring algorithm can be employed—to produce this graph, the contour routine in the graphical package IGOR (71) was used.

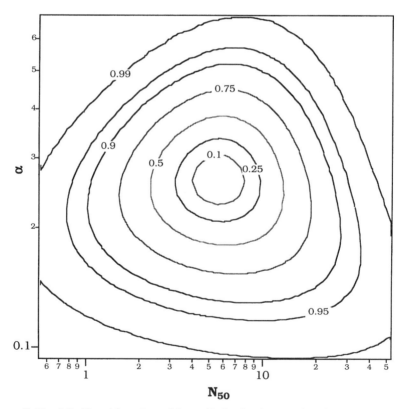

Figure 7-11 Likelihood-based confidence limits for the rotavirus beta-Poisson dose–response models. Labels indicate confidence levels.

If it is desired to estimate the confidence limits on a dose–response curve for a dose–response function with two or more parameters (i.e., similar to the dashed lines in Figure 7-9), it is necessary to solve a series of constrained minimization problems of a nature similar to equation 6-64. For a given dose d, the upper limit to the dose–response curve is determined from solving the following program (by varying the dose–response parameters, e.g., for beta-Poisson, varying α and N_{50}):

$$\max P(d;\Theta) \tag{7-37a}$$

subject to

$$Y(\Theta) - Y^* \le \chi^2 \tag{7-37b}$$

Similarly, the lower limit to the dose–response curve is found by minimizing P subject to the constraint indicated. Determination of the full upper and lower limits then requires repeating this computation at various doses and connecting the points obtained.

Figure 7-12 presents the confidence limits for the dose–response curve for rotavirus. Note that the width of the interval tends to increase at lower doses—this is characteristic; in general, the width will increase away from the central point of data. Also note that the confidence limits, as well as the best estimate, tend to become linear with a slope of 1 on a log-log plot (they will become linear on an arithmetic plot as well). This is characteristic of the exponential and beta-Poisson models (as well as all the other models that have a linear

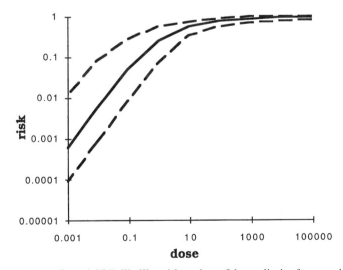

Figure 7-12 Best fit and 95% likelihood-based confidence limits for rotavirus dose–response using the beta-Poisson model.

low-dose property; note, though, that the log-probit model and the simple threshold models do not have this property).

Analysis of Suboptimal Data. Both the human rotavirus and human *Cryptosporidium* data give fairly compact and close to symmetric (at least on logarithmic scale) dose–response parameter confidence contours (Figures 7-10 and 7-11). This is a result of an experimental design that had a sufficient number of subjects and dose levels to give well-determined estimates for the dose–response parameters. However, it may be that some data, while they may yield acceptable fits to dose–response models, may give unusually shaped confidence regions. This indicates a less than optimal experimental design. Nonetheless, these data can be used for assessment of dose–response.

An analysis of experimental data with this characteristic is illustrated with reference to the following data set. McCullough and Eisele (49) administered human subjects doses of *Salmonella anatum* originally isolated from eggs. The organisms were fed in an eggnog vehicle. The response was infection. Data in Table 7-8 pertain to a single strain, designated as *S. anatum* strain I.

The first question that may be posed with this data set is whether or not it shows dose–response behavior at all. Stated formally, we would expect a tendency for the proportion of infected subjects to increase with dose. To address this in a fairly general way, preliminary to likelihood fitting, we employ the Cochran–Armitage test of trend (53). Let $x_i = \ln(d_i)$, the logarithm of the mean dose in group i. This group also has n_i total subjects and p_i positive subjects. Then Z_{CA} is computed via

$$Z_{ca} = \frac{\sum_{i=1}^{k} (x_i - \bar{x})p_i}{\sqrt{\bar{p}(1 - \bar{p}) \sum_{i=1}^{k} n_i(x_i - \bar{x})^2}} \qquad (7\text{-}38a)$$

where

$$\bar{x} = \frac{\sum_{i=1}^{k} n_i x_i}{\sum_{i=1}^{k} n_i} \qquad \bar{p} = \frac{\sum_{i=1}^{k} p_i}{\sum_{i=1}^{k} n_i} \qquad (7\text{-}38b)$$

A significant trend is asserted (i.e., a null hypothesis of lack of trend is rejected) if Z_{ca} is above the upper 5th percentile of the normal distribution (>1.644 for a one-tailed test).

Applying equation 7-36 to the data in Table 7-8, the Z_{ca} statistic is computed to be 2.19, which is above the critical value (the exact significance level is 1.4%). Hence there is a statistically significant trend, which justifies the use of a dose-response model to assess the data. This data set can be fit to the beta-Poisson model, with parameters $\alpha = 0.291$ and $N_{50} = 44,400$ and

**TABLE 7-8   Dose–Response for Infection of
Human Volunteers by *Salmonella anatum* Strain I**

Dose	Total Subjects	Positive Subjects
12,000	5	2
24,000	6	3
66,000	6	4
93,000	6	1
141,000	6	3
256,000	6	5
587,000	6	4
860,000	6	6

Source: Ref. 49.

with a residual deviance of 9.53 (19). With 6 degrees of freedom (eight doses minus two parameters), the fit cannot be rejected (p value $= 0.15$).

The confidence contours of the beta-Poisson dose–response parameters are shown in Figure 7-13. All of the contours in excess of the 0.25 curve are highly irregular in shape (both on a log-log plot as shown as well as on an arithmetic plot). The 0.95 curve is also open to the lower left. In fact, if one performs a likelihood ratio test comparing the fit of the beta-Poisson model to a "flat" model in which all response proportions are predicted to be equal to $\bar{p}$ (see equation 7-36), the null hypothesis, (of a statistically significant difference between the two models) cannot be rejected (the p value is just under 0.05). Hence the irregular, and open, confidence regions are also indicative of lack of strong evidence for dose–response.

Interestingly, this lack of rejection of a "flat" model is in contrast to the finding (from the Armitage–Cocoran test) of the existence of a trend. Hence the interpretation of this data set *in and of itself* must be treated with some caution.

Confidence Intervals and Regions: Bootstrap. The likelihood method provides a means to estimate the distribution of dose–response parameters. However, it has some significant limitations with respect to application for risk assessment when we will use Monte Carlo methods for overall risk characterization.

With respect to exposure inputs, as treated in Chapter 6, one major purpose to fitting probability distributions to these inputs was to allow for the possibility of simulating the uncertainty or variability that they depict. This is most frequently done, as will be seen later, by a Monte Carlo process. In this process, a risk computation is repeated many times by drawing a random sample from each of the underlying distributions. Since the procedures for

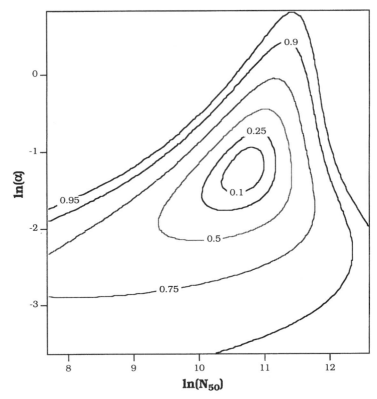

Figure 7-13   Beta-Poisson confidence regions for the *Salmonella anatum* strain I data set of McCullogh and Eisele (49).

generating random numbers from various underlying distributions are well known and many are available in commercial software, this process is relatively straightforward.

However, in the case of a dose–response distribution, particularly one with multiple parameters such as the case of the beta-Poisson, we wish to be able to generate random sets of dose response parameters (a pair, consisting of a and N_{50}, for example). Given the quite irregular nature of the sampling distribution of the dose–response parameters in some cases, such as in Figure 7-13, or even in the case of Figure 7-11 at the outer contour lines, the use of likelihood-based confidence regions to develop a method to develop random samples of the dose–response parameter pairs would be quite tedious.

An alternative approach to the problem is use of the bootstrap procedure. In particular, we present an extension of the method of bootstrapping residuals that appears to be suitable for dose–response data (17). The bootstrap is a method for developing the sampling distribution of a statistic computed from a set of data by *resampling* the data and recomputing the statistic from the resampled data. The distribution of the statistic from many resampled data sets is taken as a estimator of what the distribution of the statistic would be in the population from which our sample was drawn.

The logic of the bootstrap is illustrated in Figure 7-14. The observed dose–response data is the vector of observations $(x_1, x_2, \ldots, x_n)$, consisting of triplets (mean dose, subjects exposed, positive subjects). This is presumed to be a random sample of some unobserved population of all subjects. From the real data, we compute a statistic—the best-fit value(s) of the dose–response parameters. We want an estimate of what the uncertainty distribution of the dose–response parameters is. By constructing a series of bootstrap replicates (x^*)—in our case, the triplets (average dose, total subjects, positive subjects)—and determining (by maximum likelihood fitting) the distribution of dose–response parameters *among the bootstrap replicates,* it is inferred that the latter distribution is a good estimator of the uncertainty distribution from the total population in the real world. The key question then becomes how to construct bootstrap samples from dose–response data.

This appears not to have been a thoroughly explored subject in the statistical theory of the bootstrap. A review by Hinckley (38) notes that with respect to bootstrap methodology, when the dependent variables are discrete (as in the present case, where they are integers), "more needs to be learned." The monograph by Efron and Tibshirani (17) does not include coverage of this type of data (discrete dependent variables). We present a method extended from Efron's approach to regression analysis (termed *bootstrapping residuals*) applied to the data structure of dose–response assessment. However, further theoretical work to develop this method is required.

A central concept to this method is the definition of a residual from a model fit. Using the definition of a chi-squared residual, where π_i is the

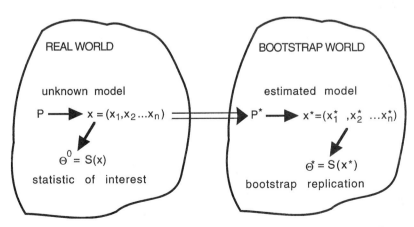

Figure 7-14 Schematic of the Logic of the Bootstrap Process. (Modified from Ref. 17.)

estimated proportion of respondents (e.g., infected individuals) at dose i from the dose–response relationship and $\pi_i^0 = p_i/n_i$ is the observed proportion, the chi-squared residual is defined by

$$\varepsilon_i = \frac{\pi_i^0 - \pi_i}{\sqrt{\pi_i(1 - \pi_i)/n_i}} \tag{7-39}$$

For dose–response data, these residuals would be expected to be (asymptotically) normally distributed with zero mean and unit variance (52).

The first step in execution of the bootstrap procedure is fitting the actual data (e.g., Table 7-8) to the dose–response model selected. This yields a set of values for π_i (from the best-fit model) and ε_i (from equation 7-39). Next, a pseudosample (bootstrap replicate) is constructed by computing

$$\pi_i^{(m)} = \pi_i^{(T)} + \delta_i \sqrt{\frac{\pi_i^{(T)} (1 - \pi_i^{(T)})}{n_i}} \tag{7-40}$$

where the superscript (m) denotes the mth set of bootstrap replicates, the subscript i refers to the particular dosage within that set, and δ_i is a randomly selected value from the set of ε values (selected with replacement). The $\pi_i^{(T)}$ values are the best-fit positive proportions to the original data. Since the proportion parameter of the binomial distribution is restricted to values between 0 and 1, the values of the bootstrap pseudoreplicate proportions are adjusted to 0 or 1 if the computations from equation 7-40 result in negative values or values in excess of 1, respectively. Finally, for each i, given n_i and $\pi_i^{(m)}$, a binomial random number (integer) is drawn to serve as the value for the number of positive responses for dose in bootstrap replication m.

Now for each pseudoreplicate, the maximum likelihood fitting process is repeated such that the best-fit parameter set, generically denoted as $\Theta^{(m)}$ (e.g., the pair $\langle \alpha^{(m)}, N_{50}^{(m)} \rangle$ for the beta-Poisson distribution), is computed. Then the parameter sets (for all m sufficiently large) represent the bootstrap estimate for the uncertainty distribution of the dose–response parameters.

Figure 7-15 presents the results of a run of 1000 bootstrap replications for the rotavirus data. The bootstrap distribution clearly surrounds the maximum likelihood estimate from the original data and qualitatively covers an area similar to the likelihood-based confidence region. These 1000 replications can then be used as an input to subsequent computations, reflecting the uncertainty distribution to dose–response.

The bootstrapped dose-response parameters can also be used to construct confidence regions to the dose–response curve alternative to likelihood-based intervals such as those shown in Figure 7-12. At a given dose, each set (say of the 1000 replicates) is used to compute the estimated response (exponential, beta-Poisson, etc). Then to obtain the confidence region (0.95, 0.99, etc.)

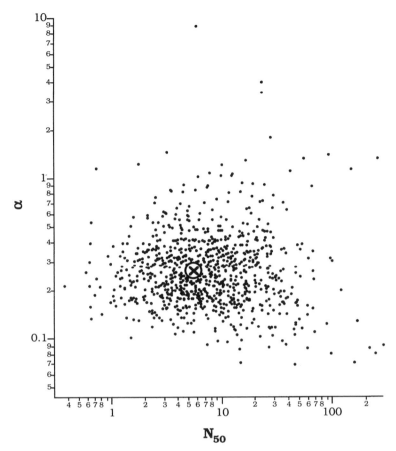

Figure 7-15 Bootstrap distribution of rotavirus parameter estimates (beta-Poisson model). Of the 1000 trials, 12 (not plotted) had $\alpha > 10$. The central marker indicates the ML estimate from the underlying data.

at a level z, the upper and lower $(1 - z)/2$ proportions of the responses (at that dose) are found; these define the confidence regions at that concentration.

Can Data Sets Be Combined? If we have several data sets, perhaps taken by different investigators, or on different microorganisms, or in different hosts (human and animal, or two animal data sets), a logical question to ask is whether the results of the two data sets can be pooled. In other words, can we characterize both data sets by the same dose–response relationship? For example, Table 7-9 summarizes human dose-response data for *Escherichia coli* It might be reasonable to ask whether there is a common dose-response relationship between all the groups (versus a null hypothesis of a different

TABLE 7-9 Experimental Data Sets for *E. coli*

Strain	Dose (no. organisms)	Affected Subjects	Unaffected Subjects	Total Subjects	Classification[a] (15)	Reference
1624 (O124)	1.00×10^4	0	5	5	EI	14
1624 (O124)	1.00×10^6	1	8	9	EI	14
1624 (O124)	1.00×10^8	3	2	5	EI	14
4608 (O143)	1.00×10^4	0	5	5	EI	14
4608 (O143)	1.00×10^6	0	5	5	EI	14
4608 (O143)	1.00×10^8	5	3	8	EI	14
B2C (O6:H16)	1.00×10^8	2	3	5	ET	14
B2C (O6:H16)	1.00×10^{10}	3	2	5	ET	14
B7A (O148:H28)	1.00×10^8	1	4	5	ET	14
B7A (O148:H28)	1.00×10^{10}	4	1	5	ET	14
H10407 (O78:H11)	2.70×10^8	9	7	16	ET	31
O111	7.00×10^6	7	4	11	EP	20
O111	5.30×10^8	8	4	12	EP	20
O111	6.50×10^9	11	0	11	EP	20
O111	9.00×10^9	12	0	12	EP	20
O55	1.40×10^8	6	2	8	EP	44
O55	1.70×10^9	5	2	7	EP	44
O55	5.00×10^9	6	2	8	EP	44
O55	1.60×10^{10}	7	1	8	EP	44

[a]EI, enteroinvasive; ET, enterotoxigenic; EP, enteropathogenic.

relationship for each group). It might be reasonable to ask whether both of the enteropathogenic strains might be combined, and so on.

Pooling data is frequently useful for two principal reasons. First, if multiple strains of microorganisms have nondistinguishable dose–response relationships, the confidence in inferring behavior of a different strain is enhanced; similarly if the same organism has the same dose response in different (animal) species, confidence in extrapolating to new species (e.g., humans) is increased. Second, the greater the number of data points (doses), the smaller the range of the confidence region—in other words, when pooling can be done, it generally acts to increase the statistical precision with which the dose–response relationship is estimated. Conversely, the inability to pool data may point to important mechanistic features. These may include differences in species sensitivity, or the possession of virulence factors (18,37,46).

The test of significance can be formulated as a test of the null hypothesis that all subsets being examined have identical dose–parameters, versus an alternative hypothesis that each subset has a different set of parameters (note that we must keep the functional forms identical with the pooled versus separated analyses; or the models must be simplified forms, such as the exponential vis-à-vis the Poisson). The individual subsets of data are denoted by capital letters; group A, B, C, and so on. The number of parameters in the individual dose–response fits is m_A, m_B, . . . , and so on. The total data set is fit by a model with m_T parameters (typically, one or two, for example).

From the individual fits, the optimal deviances from each subset are denoted Y^A, Y^B, and so on. The fit of a common model to the pooled data is denoted as having a deviance of Y^T. Then the test of the null hypothesis (that all the subsets have a common parameter set) can be constructed by computing the statistic

$$\Delta = Y^T - (Y^A + Y^B \cdots) \tag{7-41}$$

The null hypothesis is rejected if Δ is in excess of the critical χ^2 distribution with degrees of freedom $= (m_A + m_B \cdots) - m_T$. If Δ is less than the critical value, the null hypothesis (of a common dose–response model) cannot be rejected.

Example 7-4. Given the *E. coli* data in Table 7-9, determine the best-fit beta-Poisson model to O111 and O55 (the two enteropathogenic strains) and ascertain whether they can be fit by a common model.

SOLUTION. The first step in solution is the fitting of the individual data sets by the beta-Poisson model and then fitting the combined data sets by the same model. This results in three sets of parameters and three values for deviance, as shown in Table 7.10. The *p* value indicated here is the significance level for goodness of fit of that model, based on degrees of freedom equal to the number of points minus 2 (for the beta-Poisson model). Note

TABLE 7-10 Results for Example 7-4

	α	N_{50}	Y	Points	p
O111	0.2630	3.54×10^6	6.377	4	0.0412
O55	0.0869	194,057	0.481	4	0.786
Combined	0.1748	2.55×10^6	8.927	8	0.1778

that the O111 data set fails the goodness-of-fit test, and therefore we might be motivated to explore different dose–response models.

However, a test of the significance of pooling is constructed as follows:

$$\Delta = Y_{combined} - (Y_{O111} + Y_{O55})$$

$$= 8.927 - (6.377 + 0.481)$$

$$= 2.069$$

Since this is less than the upper 5th percentile of the chi-squared distribution at 2 degrees of freedom (for the difference in parameters when strains are modeled separately versus when they are modeled pooled), the difference between strains cannot be considered statistically significant. If we were to plot the confidence regions of the two organisms, we could see considerable overlap in the confidence contours.

Finally, we note that if we do a goodness-of-fit test to the combined data ($Y = 8.927$ versus 6 degrees of freedom), a null hypothesis of goodness of fit cannot be rejected. Hence the pooled model is regarded as adequate (although with the lack of fit for the O111 model handled separately, it would be desirable to conduct additional experimental work as well as additional modeling).

Explaining Lack of Fit. In some cases, none of the models above may fit a given data set to an acceptable level of goodness of fit (i.e., the residual deviance is above the critical chi-squared value). There are many possible reasons for this phenomenon. We describe some of these and suggest ways of characterizing the dose–response characteristics when the complex behavior is being manifest.

In all cases it is useful to examine the individual terms in the deviance. While the chi-squared residuals (as in equation 7-39) might be computed, an alternative approach is to compute the deviance residuals (52), defined as

$$r_i = \text{sign}(\pi_i^0 - \pi_i) \left\{ 2 \left[p_i \ln \frac{\pi_i}{\pi_i^0} + (n_i - p_i) \ln \frac{1 - \pi_i}{1 - \pi_i^0} \right] \right\} \quad (7\text{-}42)$$

where $\text{sign}(x)$ is the sign of the argument, x. Examination of these residuals can be performed in the same manner as examination of residuals from regression (5). In particular, if the model provides an adequate fit to the data, the anticipation is that the r_i values would be randomly distributed with a mean of zero.

The types of lack of fit will be illustrated using three hypothetical data sets, as described in Table 7-11. Each of these data sets is fit to the beta-Poisson model, with best-fit parameters and residual deviances as shown in Table 7-12. In all three cases, the residual deviance is sufficiently high that a lack of fit is indicated (the critical chi-squared value for $8 - 2 = 6$ degrees of freedom is 12.592). The resulting deviance residuals (as defined by equation 7-40) using the best-fit parameters to compute the values of π_i are graphed versus dose in Figure 7-16. The three hypothetical data sets show three distinct patterns of lack of fit.

The "bowing" of deviance residuals for set A suggests a systematic discrepancy from the assumed (beta-Poisson) model. Another model (perhaps incorporating an additional parameter) might be necessary to fit this data set adequately. Any trend, or systematic relationship, between deviance residual and dose (or, equivalently, the predicted proportion of response) is indicative of this type of lack of fit.

In the case of set B, there is no apparent systematic pattern in the deviance residuals. However, the residuals themselves, although apparently randomly spaced, are of larger than expected magnitude. This can occur when there is overdispersion—when there is greater diversity among replicate animals than would be predicted from the binomial distribution. This has been noted in prior (nonmicrobial) dose–response experiments (52,61) and can be described by use of a beta-binomial distribution describing host responses. There may be a mechanistic interpretation to the beta-binomial distribution in that the responses among individuals administered the same dose are (positively or negatively) correlated; perhaps all individuals were administered the same dose on the same day, and there are day-to-day correlations in sensitivity.

TABLE 7-11 Hypothetical Data Sets Indicating Lack of Fit

Dose	Total Subjects	Positive (Infected) Subjects		
		Set A	Set B	Set C
10	30	4	1	0
30	30	5	6	5
100	30	6	9	1
300	30	7	6	16
1,000	30	11	17	21
3,000	30	17	13	25
10,000	30	25	21	27
30,000	30	28	15	28

TABLE 7-12   Optimal Parameter Estimates for Hypothetical Data Sets

Parameter	Set A	Set B	Set C
N_{50}	756.7	3088.9	429.2
α	0.265	0.122	0.618
Deviance	17.954	11.684	14.501

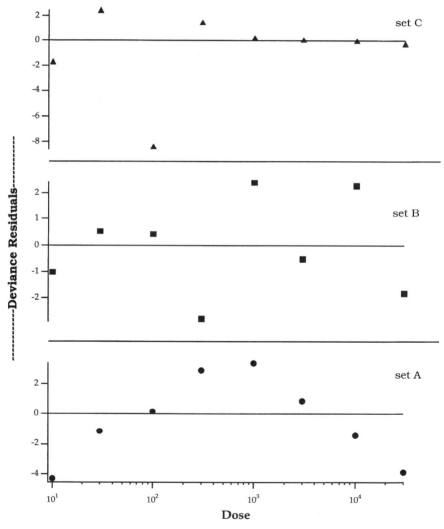

Figure 7-16   Deviance residuals for hypothetical data sets.

Mathematically, incorporation of the beta-binomial distribution for between-subject responses is obtained by the following probability of obtaining x positive (e.g., infected) subjects out of n tested, given an expected probability of response equal to π:

$$P(x|n) = \frac{n!}{x!\,(n-x)!}\,\frac{B\left(x + \dfrac{\pi}{\Theta},\; n + \dfrac{1-\pi}{\Theta} - x\right)}{B\left(\dfrac{\pi}{\Theta},\; \dfrac{1-\pi}{\Theta}\right)} \qquad (7\text{-}43)$$

where $B(a,b)$ is the mathematical beta function. In this equation the parameter Θ, which can assume any positive value, represents the degree of overdispersion (relative to the binomial). The variance in the observed proportion of positives (x/n) is given by

$$\frac{\pi(1-\pi)}{n}\left[1 + \frac{\Theta}{1+\Theta}(n-1)\right] \qquad (7\text{-}44)$$

As the parameter Θ approaches zero, equation 7-43 approaches the binomial distribution.

Figure 7-17 illustrates the beta-binomial distribution for fixed values of n and π, as a function of Θ. Note that the dispersion increases substantially. In this regard, the beta-binomial distribution generalizes the binomial distribution in much the same sense as the negative binomial distribution generalizes the Poisson distribution.

Using the beta-binomial distribution in combination with a dose–response model (predicting π as a function of dose) allows estimation of dose–response parameters, along with the value of Θ. The deviance statistic for fitting the beta-binomial variations of any dose–response model is constructed (compare to equation 7-35) from the ratio of the probability predicted from equation 7-43 to that predicted from the observed proportion of responses under assumptions of the ordinary binomial distribution. This can then be written as

$$Y = -2\sum_{i=1}^{k}\left\{\ln\frac{B\left(p_i + \dfrac{\pi_i}{\Theta},\; n_i - p_i + \dfrac{1-\pi_i}{\Theta}\right)}{B\left(\dfrac{\pi_i}{\Theta},\; \dfrac{1-\pi_i}{\Theta}\right)}\right.$$

$$\left. - [p_i \ln(\pi_i^0) + (n_i - p_i)\ln(1 - \pi_i^0)]\right\} \qquad (7\text{-}45)$$

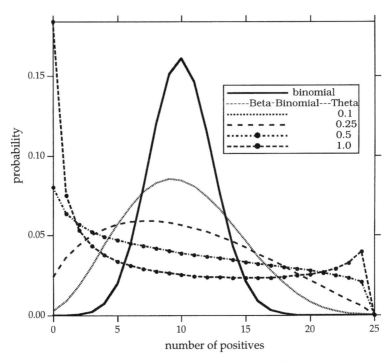

Figure 7-17   Comparison of beta-binomial and binomial distributions ($n = 25$, $\pi = 0.4$).

The latter terms represent the binomial probability under the empirical (observed) proportion of responses. This equation can then be minimized, and the values of the parameters in the dose–response model (predicting the π_i values) and Θ which produce the minimum deviance (Y) can be obtained. Goodness of fit is tested by comparing the minimum Y against the chi-squared distribution with $k - 1 - j$ degrees of freedom (where j is the number of parameters in the dose–response model). Furthermore, by comparing the minimum value of Y under the beta-binomial assumption with the minimum value computed under a binomial assumption (equation 7-35), the statistical significance of the overdispersion can be assessed.

Using the beta-binomial distribution and the beta-Poisson dose–response model,[7] the data from set B can be fit. The maximum likelihood estimates (minimum deviance) are

---

[7]We note here that there are in effect two distinct applications of the beta distribution. In one case, a distribution of individual microorganism survival probabilities is given by a beta distribution to obtain the beta-Poisson dose–response model. Then a second beta distribution is applied to modify the ordinary binomial response probability by using equation 7-43 in construction of the likelihood function.

$$N_{50} = 3081.1 \qquad \alpha = 0.122 \qquad \Theta = 0.016$$

The optimum residual likelihood is 11.038. This passes a goodness-of-fit test ($p = 5.06\%$). Note that the residual likelihood is less than that obtained from the ordinary binomial fit (Table 7-12). However, the improvement in fit (comparing the difference in likelihoods against the chi-squared distribution at 1 degree of freedom) is not statistically significant. Therefore, use of the beta-binomial distribution can be only accepted provisionally. Perhaps a larger number of doses, or more particularly a larger number of subjects per dose, would have allowed adequate statistical discrimination of the mild overdispersion. Also note that the overdispersion does not substantially alter the numerical values of the dose–response parameters (compare to Table 7-12).

The data of set C appear to show the presence of an *outlier*. By examination of the deviance residuals (Figure 7-16), the responses at a dose of 100 organisms are suspect. Ideally, such a finding would trigger reexamination of the actual experimental conditions to determine if there were an anomaly in the conditions. However, in the absence of such information (as when the data are analyzed many years after the original report), we seek systematic approaches for assessment of the dose–response in the presence of potential outliers.

The issue of determination of outliers is a complex one and has been treated in great detail in the field of regression analysis (4,6). There is very little prior work on the presence of outliers, and their implications in dose–response data such as those considered in this chapter. In general, as suggested by Barnett and Lewis (4), one may seek either to detect and remove outliers prior to analysis, or to use procedures that are robust (i.e., insensitive) to the presence of outliers. The issue of robust methods for dose–response data assessment is an important one but is not treated further here.

It should also be noted that the identification of outliers in a generalized regression context (such as maximum likelihood dose–response analysis) is with respect to a particular model. It cannot be ruled out that a more appropriate model (e.g., another dose–response relationship) would provide an adequate fit to all data, including the outlier. Particularly when the suspect observation is at the upper or lower extreme dosage, the identification and removal of an outlier should be approached with extreme caution.

With these caveats, the following procedure is proposed:

1. Use the residual deviance to identify the outlier as the observation(s) with the greatest absolute value.
2. Fit the revised data set [excluding the suspect outlier(s)] by maximum likelihood to obtain a revised deviance, Y^*.
3. If the difference between the optimal deviance for the full data set and the revised deviance, $Y - Y^*$ is greater than the chi-squared distribution with degrees of freedom equal to the number of suspect removed out-

liers, the suspect outliers should be regarded as statistically significant outliers.

For set C the dose of 100 corresponds to the maximum absolute deviance. When the revised data set is fit to the beta-Poisson model, the following are obtained:

$$N_{50} = 283.5 \qquad \alpha = 0.495 \qquad Y^* = 3.52$$

The value of $Y - Y^*$ is considerably greater than the critical chi-square distribution (at 1 degree of freedom), hence we regard the dose as being an outlier. Furthermore, we note that the revised data set yields a deviance (Y^*) that is acceptable (i.e., is less than the critical chi-squared distribution at degrees of freedom equal to the included observations, 7, minus the number of parameters). We also note that the significance level of the deleted outlier can be computed by Monte Carlo approaches and results in a similar finding.[8]

Note that the deletion of the outlier has a substantial effect on the dose–response parameters. This indicates that the outlier has a large "leverage" on the fitted parameters. Although a statistically significant outlier may not necessarily be one with large leverage, the existence of an outlier with large leverage increases the concern about investigation of possible experimental bases behind the putative outlier (since, if it cannot be discarded, it may point the way towards a new experimental finding).

POTENTIAL IMPACTS OF IMMUNE STATUS

The immune competency of people exposed to infectious organisms may influence their sensitivity to doses to which they are exposed. There are a complex set of host immune responses, as well as microbial factors, that may alter the dose–response relationship, as well as the severity of outcomes (12,21,46,63). However, due to ethical considerations, there are few direct human data to characterize the *quantitative* nature of these effects, and there are very few animal data. However, from outbreaks, [e.g., the waterborne

---

[8]To perform this computation, we assume the values of N_{50} and α from the fit to the entire data set along with the doses and numbers of subjects used. A single Monte Carlo trial consists of first computing a random number of positive response at each dose by a binomial random number (with n from the data set and π computed from the assumed dose–response parameters). This synthetic data set is fit to the beta-Poisson model using maximum likelihood, and the extreme value of the absolute deviance is obtained. By repeating for a large number of Monte Carlo trials, the distribution of the maximum absolute deviance is obtained, and it is found that the maximum absolute deviance in the observed data set is greater than the upper 0.1% of the Monte Carlo distribution and is therefore regarded as a statistically significant outlier (since it is more extreme than the upper 5%).

Cryptosporidium outbreaks in Milwaukee (47,48) and Las Vegas (30)], it is clear that greater severity is often observed in elderly, young, and immuno-compromised populations. However, not all agents show enhanced severity in the immunocompromised. In particular, neither Norwalk virus nor hepatitis A appear to be associated with greater severity in immunocompromised populations (27).

Gerba et al. (27) have reviewed factors responsible for differential sensitivity in exposed populations. For many enteric viruses, the morbidity ratio in children is less than in normal healthy adults. Coxsackie virus B has been implicated as a potential agent of spontaneous abortion. In the elderly, the mortality ratio for a variety of agents is greater than for the general population.

One outcome of infection (either symptomatic or asymptomatic) may be the development of complete or partial immunity to subsequent exposure. Data to quantitatively evaluate these effects are not available for water- or foodborne agents. However, it should also be noted that there are some circumstances in which prior exposure may result in hypersensitization to subsequent exposure (36). Discussion of quantitative approaches to assess risk from multiple exposures is included in Chapter 8.

RELATIONSHIP BETWEEN DOSE AND SEVERITY (MORBIDITY AND MORTALITY)

The focus of the chapter so far has been on the description of the relationship between dose and probability of infection. However, as noted in Figures 7-1 and 7-2, a variety of outcomes of different severity may occur from microbial exposure. These may be described by a series of ratios (probabilities) defined as the frequency of the more severe outcome divided by the frequency of the predecessor outcome of lesser severity, such as $P_{D:I}$ and $P_{M:D}$ in Figure 7-1. A key question in the quantification of the likelihood of more severe outcomes is whether or not these ratios (morbidity ratio, mortality ratio, for example) are constant or dose dependent (for a given organism).

Morbidity Ratio

From a retrospective review of outbreaks of *Salmonella,* Glynn and Bradley (28) assessed the correlation between dose and severity in water- and foodborne episodes. There was no relationship between dose and severity for outbreaks of *Salmonella typhi,* but there was a relationship for *S. typhimurium, enteriditis, infantis, newport,* and *thompson.*

In the case of *Giardia,* review of outbreak attack rates in comparison with infectivity dose–response relationships has permitted an indirect assessment of morbidity ratio to be made (57). From this analysis it has been estimated that the *Giardia* morbidity ratio does not appear to be dose dependent and is about 50%. For *Cryptosporidium,* based on dose–response studies in healthy adults (13), the morbidity ratio appears to be constant with a best estimate of 39% (34).

The description of the quantitative dose–response relationships for dose-dependent morbidity ratios has not been well studied. However, with very simple assumption,[9] it can be shown that a relationship of identical form to the beta-Poisson model can be derived (66). This can be written as

$$P_{D:I}(d) = 1 - \left[1 + \frac{d}{N_{50}^*} (2^{1/\alpha^*} - 1) \right]^{-\alpha^*} \tag{7-46}$$

Note the use of the asterisked parameters to refer to morbidity ratio as the response (rather than infectivity).

The statistical significance of dose-dependent morbidity can be assessed by using maximum likelihood analysis to find the best values of N_{50}^* and α^* that provide the minimum deviance when the morbidity ratio is used as a response. This residual deviance is compared with the deviance by assuming a constant morbidity ratio.[10] If the reduction in deviance is greater than the critical chi-squared distribution at 1 degree of freedom (two parameters in equation 7-46 minus one parameter for a constant morbidity ratio), the dose dependence using the beta-Poisson model for morbidity ratio is statistically significant. Similarly, the goodness of fit can be assessed by comparing the residual deviance (from either the beta-Poisson or the constant-morbidity-ratio models) with the chi-squared distribution at $k - 2$ (beta-Poisson) or $k - 1$ (constant ratio) models, rejecting the model if the residual deviance exceeds the critical values. Similarly, morbidity ratio data can be assessed by the Cochoran–Armitage test of trend.

The approach to assessment of morbidity ratio data is illustrated by a series of examples using data on human exposure to *Vibrio.* Honick et al. (39) exposed healthy human volunteers to various doses of *V. cholerae* Inaba strains with simultaneous ingestion of sodium bicarbonate. The latter agent served to buffer intestinal tract pH and protect against acid inactivation of the microorganisms. In this study, the number of respondents was measured using three distinct endpoints:

1. Presence of *V. cholerae* in stools, or presence of positive antibody response
2. Occurrence of diarrhea, with microorganisms present but of severity not requiring rehydration
3. Severe watery diarrhea requiring intravenous rehydration

---

[9]Specifically, that the clearance time for microbial infection is given by a gamma distribution with scale parameter linearly proportional to mean dose.
[10]Which, according to the same analysis indicated earlier, can be estimated directly as the total number of illnesses divided by the total number of infected persons.

For the purpose of the present analysis, all persons showing any of the foregoing responses are regarded as positive (infected) cases. People showing none of the foregoing responses are regarded as negative cases. Positive cases showing only positive stool or antibody reaction, without diarrhea, are termed *asymptomatic,* while all other cases are termed *symptomatic* (ill). Table 7-13 summarizes the results presented by Hornick et al. (39) using the foregoing definitions.

Example 7-5. For the *Vibrio* data, determine the adequacy of an assumption of constant morbidity ratio.

SOLUTION. The maximum likelihood estimator of the morbidity ratio (assuming a dose-independent formulation) is given by the total number of ill (symptomatic) cases divided by the total infected (symptomatic + asymptomatic). From Table 7-13, this yields

$$P_{D:I} = \frac{37}{44} = 0.841$$

The residual deviance is determined by application of equation 7-35 with π_i^0 given by the number of symptomatic cases at dose i to the total infected at dose i. For example, at a dose of 10,000 organisms, this would be 9/11, or 0.909. The value of π_i is given by the estimated constant morbidity ratio, $P_{D:I}$. The quantities n_i and p_i are given by the total infected and the number of symptomatic cases at dose i, respectively. The deviance is evaluated for the doses at which at least one person is infected, even asymptomatically (i.e., the dose of 10 organisms is not used). By equation 7-35 as modified, the residual deviance is computed to be $Y = 14.344$.

For 4 degrees of freedom (five doses with at least one infected case, minus 1 degree of freedom for the estimated morbidity ratio), this has a p value of 0.6% (i.e., 99.5% of the area under a chi-squared distribution is below this level), and therefore a null hypothesis of adequacy of fit is rejected. Therefore,

TABLE 7-13 Experimental Data of Hornick et al. (39): Response to *Vibrio cholerae* Inaba 569B with Simultaneous Ingestion of Bicarbonate

			Infected Subjects	
Dose	Total Subjects	No response	With Illness	Asymptomatic Cases
10	2	2	0	0
1,000	4	1	0	3
10,000	13	2	9	2
100,000	8	1	6	1
1,000,000	23	2	20	1
100,000,000	2	0	2	0

the use of a constant morbidity ratio to characterize these data is inappropriate.

In principle, rejection of an assumption of constant morbidity ratio may be due to the existence of systematic relationships between dose and morbidity. Alternatively, it may be due to variability in excess of binomial. The latter case will not be considered further here; however, the use of a beta-binomial distribution may be sufficient to deal with this problem. The existence of systematic trends in morbidity can be detected using the Cochoran–Armitage test.

Example 7-6. For the data on *Vibrio* in Table 7-13, use the Cochoran–Armitage test to determine the significance of a trend in morbidity ratio.

SOLUTION. Equation 7-36 is applied to the data, with n_i equal to the total (symptomatic+asymptomatic) infected subjects at dose i, and p_i equal to the symptomatic subjects. The intermediate terms are as follows:

$$\bar{x} = 12.036 \qquad \bar{p} = 0.841$$

Then a value of Z_{CA} is computed to be 3.039. Since this is outside the critical value (>1.64), a null hypothesis of lack of trend is rejected.

Having rejected a constant morbidity ratio and ascertained the existence of a statistically significant trend, we can now determine the suitability of particular parametric models for a description of this trend. The following example illustrates the use of the beta-Poisson model (equation 7-44).

Example 7-7. Determine the best-fit parameters of the beta-Poisson morbidity model and the adequacy of this model to characterize the *Vibrio* data in Table 7-13.

SOLUTION. The computations are set up using maximum likelihood. From iterative solution, the parameters that minimize the residual deviance are found to be

$$N_{50}^* = 3364 \qquad \alpha^* = 0.495$$

Table 7-14 provides the deviance values and the mortality ratios predicted for this parameter combination. The residual deviance is given by the sum of the last column, and is therefore equal to 3.17. To determine goodness of fit, this should be compared against the chi-squared distribution with 3 ($= 5 - 2$) degrees of freedom. This is below the upper 5% critical value ($p = 0.366$), and hence the null hypothesis of adequacy of fit cannot be rejected. Therefore, the beta-Poisson morbidity model is considered adequate for the description of these data.

TABLE 7-14   Results for Example 7-7

Dose, d_i	Infected Subjects, n_i	Symptomatic Subjects, p_i	π_i^0	π_i	Y_i
1,000	3	0	0	0.2737	1.9195
10,000	11	9	0.8181	0.6813	1.0493
100,000	7	6	0.8571	0.8932	0.0874
1,000,000	21	20	0.9523	0.9657	0.09997
100,000,000	2	2	1	0.9965	0.0141

Mortality Ratio

The final outcome in the infection–disease process is mortality of a fraction of people who become ill. As indicated in Figure 7-1, the quantity $P_{M:D}$ is the fraction of ill people who die of the illness. There are few data on the potential dose dependency of this quantity. In fact, actual estimates of the true mortality ratio are quite uncertain, since most such estimates represent the ratio of deaths (diagnosed as caused by the infection) to the number of identified (and usually severe, often seeking hospitalization) ill individuals. It is likely that for many agents, particularly for those causing relatively mild illness, the true mortality ratio is much lower, since many ill persons may not be hospitalized or even seek medical attention. The quantity that is difficult to estimate is the fraction of people who seek medical attention (or hospitalization) among those ill. Based on the 1993 Milwaukee *Cryptosporidium* outbreak, this proportion may be substantially less than 1%. Frequently, the mortality ratio in a widespread outbreak may also be a strong function of the proportion of sensitive persons among those who contract the illnesses. Notwithstanding these difficulties, the mortality ratio (generally based on hospitalized or severe cases) for a variety of agents has been reviewed (26,27). Table 7-15 summarizes these findings.

REALITY CHECKING: VALIDATION

An important part of dose–response assessment is validation. Although, in general, dose–response curves for many pathogens can be estimated from human feeding trials, and therefore issues of interspecies extrapolation that are present in chemical risk assessment (7,68) do not exist,[11] it is important to assess whether controlled human feeding trials provide good estimates of sensitivity in real populations. This process of validation can be performed by comparing the risk computed from a dose–response relationship to that in

---

[11]In fact, there are cases in which we must rely on animal dose–response data and hence where we do need to face the issue of interspecies extrapolation.

TABLE 7-15 Summary of Mortality Ratios for Various Pathogens

Agent	Mortality Ratio (%)
Viruses	
Hepatitis A	0.6
Coxsackie A viruses	0.12–0.5
Coxsackie B viruses	0.6 –0.94
Echoviruses	0.27–0.29
Bacteria	
Salmonella	0.1
Shigella	0.2
Protozoa	
Giardia and *Cryptosporidium*	~0.1[a]

[a]Estimated from recent outbreaks.
Source: Data from Refs. 26 and 27.

a well-defined outbreak. We illustrate this process with a case study. Several additional cases of validation are presented in Chapter 8.

One result, developed further in Chapter 8, that will be needed for validation is proper consideration of multiple exposures. If multiple exposures (e.g., on different days) each carry a certain probability of response (whether infection, illness, or mortality) denoted as $p_1, p_2, \ldots, p_j$, the overall probability that a person responds one or more times, denoted p_T, can be given by the theorem of independence from probability by

$$p_T = 1 - \prod_{i=1}^{j} (1 - p_i) \qquad (7\text{-}47)$$

In other words, the probability of one or more occurrences is the complement of the probability of no responses from any individual exposure.

Validation: 1993 Milwaukee Outbreak

Based on investigation of the Milwaukee outbreak, the following information can be gleaned (47): (1) the most likely duration of contamination (t) appears to have been about 21 days, with a possible range of 15 to 30 days; and (2) the attack rate determined from an epidemiological survey was 0.21. During the course of the outbreak, a number of samples were taken for protozoan analysis. Samples of ice manufactured during the outbreak were analyzed by membrane filter concentration and found to have a geometric mean of 0.079 L^{-1} (55); however, it is believed that as much as a 90% loss could have occurred in oocyst concentration from freezing and thawing (as well as from the oocyst concentration procedure itself) (56). Applying this correction, the geometric mean[12] oocyst concentration could have been 0.79 L^{-1}.

To complete the analysis, information is needed on the water ingestion rates of the exposed population. In the absence of site-specific information, the distribution of Roseberry and Burmaster (58) is used in which the daily water ingestion rate (q) (in mL) is lognormally distributed with a mean of 1948 mL.

From the water consumption and oocyst concentration, a daily mean dose of 1.948 L $\times$ 0.79 L^{-1} = 1.54 oocysts is computed. The dose–response relationship for *Cryptosporidium* infection has been estimated from the data of DuPont (Example 7-3). Using exponential best-fit dose–response parameters, the risk of infection from a single exposure is computed as

$$P_I = 1 - \exp[-(0.00419)(1.54)] = 0.0064$$

Since the risk of illness is dose independent and estimated as 0.39 (34), the single exposure risk of illness is 0.0064 $\times$ 0.39 = 0.0025.

If each day of the 21-day exposure period represents an independent and identical risk; the overall proportion of persons ill at any point is given by application of equation 7-46:

$$P_M = 1 - (1 - 0.0025)^{21} = 0.052$$

Hence we would estimate an illness ratio of 5.2% given the estimated exposure and exposure duration. It is not surprising that this is not identical to the ratio found (21%). Qualitatively, given the many assumptions and possible sources of uncertainty in this computation, however, the estimated attack ratio and the observed ratio are comfortably similar. A more rigorous comparison may be conducted by using the methods of Chapter 8 to assess the confidence regions to the predicted attack ratio and comparing it to the observed attack ratio (along with uncertainty in estimating the observed attack ratio). From this more exact computation, it has been concluded that the dose–response of *Cryptosporidium* is validated by the Milwaukee outbreak data (35).

APPENDIX

The best-fit dose–response parameters can be determined by unconstrained optimization. This computation can be performed in a spreadsheet such as Microsoft EXCEL with the Solver add-in using an approach similar to that

---

[12]We use geometric mean at this point since for the usual skew distribution of organisms it is a close estimator of central tendency. If our objective was to estimate the average population risk, rather than a single point estimate, the methods in Chapter 8 should be used.

described in Chapter 6. In this appendix we illustrate the use of a spreadsheet to ascertain confidence limits to a dose–response relationship.

We consider the rotavirus data set presented in Table 7-4. We wish to compute the upper 95% confidence limit to the dose-response curve at a mean dose of 0.0001 virus. In other words, what is the upper 95% confidence limit to the estimated risk associated with exposure to a mean dose of 0.0001 virus? Recall that the beta-Poisson model fit this data set with a residual deviance of 6.82.

The first step is to compute the critical value for the deviance; that is, what is the maximum acceptable value of deviance such that a parameter set (and hence a derived risk estimate) will be within the 95% confidence region? There are two parameters that will be varied (α and N_{50}); hence there are 2 degrees of freedom. The upper 95th percentile of the chi-squared distribution at 2 degrees of freedom equals 5.992; hence any parameter set yielding a deviance less than or equal to $6.82 + 5.991 = 12.807$ is within the confidence region. Therefore, we seek the parameter set that maximizes the estimated risk at a dose of 10^{-4} subject ot the constraint that the deviance (with respect to the original dose–response data) is less than or equal to 12.807.

Figures 7-18 and 7-19 provide the spreadsheets for this computation. Figure 7-18 is a numeric view, and Figure 7-19 indicates the formulas within the individual cells. In columns B, C, and D (rows 2 to 9), the original dose response data are indicated. Cells D18 and D19 contain assumed values of the log transforms of the dose–response parameters, and cells C18 and C19 contain the arithmetic values of these parameters. Cells E2 through E9 contain the observed response proportions (π_i^0). In cells F2 through F9, the predicted response (from the beta-Poisson model) using the assumed dose–response parameters is computed. Cells G2 through G9 contain the deviance contributions from each row. Note that in setting up the formulas in cells G2 through G9, a conditional ("IF" statement) statement is used to avoid taking the logarithm of zero or infinity.

Cell B14 contains the numerical value of the target dose for which we seek the confidence limit. Cell F14 contains the risk computed from this value (by the identical formula as in cells F2 through F9). In cell G19, the significance level for the confidence limit is placed. From this value and the degrees of freedom, cell G20 computes the critical chi-squared value. Cell C20 contains the deviance obtained by summing the values in cells G2 through G9. In cell G21, the numerical value of the optimum deviance (from the original optimization) is placed. Therefore, the "excess" deviance is obtained by difference in cell C23.

Figure 7-20 contains the panel for invocation of the solver. We seek to maximize the predicted response at the target dose (cell F14). This is to be done by changing the numerical values of the log dose–response parameters (cells D18 and D19). The maximization is to be performed subject to the constraint that the excess deviance (cell C23) is less than or equal to the critical chi-squared value (cell G20). To obtain the lower limit to the dose–

	A	B	C	D	E	F	G
1		dose	pos	total	f obs	f beta	2 ln L
2		90000	3	3	1.000	0.896	0.656
3		9000	5	7	0.714	0.856	0.932
4		900	7	8	0.875	0.800	0.315
5		90	8	9	0.889	0.722	1.491
6		9	8	11	0.727	0.614	0.626
7		0.9	1	7	0.143	0.464	3.284
8		0.09	0	7	0.000	0.267	4.350
9		0.009	0	7	0.000	0.079	1.153
10							
11							
12							
13		target dose				risk	
14		1.00E-04				0.001232	
15							
16							
17					ln transforms		
18		N50	1.464	0.382			
19		alpha	0.143	-1.946		p value	0.05
20		Y( )=2 ln L	12.807			chi sq 2 df,	5.991
21		Y*	6.815				
22							
23		Y( )-Y*	5.991				

Figure 7-18 Spreadsheet for computation of dose–response confidence limits.

response curve, the button "Min" in the Solver window (Figure 7-20) can be checked. Using the spreadsheet, the full two confidence bands (as in the lines in Figure 7-12) can be obtained by repeating this process at different values of dose (cell B14) and connecting the resulting risk values.

PROBLEMS

7-1. Take the beta-Poisson model in equation 7-21 to the limit as $\alpha \to \infty$. What does this result say with respect to the relationship between the beta-Poisson parameters (at the limit) and the exponential r in equation 7-7?

7-2. Examine the fit of the rotavirus data analyzed in Example 7-1 to the log-logit and Weibull dose–response models.

7-3. Using the parameters of the beta-Poisson model fit to the data in Example 7-2, plot the beta distribution corresponding to these parameters. Discuss the biological plausibility of your observations.

	B	C	D	E	F	G
1	dose	pos	total	f obs	f beta	2 ln L
2	90000	3	3	=C2/D2	=1-(1+B2*(2^(1/$C$19)-1)/$C$18)^(-$C$19)	=-2*(C2*IF(C2>0,LN(F2/E2),0)+(D2-C2)*IF(E2<1,LN((1-F2)/(1-E
3	9000	5	7	=C3/D3	=1-(1+B3*(2^(1/$C$19)-1)/$C$18)^(-$C$19)	=-2*(C3*IF(C3>0,LN(F3/E3),0)+(D3-C3)*IF(E3<1,LN((1-F3)/(1-E
4	900	7	8	=C4/D4	=1-(1+B4*(2^(1/$C$19)-1)/$C$18)^(-$C$19)	=-2*(C4*IF(C4>0,LN(F4/E4),0)+(D4-C4)*IF(E4<1,LN((1-F4)/(1-E
5	90	8	9	=C5/D5	=1-(1+B5*(2^(1/$C$19)-1)/$C$18)^(-$C$19)	=-2*(C5*IF(C5>0,LN(F5/E5),0)+(D5-C5)*IF(E5<1,LN((1-F5)/(1-E
6	9	8	11	=C6/D6	=1-(1+B6*(2^(1/$C$19)-1)/$C$18)^(-$C$19)	=-2*(C6*IF(C6>0,LN(F6/E6),0)+(D6-C6)*IF(E6<1,LN((1-F6)/(1-E
7	0.9	1	7	=C7/D7	=1-(1+B7*(2^(1/$C$19)-1)/$C$18)^(-$C$19)	=-2*(C7*IF(C7>0,LN(F7/E7),0)+(D7-C7)*IF(E7<1,LN((1-F7)/(1-E
8	0.09	0	7	=C8/D8	=1-(1+B8*(2^(1/$C$19)-1)/$C$18)^(-$C$19)	=-2*(C8*IF(C8>0,LN(F8/E8),0)+(D8-C8)*IF(E8<1,LN((1-F8)/(1-E
9	0.009	0	7	=C9/D9	=1-(1+B9*(2^(1/$C$19)-1)/$C$18)^(-$C$19)	=-2*(C9*IF(C9>0,LN(F9/E9),0)+(D9-C9)*IF(E9<1,LN((1-F9)/(1-E
10						
11						
12						
13	target dose				risk	
14	0.0001				=1-(1+B14*(2^(1/$C$19)-1)/$C$18)^(-$C$19)	
15						
16						
17			ln transforms			
18	N50	=+EXP(D18)	0.381504			
19	alpha	=+EXP(D19)	-1.9457125		p value	0.05
20	Y()=2 ln L	=SUM(G2:G9)			chi sq 2 df, 95 %	=CHIINV(G19,2)
21	Y*	6.81529				
22						
23	Y()-Y*	=+C20-C21				

Figure 7-19 Formulas in spreadsheet for computing dose–response confidence limits.

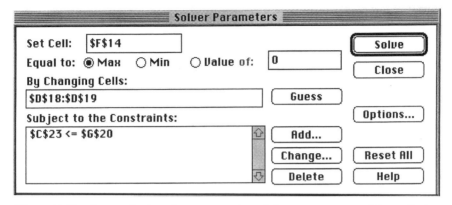

Figure 7-20   Setup of solver for computation of dose–response confidence limits.

7-4. The exponential dose–response relationship can be considered as a special case of the Weibull dose–response relationship. From the answer to Problem 7-2, determine if the Weibull provides a statistically significant improvement in fit as compared to the exponential model.

7-5. Apply the Cochoran–Armitage test of trend to the *Cryptosporidium* data in Table 7-7.

7-6. Fit the *Salmonella anatum* data (Table 7-8) to the exponential model, and determine the statistical significance of the improvement in fit for the beta-Poisson model relative to the exponential model.

7-7. Determine the 95% confidence limits for the *S. anatum* (Table 7-8) beta-Poisson dose–response curve.

7-8. Fit the *S. anatum* data to the log-probit model, and construct 95% confidence limits to the dose–response curve. Compare the fit and the dose–response curve confidence limits to those from the beta-Poisson model.

REFERENCES

1. Abramowitz, M., and I. A. Stegun, eds. 1965. Handbook of mathematical functions. Dover Publications, New York.

2. Armitage, P., G. G. Meynell, and T. Williams. 1965. Birth–death and other models for microbial infection. Nature 207:570–572.

3. Armitage, P., and C. C. Spicer. 1956. The detection of variation in host susceptibility in dilution counting experiments. J. Hyg. 54:401–414.

4. Barnett, V., and T. Lewis. 1994. Outliers in statistical data, 3rd ed. Wiley, New York.

5. Bates, D. M., and D. G. Watts. 1988. Nonlinear regression analysis and its applications. Wiley, New York.

6. Beckman, R. J., and R. D. Cook. 1983. Outlier........s. Technometrics 25(2):119–163.

7. Chappell, W. R. 1992. Scaling toxicity data across species. Environ. Geochem. Health 14(5):71–80.

8. Christensen, E. R. 1984. Dose response functions in aquatic toxicity testing and the Weibull model. Water Res. 18:213–221.

9. Christensen, E. R., and C.-Y. Chen. 1985. A general noninteractive multiple toxicity model including probit, logit and Weibull transformations. Biometrics 41:711–725.

10. Council for Agricultural Science and Technology. 1994. Foodborne pathogens: risks and consequences. Task Force Report 122. CAST, Ames, IA.

11. Crump, K. S., H. A. Guess, and K. L. Deal. 1977. Confidence intervals and test of hypotheses concerning dose response relations inferred from animal carcinogenicity data. Biometrics 33:437–451.

12. Duncan, H. E., and S. C. Edberg. 1995. Host–microbe interaction in the gastrointestinal tract. Crit. Rev. Microbiol. 21(2):85–100.

13. Dupont, H., C. Chappell, C. Sterling, P. Okhuysen, J. Rose, and W. Jakubowski. 1995. Infectivity of *Cryptosporidium parvum* in healthy volunteers. N. Engl. J. Med. 332(13):855.

14. Dupont, H. L., S. B. Formai, R. B. Hornick, M. J. Snyder, J. P. Libonati, D. G. Sheahan, E. H. Labrec, and J. P. Kalas. 1971. Pathogenesis of *Escherichia coli* diarrhea. N. Engl. J. Med. 285(1):1–9.

15. Dupont, H. L., and J. J. Mathewson. 1991. *Escherichia coli* diarrhea, pp. 239–254. *In* A. S. Evans and P. S. Brachman, eds., Bacterial infections of humans: epidemiology and control, 2nd ed. Plenum Medical Book Co., New York.

16. Edberg, S. C. 1996. Assessing health risk in drinking water from naturally occurring microbes. J. Environ. Health 58(6):18–24.

17. Efron, B., and R. J. Tibshirani. 1993. An introduction to the bootstrap, monographs on statistics and applied probability. Chapman & Hall, New York.

18. Ewald, P. W. 1991. Waterborne transmission and the evolution of virulence among gastrointestinal bacteria. Epidemiol. Infect. 106:83–119.

19. Fazil, A. M. 1996. M.S. thesis. Drexel University.

20. Ferguson, W. W., and R. C. June. 1952. Experiments on feeding adult volunteers with *Escherichia coli* 111 B$_4$: a coliform organism associated with infant diarrhea. Am. J. Hyg. 55:155–169.

21. Finlay, B. B., and S. Falkow. 1989. Common themes in microbial pathogenicity. Microbiol. Rev. 53(2):210–230.

22. Finney, D. J. 1971. Probit analysis, 3rd ed. Cambridge University Press, Cambridge.

23. Furumoto, W. A., and R. Mickey. 1967. A mathematical model for the infectivity–dilution curve of tobacco mosaic virus: experimental tests. Virology 32:224.

24. Furumoto, W. A., and R. Mickey. 1967. A mathematical model for the infectivity–dilution curve of tobacco mosaic virus: theoretical considerations. Virology 32:216.

25. Gaddum, J. H. 1933. Reports on biological standards. III. Methods of biological assay depending upon a quantal response. His Majesty's Stationery Office, London.

26. Gerba, C. P., and C. N. Haas. 1988. Assessment of risks associated with enteric viruses in contaminated drinking water. ASTM Spec. Tech. Publ. 976:489–494.

27. Gerba, C. P., J. B. Rose, and C. N. Haas. 1996. Sensitive populations: who is at the greatest risk? Int. J. Food Microbiol. 30(1–2):113–123.

28. Glynn, J. R., and D. J. Bradley. 1992. The relationship between infecting dose and severity of disease in reported outbreaks of *Salmonella* infections. Epidemiol. Infect. 109:371–386.

29. Goldberg, L. J., H. M. S. Watkins, M. S. Dolmatz, and N. A. Schlamm. 1954. Studies on the experimental epidemiology of respiratory infections. VI. The relationship between dose of microorganisms and subsequent infection or death of a host. J. Infect. Dis. 94:9–21.

30. Goldstein, S. T., D. D. Juranek, O. Ravenholt, A. W. Hightower, D. G. Martin, J. L. Mesnick, S. D. Griffiths, A. J. Bryant, R. R. Reich, and B. L. Herwaldt. 1996. Cryptosporidiosis: an outbreak associated with drinking water despite state of the art water treatment. Ann. Intern. Med. 124(5):459–468.

31. Graham, D. Y., M. K. Estes, and L. O. Gentry. 1983. Double-blind comparison of bismuth subsalicylate and placebo in the prevention and treatment of enterotoxigenic *Escherichia coli* induced diarrhea in volunteers. Gastroenterology 85:1017–1022.

32. Haas, C. N. 1994. Dose–response analysis using spreadsheets. Risk Anal. 14(6):1097–1100.

33. Haas, C. N. 1983. Estimation of risk due to low doses of microorganisms: a comparison of alternative methodologies. Am. J. Epidemiol. 118(4):573–582.

34. Haas, C. N., C. Crockett, J. B. Rose, C. Gerba, and A. Fazil. 1996. Infectivity of *Cryptosporidium parvum* oocysts. J. Am. Water Works Assoc. 88(9):131–136.

35. Haas, C. N., and J. B. Rose. 1994. Presented at the Annual Conference of the American Water Works Association, New York.

36. Halstead, S. B. 1982. Immune enhancement of viral infection. Prog. Allergy 31:301–364.

37. Haraldo, C., and S. C. Edberg. 1997. *Pseudomonas aeruginosa:* assessment of risk from drinking water. Crit. Rev. Microbiol. 23(1):47–75.

38. Hinckley, D. V. 1988. Bootstrap methods. J. R. Stat. Soc. B 50(3):321–337.

39. Hornick, R. B., S. I. Music, R. Wenzel, R. Cash, J. P. Libonati, and T. E. Woodward. 1971. The Broad Street Pump revisited: response of volunteers to ingested cholera vibrios. Bull. N. Y. Acad. Med. 47(10):1181–1191.

40. Huq, A., R. R. Colwell, R. Rahman, A. Ali, and M. A. R. Chowdhury. 1990. Detection of *Vibrio cholerae* O1 in the aquatic environment by fluorescent-monoclonal antibody and culture methods. Appl. Environ. Microbiol. 56(8):2370–2373.

41. Johnson, N. L., S. Kotz, and N. Balakrishnan. 1994. Continuous univariate distributions, Vol. 1, 2nd ed. Wiley-Interscience, New York.

42. Johnson, N. L., S. Kotz, and N. Balakrishnan. 1995. Continuous univariate distributions, Vol. 2, 2nd ed. Wiley, New York.

43. Johnson, N. L., S. Kotz, and N. Balakrishnan. 1994. Discrete univariate distributions, 2nd ed. Wiley-Interscience, New York.

44. June, R. C., W. W. Ferguson, and M. T. Worfel. 1953. Experiments in feeding adult volunteers with *Escherichia coli* 55 B$_5$: a coliform organism associated with infant diarrhea. Am. J. Hyg. 57:222–236.

45. Lauffer, M. A., and W. C. Price. 1945. Infection by viruses. Arch. Biochem. 8: 449–469.

46. Levin, B. R. 1996. The evolution and maintenance of virulence in microparasites. Emerg. Infect. Dis. 2(2):93–102.

47. Mac Kenzie, W. R., N. J. Hoxie, M. E. Proctor, M. S. Gradus, K. A. Blair, D. E. Peterson, J. J. Kazmierczak, K. R. Fox, D. G. Addias, J. B. Rose, and J. P. Davis. 1994. Massive waterborne outbreak of *Cryptosporidium* infection associated with a filtered public water supply, Milwaukee, Wisconsin, March and April 1993. N. Engl. J. Med. 331(3):161–167.

48. Mac Kenzie, W. R., W. L. Schell, B. A. Blair, D. G. Addiss, D. E. Peterson, N. J. Hozie, J. J. Kazmierczak, and J. P. Davis. 1995. Massive outbreak of waterborne *Cryptosporidium* infection in Milwaukee, Wisconsin: recurrence of illness and risk of secondary transmission. Clin. Infect. Dis. 21:57–62.

49. McCullough, N. B., and C. W. Eisele. 1951. Experimental human salmonellosis. I. Pathogenicity of strains of *Salmonella meleagridis* and *Salmonella anatum* obtained from spray dried whole egg. J. Infect. Dis. 88:278–289.

50. Meynell, G. G., and B. A. D. Stocker. 1957. Some hypotheses on the aetiology of fatal infections in partially resistant hosts and their application to mice challenged with *Salmonella paratyphi-B* or *Salmonella typhimurium* by intraperitoneal injection. J. Gen. Microbiol. 16:38–58.

51. Moran, P. A. P. 1954. The dilution assay of viruses. J. Hyg. 52:189–193.

52. Morgan, B. J. T. 1992. Analysis of quantal response data. Chapman & Hall, London.

53. Piegorsch, W. W. 1994. Environmental biometry: assessing impacts of environmental stimuli via animal and microbial laboratory studies, pp. 535–559. *In* G. P. Patil and C. R. Rao, eds., Handbook of statistics, Vol. 12. Elsevier Science, New York.

54. Prentice, R. L. 1976. A generalization of the probit and logit methods for dose response curves. Biometrics 32:761–768.

55. Rose, J. B. 1993. Results of the samples collected from Milwaukee associated with a waterborne outbreak of *Cryptosporidium*. Personal communication.

56. Rose, J. B., and C. N. Haas. 1994. Presented at the Annual Conference of the American Water Works Association, New York City.

57. Rose, J. B., C. N. Haas, and S. Regli. 1991. Risk assessment and the control of waterborne giardiasis. Am. J. Public Health 81:709–713.

58. Roseberry, A. M., and D. E. Burmaster. 1992. Log-normal distributions for water intake by children and adults. Risk Anal. 12(1):99–104.

59. Roszak, D. B., and R. R. Colwell. 1987. Survival strategies of bacteria in the natural environment. Microbiol. Rev. 51(3):365–379.

60. Rubin, L. G. 1987. Bacterial colonization and infection resulting from multiplication of a single organism. Rev. Infect. Dis. 9(1):488–493.

61. Ryan, L. 1992. The use of generalized estimating equations for risk assessment in developmental toxicity. Risk Anal. 12(3):439–447.

62. Singh, A., R. Yeager, and G. McFeters. 1986. Assessment of in vivo revival, growth and pathogenicity of *Escherichia coli* strains after copper and chlorine induced injury. Appl. Environ. Microbiol. 52(4):832.

63. Smith, H. 1990. Pathogenicity and the microbe in vivo. J. Gen. Microbiol. 136: 377–383.

64. Stuart, A., and J. K. Ord. 1987. Kendall's advanced theory of statistics, 5th ed., Vol. 2: Classical inference and relationship. Oxford University Press, New York.

65. Teunis, P. F. M. Infectious gastroenteritis: opportunities for dose–response modeling. Unpublished manuscript. Rijkinstituut Voor Volksgezondheid en Milieu.

66. Teunis, P. F. M., N. J. D. Nagelkerke, and C. N. Haas. Dose response models for infectious gastroenteritis. Unpublished.

67. The MathWorks Inc. 1994. MATLAB, 4.1 ed. MathWorks, Natick, MA.

68. Travis, C. C., and J. M. Morris. 1992. On the use of 0.75 as an interspecies scaling factor. Risk Anal. 12(2):311–317.

69. Turner, M. E., Jr. 1975. Some classes of hit-theory models. Math. Biosci. 23:219–235.

70. Ward, R. L., D. L. Bernstein, E. C. Young, J. R. Sherwood, D. R. Knowlton, and G. M. Schiff. 1986. Human rotavirus studies in volunteers: determination of infectious dose and serological response to infection. J. Infect. Dis. 154(5):871.

71. WaveMetrics Inc. 1996. Igor Pro, 3.0 ed. WaveMetrics, Lake Oswego, OR.

72. Williams, T. 1965. The basic birth–death model for microbial infections. J. R. Stat. Soc. B 27:338–360.

CHAPTER 8

CONDUCTING THE RISK CHARACTERIZATION

Risk characterization "integrates the results of [dose response and exposure assessment] into a risk statement that includes one or more quantitative estimates of risk" (16). One decision to be made in framing a risk estimate is what outcome(s) are most appropriate to the needs of decision-makers and other stakeholders. For example, the following outcome measurements might be relevant for microbial risk assessment (each of the measurements would be evaluated for a spectrum of policy choices to ascertain the relative consequences of a given decision):

- Expected risk of infection to a "typical" person
- Expected number of illnesses in a community
- Upper confidence limit to expected number of illnesses
- Upper confidence limit for illness to a "highly exposed" person
- Maximum number of illnesses existing in a community at any one time

Each of these measurements (the list is by no means exhaustive) consists of a single numerical value (for a given policy alternative) and is termed a *point estimate*. The first two quantities attempt to depict a central measure of a consequence. The next two quantities attempt to depict some upper value or "conservative" measure of consequence. The final quantity required information of the dynamics (incubation, duration) of disease to assess the case burden as a function of exposure. What none of these (and other *point estimates*) depict is a range of values that denote the uncertainty and variability of the input quantities and assumptions of the characterized risk.

In contrast, an interval estimate of risk is presented as either a confidence region or a full probability distribution of the resulting risk. The assessment of uncertainty and variability in a risk characterization is important since it has been argued that "a decision made without taking uncertainty into account is barely worth calling a decision" (59). Indirectly, the third and fourth outcomes enumerated in the preceding list attempt to account for underlying uncertainty and variability, but as will be noted, do so in an unclear manner.

The potential importance of knowing the full probability distribution of a risk may be illustrated by an example. Suppose that three policy alternatives (designated as A, B, and C) are evaluated. Using point estimates, the risk characterizations given by the solid lines in Figure 8-1 are obtained. The full probability density functions of estimated risk are given by the curves. For all three policy options, a central point estimate of risk might be computed as the solid lines (these are intended, for example, to reflect the arithmetic means of the risk).

Clearly however, for option A, the central estimate has a lower probability of being exceeded (or undershot) than for the other options, and the probability distribution is relatively symmetric about the central estimate. In cases B and C, however, there is a greater chance of falling below or exceeding, respectively, the computed central estimate. Hence, if the decision maker has particular sensitivity to the possibility of a risk being unexpectedly large (the right-hand "tail" of a distribution), then despite equal central tendencies, policy option C would be the least favorable alternative. Therefore, the distribution function (or for an interval estimate, in general, conveys additional information above that of any single point estimate).

A second important use of interval estimation techniques is in the assessment of the inputs to a risk characterization that most influence the width of a confidence interval. For example, with reference to Figure 8-1, one could pose the question of what input(s) to a risk computation (with respect to their uncertainty or variability) may most substantially influence the length of the right-hand tail of case C. It might then be desirable, if possible, to reduce this source of uncertainty or variability (e.g., by performing additional measurements) to give a decision maker greater confidence in the alternative selected.

We first outline the process of reaching a point estimate. This approach will then be generalized to producing a full interval estimate of a microbial risk.

POINT ESTIMATES OF RISK

For a numerical point estimate of risk from exposure to microorganisms, a point estimate of exposure may be directly substituted into the dose–response equation using the point estimate of relevant parameters to obtain the risk

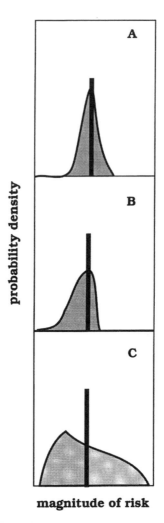

Figure 8-1   Point versus interval estimates of risk.

from a single exposure. For example, suppose that we estimate that in a given drinking water exposure scenario, a point estimate of exposure is 10^{-3} virus per exposure. From an analysis of dose–response experiments (e.g., the data shown in Table 7-4), it is determined that the dose–response relationship is a beta-Poisson equation with point estimates for the parameters of $\alpha = 0.265$ and $N_{50} = 5.597$ (Example 7-1). From this, the risk for a single exposure may be computed as

$$\pi = 1 - \left[1 + \frac{10^{-3}}{5.597} (2^{1/0.265} - 1) \right]^{-0.265} = 6 \times 10^{-4} \qquad (8\text{-}1)$$

By extension, if we further assume that a number of people are subjected to this exposure, the expected number of infections may be computed by multiplication. Continuing the example, if 10,000 people were exposed to the indicated dose, it would be anticipated that $(6 \times 10^{-4})(10,000) = 6$ infections would result. Point estimates for risk can require more than two or three inputs for computation, as indicated in the following example.

Example 8-1. A certain sample of raw chicken contains 10 *Salmonella anatum* bacteria per gram. The dose–response curve is beta-Poisson with parameters $\alpha = 0.291$ and $N_{50} = 44{,}400$ (see Table 7-8). People consume 200 g per meal. In cooking, there is a reduction of 99.5% in *Salmonella* levels; however, storage after cooking permits increase in *Salmonella* levels by 50%. Compute the risk (per person) from consumption of this product.[1]

SOLUTION. The density immediately after cooking is given by

$$\left(10 \; \frac{\text{organisms}}{\text{gram}}\right)(1 - 0.995) = 0.05 \; \frac{\text{organism}}{\text{gram}}$$

With regrowth, the density of organisms as consumed is

$$\left(0.05 \; \frac{\text{organism}}{\text{gram}}\right)(1.5) = 0.075 \; \frac{\text{organism}}{\text{gram}}$$

Therefore, for each meal, there is a consumption of 15 organisms. Using this as the dose in the beta-Poisson model produces a risk estimate (for infection) of

$$\pi = 1 - \left[1 + \frac{15}{44{,}000} \, (2^{1/0.291} - 1)\right]^{-0.291} = 9.7 \times 10^{-4}$$

Note that there are five inputs to this risk computation: (1) organism density in raw chicken, (2) two beta-Poisson dose parameters (α and N_{50}), (3) reduction in cooking, (4) increase subsequent to cooking, and (5) amount of portion consumed.

The use of point estimates for risk, either the risk per person or the expected number of adverse outcomes in a population, has the advantage of being simple to compute and being information relatively simple to convey to risk managers. However, it is disadvantageous in that for some (perhaps most) situations it conveys a false sense of certainty in the computed number.

---

[1] The numbers in this example are meant to be illustrative only and may not reflect particulars of specific cooking or storage practices.

Furthermore, many would presume that if central values are used for each of the inputs to a risk characterization, a central value for the result would occur; similarly, that if upper confidence limits for each of the inputs are used, an upper confidence limit (at a similar confidence level) for the output would be computed. Neither of these presumptions are universally correct. The following illustration addresses the first presumption; a subsequent illustration will address the second presumption.

Consider a risk characterization based on two variables, which we shall call x and y (perhaps one denotes exposure and the other denotes a potency). The overall risk is estimated as x/y. The input x may take one of five values $\langle 3 \times 10^{-4}, 4 \times 10^{-4}, 5 \times 10^{-4}, 0.001, 0.008 \rangle$. The input y may take one of three values (independent of x) $\langle 0.1, 0.5, 0.7 \rangle$. The means of x and y are 0.00204 and 0.433, respectively. The medians of x and y are 5×10^{-4} and 0.5, respectively.

If the quantity x/y is evaluated at the median of the two inputs, it is computed to be 0.001, while evaluated at the mean of the two inputs it is computed to be 0.00471. Now, if the function x/y is evaluated at each of the 15 combinations of x and y, the median and mean of the resulting values can be determined to be 0.002 and 0.00913, respectively. This illustrates that, in general, for a function of random variables (those which can be described as probability distributions), the expected value (mean) is not always equal to the function evaluated at the mean of the inputs, nor is the median equal to the function evaluated at the median of the inputs (nor, in general, is the nth percentile of the result necessarily given by substituting the nth percentile of all the inputs). The extent of such differences can be computed by the theory of propagation of errors (39). However, it is presently in general easier to approach the problem by use of the numerical Monte Carlo method, to be discussed subsequently.

Multiple Exposures

Before proceeding further, the issue of multiple exposures needs to be addressed. In developing the dose–response models in Chapter 7, data used were invariably formulated for exposure to a single dose (or bolus) of microorganisms. In exposure to infectious agents in most situations, there is the potential for multiple, or continual, exposures. People consume water every day, they consume food at every meal, and so on, and therefore incur a chance of exposure on a frequent or virtually continuous basis. There is thus an issue of how dose–response curves obtained for well-controlled bolus doses should be interpreted for multiple or continuous exposures.

There are virtually no experimental data in well-controlled settings to provide underlying support for a theory of multiple exposures. Hence, several empirical assumptions will be made, which are believed to be generally the most biologically plausible. However, in the future, it is possible that additional data will accumulate requiring revisions in this framework. Hence some approaches for incorporating such additional information will also be suggested.

In evaluating the risk from organisms in food, generally a single meal will be assumed to constitute an exposure. In evaluating the risk from organisms in water or air, each day will be assumed to constitute an exposure (i.e., the total organisms that constitute an exposure in a single day is construed to be d for substitution into a relevant dose–response equation).

There are two approaches to dealing with multiple exposure. The first approach is to consider the risk posed by any other exposure to be statistically independent of the risk posed by any other single exposure, following the approach of Bliss for chemical toxins (11). In this circumstance, if π_i is the risk from exposure to dose d_i, the overall risk (i.e., the probability of having one or more "positive" effects–infection, illness, etc.) from all the exposures, π_t, can be given by

$$\pi_t = 1 - \prod_{j=1}^{i} (1 - \pi_i) \tag{8-2}$$

This result, which follows from the binomial theorem, essentially states that the probability of having an effect is given by the complement of the product of probabilities of having no effect from any and all exposures.

A second approach, which in the chemical field has been termed the *approach of linear isoboles* (40), has been expounded quantitatively by Berenbaum (7–9). This same method, in the guise of hazard indices, has also been used in ecological risk assessment and in the assessment of risk from non-carcinogenic chemicals (41,54). In the microbial context, this might be termed "application of a principle of dose accumulation, for a reason that will be obvious. The microbial dose–response equation can be depicted as a function of dose, $f(d)$, that results in the risk. This function may be inverted, and hence we can write $f^{-1}(\pi)$, which computes the dose that resuls in a defined risk, π. With this nomenclature, the overall risk is given by the equation

$$\sum_{j=1}^{i} \frac{d_j}{f^{-1}(\pi_t)} = 1 \tag{8-3}$$

However, since the denominator in each term in the summation is identical, this equation can be written as

$$f^{-1}(\pi_t) = \sum_{j=1}^{i} d_j \tag{8-4}$$

The left-hand side can now be inverted to yield the result:

$$\pi_t = f\left(\sum_{j=1}^{i} d_j\right) \tag{8-5}$$

This equation indicates that the overall risk from multiple exposures is obtained by adding the individual doses and substituting the summed dose into the applicable dose–response equation; in other words, under this approach, the order and schedule of exposure has no effect on the response.

Parenthetically, note that both of these approaches give the same results regardless of the order with which multiple exposures are received. In other words, a large exposure followed by a small exposure is predicted to yield the same response as a small exposure followed by a large exposure. This may not necessarily be biologically plausible, and suggests the need for additional research. For example, initial exposure to microorganisms may produce an immune response which has the effect of reducing susceptibility to subsequent exposures. Alternatively, there have been some reports (albeit not with organisms generally regarded as transmissible from food, water, or other environmental exposure) of initial exposure to certain infectious agents causing hypersensitization to subsequent exposures (31).

In the case of the exponential dose–response model, the results of multiple exposure via equations 8-2 and 8-3 are identical. This can be seen by substitution of individual exponential terms in equation 8-2 to yield

$$\pi_t = 1 - \prod_{j=1}^{i} \{1 - [1 - \exp(-rd_j)]\}$$

$$= 1 - \prod_{j=1}^{i} \exp(-rd_j) = 1 - \exp\left(-r\sum_{j=1}^{i} d_j\right) \tag{8-6}$$

which is in fact identical in form to equation 8-5. In fact, it can be shown that the exponential model is the only such dose–response model for which the approach of equations 8-2 and 8-3 give formally identical results (M. J. Frank, personal communication to C. N. Haas).

For any other dose–response model that is approximately linear at low doses, such as the beta-Poisson (but not, for example, the log-probit or Weibull), the approaches of equations 8-2 and 8-3 give numerically similar answers provided that the overall risk itself is low. First, recognize that we can approximate equation 8-2 as follows:

$$\pi_t = 1 - \prod_{j=1}^{i} (1 - \pi_i) \approx \sum_{j=1}^{i} \pi_i \tag{8-7}$$

This approximation holds true provided that $\pi_u \pi_v \ll 1$ for all $u \neq v$. Now, let the individual risks be given as:

$$\pi_i = ad_i \qquad\qquad (8\text{-}8)$$

Substitution of equation 8-8 in equation 8-7 proves the assertion.

However, for other circumstances, there may be a substantial difference in the estimated risk from multiple exposures under an independence assumption versus one of dose accumulation. This is illustrated in the following example.

Example 8-2. Consider exposure to *Salmonella anatum* via food. It is considered that there are two exposures, at each of two meals. One meal has an exposure of 10,000 organisms. The second meal has an exposure of 25,000 organisms. What is the combined risk of infection from these two exposures under the assumptions of independence or of dose accumulation?

SOLUTION. For the doses of 10,000 and 25,000, the computed risks from application of the beta-Poisson dose–response equation equal 0.288 and 0.421, respectively. Hence, for the independence model, the total risk is estimated to be

$$\pi_t = 1 - (1 - 0.288)(1 - 0.421)$$

$$= 0.468$$

The summation of doses is equal to 35,000. Substituting this into the beta-Poisson model yields a risk estimate of 0.588. Hence there is a roughly 20% difference between the two assumptions.

Both the independence model and the dose accumulation model may be modified (by the inclusion of additional parameters) to reflect interactions, either in a synergistic or antagonistic sense, between dosages (28–30). However, data do not appear to exist to assess the necessity of more complex mixture models (which might be appropriate for consideration of immunity or hypersensitivity phenomena).

INTERVAL ESTIMATES

The objective of an interval estimate is to provide a range, or a probability distribution (or a set of probability distributions), that provides a sense of the precision with which a risk has been estimated. Furthermore, by using a sensitivity analysis in conjunction with interval estimation techniques, it is possible to ascertain the inputs whose lack of precise specification contribute to the imprecision with which risk can be estimated.

To perform such interval estimates for a risk, we need first to characterize the inputs with respect to whether or not they are known precisely (as single point values) or whether there is a distribution or range of values that they may take. In doing this assessment, it may be useful to distinguish between the concepts of uncertainty (and its types) and variability.

Uncertainty versus Variability

Inputs to a risk estimate are uncertain if they are the result of ignorance from "the partial incertitude that arises because of limits on empirical study or mensurational precision" (22). Empirically, the magnitude of uncertainty (i.e., the standard deviation of the distribution used to characterize it), can be diminished by additional effort devoted to measuring that input.

In contrast, variability in an input quantity results from an intrinsic heterogeneity in the input. For example, if we were doing a risk estimate (for exposure to infectious agents) among persons going to a beach, we would naturally get a variable exposure (to infectious agents in the water) arising from differing amounts of swimming activity, those who engage in active swimming being more highly exposed than those who do not engage in any water contact, for example. By describing the relative proportions of those engaged in activities leading to different intensities of exposure (again, perhaps, by a probability distribution), we could characterize variability. No amount of additional effort (at measurement) could reduce the magnitude of heterogeneity arising from this differential exposure.

Hence, although both uncertain and variable inputs can each be described mathematically by probability distributions, it is important to realize that they are addressing fundamentally different phenomena. Hence some have advocated dissecting these sources of heterogeneity (in a risk estimate) in the final characterization, which we shall discuss. In fact, a small number of practitioners of risk assessment have argued that probability distributions should not be used *at all* to describe uncertainty, but rather, interval arithmetic (or fuzzy arithmetic) should be used (22). However, we shall follow a probabilistic framework that appears to have more adherents than the approach using mixed probability distributions and interval arithmetic.

Sources of Uncertainty. There are many sources of uncertainty that may arise in inputs to a risk assessment. These have been categorized into a taxonomy by Finkel (23). These are arranged schematically in Figure 8-2.

Parameter uncertainty arises from measurement of discrete quantities. In the case of measurement error, physical limitations prevent more precise determinations. For example, in the assessment of exposure to drinking water, reliance is of necessity placed on information on water consumption obtained from questionnaires. There may be imprecision (as well as systematic bias) in the evaluation of this quantity.

Random errors leading to parameter uncertainty typically arise from the use of small data, sets. The use of only a small number of MPN assays to characterize microbial density as a component of exposure would be a frequent type of random error. Additionally, the use of small numbers of subjects

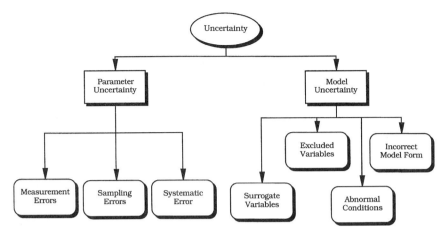

Figure 8-2 Taxonomy of uncertainty. (After Ref. 23.)

for dose–response assessment would lead to parameter uncertainty in the dose–response parameters.

Systematic errors arise from inherent flaws in data collection leading to bias in determination of information. In a recreational water situation, if exposure were determined based on microbial densities taken only on samples collected (for example) at 3:00 P.M., there might be a systematic error in estimation with respect to risk to all swimmers if systematic diurnal variability in microbial levels (perhaps from tidal flushing, or even inputs from swimmers themselves) existed. Another potential source of systematic error may be the failure to consider the potential for subpopulations that may be more sensitive to, or more adversely affected by, microbial infections (26).

Model uncertainty arises from the structural equations or relations used to develop exposure assessment or dose–response assessment. There are several components of model uncertainty that can be further differentiated. The use of surrogate variables contributes to model uncertainty when there is a potential difference (either random or systematic) between the quantity of interest and the quantity being modeled. A subtle type of use of surrogate variables arises in microbial risk assessment from the different nature of organism enumeration methods employed in dose–response measurement and in environmental sampling. In the development of dose–response relationships, typically pure culture of organisms of high viability are used. In many microbial measurements from environmental media or food, the assessment of viability is impossible, impractical, or very imprecise. Therefore, the use of "measurable" pathogen levels as surrogates for actual infectious levels presents a source of uncertainty of this type.

Errors from excluded variables arise when there may be factors, not considered in a model, that are in fact significant. For example, if the survival of an organism in food is modeled as a function of pH, water activity, and temperature, this may be used to predict pathogen levels at the point of an exposure (10,12). However, if, in fact, changes in redox potential influence survival, this have been excluded from the model and would represent a potential source of error. This chance of surprise is difficult to assess and has not been widely considered from a quantitative point of view.

The presence of abnormal conditions is related to the issue of excluded variables. If we fail to consider the potential for rare but catastrophic failure, we may fail to account for some aspects contributing to a risk. For example, if we are modeling the risk of microbial illness from drinking water produced from a given water treatment plant, there is a chance (perhaps minor, depending on locale) of treatment failure due to simultaneous failure of a chlorine application system (for disinfection) and the chlorine analyzer (which presumably would alert plant personnel to a performance failure). Although we expect that such simultaneous conditions would be extremely rare, they might never be zero, and hence unless we consider such possibilities, we have excluded the contribution of this situation to the resulting total risk.

Uncertainty due to model form may arise in a number of ways. In an exposure assessment, we may have several forms of a relationship characterizing transport. In the description of virus removal by lime treatment in the example in Figure 6-17, a lognormal distribution of the survival ratio appears to be supported by the data; however, this may not be substantially distinguishable from a gamma or Weibull distribution, and the alternative distributions may have very different behaviors at the extreme tails (and thus give different projected risks at high or low recurrence probability). In the assessment of dose–response, we may have a number of plausible dose–response models whose fits are consistent with the data (and again have different behaviors with respect to low-dose extrapolation). Some means to incorporate this possible source of uncertainty may be desirable.

Sources of Variability. Sources of variability arise from identifiable characteristics which result in differential exposure or differential dose–response characteristics. For example, as shown in Table 6-35, there are systematic differences in food consumption patterns with ethnic group and with region. Hence we would expect variability due to ethnic group and/or region with respect to microbial risk from particular food items that show such heterogeneity. Similarly, there are differences in drinking water consumption amounts with age and gender (50).

Variability in dose–response sensitivity may exist, although it is less well documented. Particularly with respect to severity of illnesses, it is known (26) that there are differential sensitivities to certain organisms in the elderly or the very young. Similarly, the variation in competency of different components of the immunological system as a function of nutrition or other components of health status (such as AIDS, or use of immunosuppressant drugs) may alter the intrinsic dose–response or severity responses (18).

Variability That Is Uncertain. Certain inputs that describe variability among a population exposed may also not be known with certainty, and hence we can have variability that is uncertain. This may be characterized by two principal aspects (analogous to Figure 8-2) or their combination.

We may have variability that is uncertain as to form. For example, the water consumption distribution has ordinarily been described as a lognormal distribution (50), but perhaps in a particular situation it may be described adequately by several alternative distributional forms. Variability may also be uncertain with respect to the parameters that characterize the distribution describing variability. For example, consider the food consumption distributions in Table 6-35. The entries (consumption of eggs, etc.) describe averages from a survey. However, due to a finite sample size, there is of necessity sampling error that must be accounted for (perhaps by a normal distribution centered on these averages, with a standard deviation given from further detailed analysis of the survey results). Whether this uncertainty in variability is substantial enough to affect the results will depend on the size of the survey,

Several authors have described the use of a second-order random variable to convey information about variability that is uncertain (15). At this point, the application of such methods is not well reduced to easy utility in terms of available software. However, a brief illustration of the concept is presented later.

Available Tools

The task before a risk analyst is to characterize the distribution of a measure of risk, such as total cases, or probability of risk to a single class of exposed persons, given known or assumed forms for the input distributions (whether due to variability or uncertainty or the combination of these factors). The most widely applied tool for such a problem is the Monte Carlo method (14), which we describe in detail below.

There are alternative tools that, in principle, can be used instead of Monte Carlo. For example, it is always possible to perform a (perhaps multiple) integration over the input probability distributions to generate the probability distribution of a selected output quantity (53). However, the evaluation of such integrals analytically may not be possible, and the resulting formal numerical integration may not necessarily be efficient or simple to perform (compared with Monte Carlo). A second method would rely on the use of propagation of error formulas for functions of random variables to assess the moments of the resulting output quantity (39). However, this approach may not be computationally simple, and furthermore, it does not lead to an efficient way to compute the tails or extreme values of an output quantity. Hence, particularly with the growing availability of efficient computer resources, use of the Monte Carlo approach has become the method of choice in developing a distribution of a derived risk quantity.

The term *Monte Carlo* derives from the extensive use of random numbers in the computational process. Let the quantity Z be the desired output of a risk assessment. This is determined as a function of a number of input quantities (X_i, Y_j). In other words, $Z = f(X_i, Y_j)$. There are m quantities X_i (i.e., $i = 1$ to m) and n quantities Y_j ($j = 1$ to n). The quantities X and Y may represent intake quantities (such as water consumption), microbial density measurements (densities in a dose), dose–response parameters, or any other factors that connect the output Z to the inputs. The quantities X_i may have some variability or uncertainty (i.e., they may be characterized by probability distributions of stipulated form with certain fixed parameters), while the quantities Y_j are regarded as fixed values (without either variability or uncertainty). Furthermore, this structure is general enough so that the distribution of some of the X_i may be dependent on the numerical values of other X_i (as when the variability itself contains a component of uncertainty), as long as evaluation of the relationship yielding Z proceeds in a manner such that the X_i values which comprise pure uncertainty or variability are evaluated first.

In the Monte Carlo method, a series of trials are performed. In each trial, random values of the X's are selected. For example, in the kth trial, the set of random quantities $X_1^k, X_2^k, \ldots, X_i^k$ are selected. Then from the defining equation for Z, the output is computed from

$$Z^k = f(X_i^k, Y_j) \tag{8-9}$$

By repeating this computation for a large number of trials (k large, say 1000), the resultant set of outputs, Z^k, provides an estimate of the output distribution given the distributions in the variable and/or uncertain inputs. The application of the Monte Carlo method to a simple problem is illustrated by the following example.

Example 8-3.   Consider the raw water supply described in Example 6-16, where it was found that the influent *Giardia* density (number per 100 L) could be described as a gamma distribution with parameters $\alpha = 0.326$ and $\beta = 134$. A regulation is contemplated in which a water supply is deemed inadequate if any three consecutive samples of 2 L contain more than a total of 10 cysts. What is the likelihood that this water supply will be deemed inadequate?

SOLUTION. This problem really involves two separate distributions. First, the density in any given sample is given by gamma-distributed random numbers with the indicated parameters. Second, *given that density,* the actual number of cysts is given by a Poisson random number with parameter equal to the density in that sample. Therefore, a two-step process must be conducted, as illustrated schematically in Table 8-1. The second column are gamma-distributed densities obtained (as random numbers) from a distribution

TABLE 8-1   Trials for Example 8-3

Trial	Sample Density (no./100 L)	Mean Number in 2 L	Cysts (no.) in 2-L Portion	Sum of Three Portions
1	39.38	0.79	0 ⎫	
2	4.32	0.09	0 ⎬	4
3	170.6	3.41	4 ⎭	
4	72.32	1.45	2 ⎫	
5	34.34	0.69	0 ⎬	9
6	309.4	6.19	7 ⎭	

using the underlying parameters. The third column represents the mean number of cysts that would be anticipated in 2 L (obtained by multiplying the second column by 2 L/100 L. The fourth column consists of random integers sampled from a Poisson distribution whose mean is given in the corresponding row of the third column. The final column represents the summation of three consecutive portions (with respect to total number of cysts).

Repeating this computation a large number of times yields a table similar to Table 8-2 (for 1000 sets of three samples). There are 36 sets of (three) samples with 11 or more cysts. Therefore, we estimate the probability of exceedence (of a maximum acceptable level of 10 cysts) to be 3.6%.

While the preceding example is a relatively simple one (and in fact, can be computed analytically[2]), more complex computations are not as amenable to simple analytical computations. It is in the latter situations that Monte Carlo method's show their application.

Unfortunately, most Monte Carlo problems cannot be computed using simple spreadsheets. However, several add-in products to spreadsheets are available which make the application of Monte Carlo methods to risk analysis problems more user friendly. These include the following:

Crystal Ball
> by Decisioneering Inc.
>> 1515 Arapahoe Street, Suite 1311
>> Denver, CO 80202
>> phone: 800-289-2550 or 303-534-1515

---

[2] A gamma mixture of Poisson distributions is a negative binomial, as shown in Chapter 6. Furthermore, a sum of negative binomial observations (independently and identically distributed) is also a negative binomial (with parameters obtained from the underlying distribution). Hence this problem can be solved analytically by investigating the cumulative negative binomial distribution.

TABLE 8-2 Results for Example 8-3

Cysts in Three Consecutive Samples	Number of Trials (out of 1000)
0	275
1	198
2	152
3	101
4	79
5	49
6	38
7	24
8	22
9	16
10	10
11 or more	36

URL: http://www.decisioneering.com

@RISK

by Palisade Corporation
        31 Decker Road
        Newfield, NY 14867
        phone: 800-432-7475 or 607-277-8000
        URL: http://www.palisade.com/

A good general-purpose introduction to the use of these types of programs in a risk assessment context is provided by Vose (55).

In addition, several stand-alone general purpose mathematical packages are quite suitable for use in risk assessment applications, although typically these require somewhat more programming (and are less menu oriented in nature) than either of the two programs cited above. In particular, the senior author has had good success with the following package:

MATLAB

by   The MathWorks, Inc.
        24 Prime Park Way
        Natick, MA 01760-1500
        phone: 508-647-7000
        URL: http://www.mathworks.com/

Guidelines for Performing Monte Carlo Analysis

An important aspect of performing a Monte Carlo risk estimation is the assessment of how many *trials* to perform to reach a stable result. This depends on the particular aspect of the risk computation that is of interest (i.e., it typically requires far fewer trials to obtain a stable estimate of the mean or median than of an upper percentile), the intrinsic heterogeneity of the outcome (variability plus uncertainty), and the nature of the sampling scheme[3]. The best answer to assessing how many runs should be made is obtained by numerical experimentation.

This is illustrated by example with reference to the problem posed in Example 8-3. For various numbers of trials (e.g., 50, 100, 200), six runs were made to obtain an estimate of the fraction of triplets exceeding 10 cysts. From each set of six runs at a given sample size, the mean (of the exceedence fraction in a run) and the standard deviation were computed. Figure 8-3 plots the resulting mean and standard deviation as a function of sample size, ranging from 20 trials per run to 10,000 trials per run. This analysis shows that there is greater instability in the estimated exceedance fraction at smaller sample sizes and also that the standard deviation decreases systematically as sample size increases. In fact, in this case, the decline is approximately inversely proportional to the square root of sample size (number of trials). There are reasonable theoretical reasons why such inverse-square-root behavior should be expected. In any event, this illustrates the fact that by increasing the size of a Monte Carlo trial, any (arbitrarily small) degree of precision of an estimator can be achieved given sufficient computer resources. For example, for this problem, if a precision (defined as standard deviation divided by mean) of 10% is acceptable, a sample size of about 2000 appears adequate.

While the precise relationship between sample size and precision of a Monte Carlo output calculation may vary (with the input distributions, the nature of the output, and the nature of the random number generator), a risk analyst can explore the relationship between sample size and precision of an estimate by performing the calculations in the manner of Figure 8-3. In this manner it is possible to develop a sense of security with respect to knowing the level of precision afforded by a Monte Carlo computation of a given sample size.

The use of Monte Carlo methods for risk assessment has a number of aspects that are not completely objective and offer flexibility for interpretation by different analysts. Burmaster and Anderson (14) have enumerated a number of principles of good practice, which are summarized in Table 8-3, some of which will be expanded upon subsequently.

---

[3] *Latin hypercube* sampling frequently results in fewer necessary iterations than ordinary Monte Carlo sampling—the former is a method in which the random numbers are forced to lie in all intervals of the sampling space. There are also developing methods for sampling, such as the *Gibbs sampler,* or *importance sampling,* which may speed up run times when tail probabilities are sought. However, these algorithms have yet to be included in the most commonly available software packages.

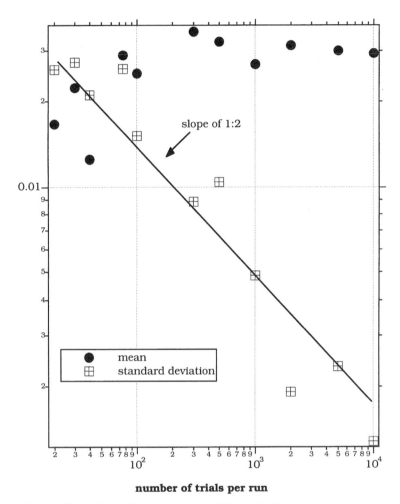

Figure 8-3   Effect of number of trials per Monte Carlo run on mean output and standard deviation of output (eight runs per sample size) for Example 8-3.

Assigning Input Distributions. The particular task of assigning distributional forms to input quantities in a risk assessment, and estimating their values, is a critical step in the process. Aspects of the problem with respect to exposure assessment and dose-response assessment were discussed in Chapters 6 and 7, respectively,

Moore (45) has enumerated a hierarchical decision process for developing input distributions to risk assessments:

1. Will the variable have an important influence on the output? If not, don't worry about it (in terms of assigning an input distribution).

TABLE 8-3   Good Risk Assessment Principles

1. Show all formulas used to compute exposure, potencies, and endpoints, either in the text, as spreadsheets, or in appendices to a report.
2. Calculate both point estimates as well as interval (Monte Carlo) estimates. Point estimation is a desirable first step.
3. Conduct sensitivity analyses of the point estimates to ascertain the most significant inputs to consider in the Monte Carlo steps.
4. Restrict application of Monte Carlo techniques to the most important pathways, routes of exposure, and endpoints (most important to potential risk managers).
5. Document input distributions with respect to means, medians, minimum and maximum (for truncated distributions), and 95th percentiles. Justify selection of distribution either from data, expert judgment, or mechanistic considerations.
6. Document the contribution and extent of variability and uncertainty to inputs to individual distributions and their prameters.
7. When possible, use actual data to select distributional forms and their parameters.
8. Document goodness-of-fit statistics used to obtain parameters for input distributions.
9. Discuss the presence or absence of moderate-to-strong correlations (rank correlation absolute value in excess of 0.6) and the potential impact on computed results.
10. Provide detailed information in graphical and numerical form for all output distributions.
11. Perform probabilistic sensitivity analyses of the Monte Carlo results.
12. Document the numerical stability of the output risk distribution with respect to the number of trials used in the simulation.
13. Document the quality of the random number generator.
14. Provide a qualitative discussion of limitations of the methods, biases, and potential factors not considered in the analysis.

Source: Adapted from Ref. 14.

2. Is the distribution known for the input variable?
3. If not, are there sound theoretical reasons for assigning a specific distribution to the input variable?
4. If not, are the data adequate for fitting a distribution?
5. If not, do appropriate surrogates exist? If yes, repeat steps 2 through 4.
6. If not, do data exist addressing components of the variable? If yes, repeat steps 2 through 4.
7. If not, solicit expert opinion (20,46).

Obviously, as one proceeds to later steps, the assignment of distributional form and parameters becomes more subjective, and hence particular caution should be observed if the less well defined inputs play a substantial role in defining the uncertainty and variability in the outcome risk measure(s).

A Simple Example

The importance of doing interval estimates of risk, using either a Monte Carlo method or another approach, is illustrated with a simple example. We modify the problem posed in Example 8-1 to include elements of variability as noted in Table 8-4. The choice of distribution forms in the table would be guided by experimental data and by expert judgment. For example, the density of *Salmonella* could be estimated by random sampling of carcasses, the portion consumed could be estimated by consumer surveys, and the reductions in cooking and increases in holding times could be estimated from surveys on storage practices (time and temperature) combined with predictive microbiology models.

The mean values in Table 8-4 are computed using the distributional parameters and analytical expressions for the distribution mean. The upper 95% confidence limits are computed from analytical expressions for the cumulative distribution function. In the case of reduction due to cooking, the lower 95% confidence limit is computed (i.e., the value of x_3 that is exceeded by 95% of the distribution), since a low value of x_3 increases exposure (as opposed to the other inputs, where a high value increases exposure).

In a manner similar to Example 8-1, the average[4] number of organisms consumed expected to be ingested (y) could be computed from the inputs in Table 8-4 using the following relationship:

$$y = x_1 x_2 (1 - x_3)(1 + x_4) \qquad (8\text{-}10)$$

The value of y can be computed using the mean values of the parameters or using the 95% confidence limits of the parameters (or any other combination, such as the median, the 75% confidence limits, etc.). For these two situations, application of equation 8-10 yields

$$y_{\text{mean}} = 10(216.67)(1 - 0.995)(1 + 0.4431) = 15.63 \qquad (8\text{-}11)$$

$$y_{0.95} = 38.69(340.73)(1 - 0.9725)(1 + 0.8654) = 676.26 \qquad (8\text{-}12)$$

Considering the distributions in Table 8-4, a Monte Carlo analysis is conducted. For each trial, a random number is drawn from each distribution, and using these values (for $x_1, \ldots, x_4$), the resulting value of y is computed via equation 8-10. For a run with 10,000 trials, Figure 8-4 shows the distribution of average exposures (y). Figure 8.4b provides the *cumulative distribution,* the probability that the average exposure is less than or equal to the indicated value). The upper panel provides the *complementary cumulative distribution*

---

[4] We use average here since if we were actually desirous of computing the *number* of organisms consumed, this average would form the input to a Poisson distribution.

TABLE 8-4 Modifications of the Chicken Problem to Include Variability

		Mean	Upper 95% Confidence Limit
Density of *Salmonella* in chicken x_1 (no./g)	Lognormal Mean = 10 standard deviation = 30 ζ = 1.151 δ = 1.517	10	38.69
Portion consumed, x_2 (g)	Triangular Minimum = 50 Mode = 200 Maximum = 400	216.67	340.73
Reduction by cooking, x_3 (proportion)	Beta α = 10 β = 0.05	0.995	0.9725[a]
Increase by holding, x_4	Weibull a = 0.5 c = 2	0.4431	0.8654

[a] Lower confidence limit.

(CCD), the probability that the average exposure is greater than or equal to the value indicated. The CCD plot is particularly useful for examining the details of the upper tail of a distribution; for example, it is clear that there is about a 10% chance of the average exposure exceeding 100 organisms.

Taking the average of the 10,000 trials, the mean of the distribution of average exposures is estimated to be 54.57 organisms. The upper 95% confidence limit to the distribution of average exposures is estimated (graphically from examining Figure 8-4a at 5% or numerically from the 10,000 sorted values) as 199.8 organisms. Comparing these two numbers to the results from equations 8-10 and 8-11, it is found that the mean of average exposures is greater than that computed simply by substituting the mean of all inputs into the computation. The upper 95th percentile of average exposures is less than that which would be obtained from a naive substitution of the individual extreme percentile values into the computation. The latter finding, particularly, indicates a substantial value of the Monte Carlo approach, in that a more accurate estimate of the likelihood of extreme occurrences can be made rather than in the simple substitution method.

As a matter of general principle it can be stated that (except for very simple situations) the mean of a derived quantity of random variables is not the simple result of substituting the means of the individual random variables into the defining equation for the quantity derived. Similarly, the qth percentile of a derived quantity is not in general obtained by substituting the qth percentile of each of the inputs into the defining equation.

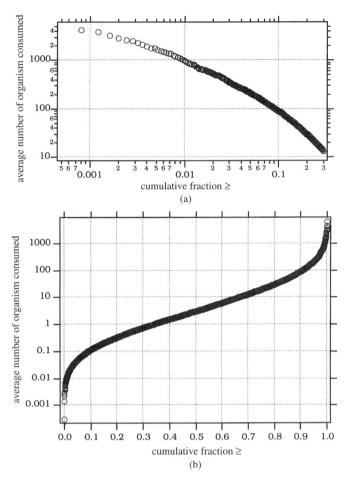

Figure 8-4 Distribution of average microbial exposure from Monte Carlo trials: (*a*) complementary cumulative distribution; (*b*) full distribution.

One output of a Monte Carlo computation is a sensitivity analysis. In other words, the relative impact of the changes in the various inputs on the computed output can be determined. Each Monte Carlo trial has a particular combination of inputs. By examining the correlation between the inputs and the output, the relative importance of individual components can be ascertained. This may be done using a correlation[5] or an analysis-of-variance approach.

---

[5] The most common procedure is to examine the rank correlation, or Spearman, coefficient. This is equivalent to the ordinary (Pearson) correlation when the inputs and the output are each replaced by their ranks (among their respective sets of inputs and outputs). The Spearman coefficient can account for nonlinear but monotonic relationships (such as exponential, logarithmic, or parabolic), while the ordinary correlation coefficient measures only the linear component of a relationship.

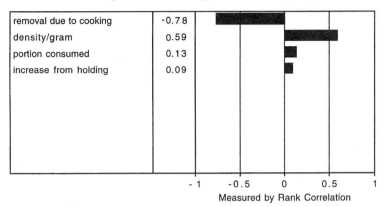

Figure 8-5 Rank correlation coefficient of inputs to the estimation of average microbial dose for the *Salmonella*-on-chicken problem.

Figure 8-5, an actual output from the CrystalBall analysis of the problem, summarizes the rank correlation coefficients for the four inputs. It is clear that the removal during cooking (x_3) and the initial density (x_1) have a much stronger influence on the average organism consumption than do the other inputs. Hence a reduction in the range of removals due to cooking, or of initial density, would have the greatest impact on reducing the range of the output. We return to this issue below with respect to differentiating uncertainty and variability for the purpose of focusing additional data-gathering efforts.

Correlated Inputs

The example just presented assumed that all of the input distributions were statistically independent. Stated simply, the probability of observing a particular value of one of the inputs is independent of the particular value (in) that sample) of all the other inputs. This may or may not be reasonable in all cases. For example, with reference to the example, the removal due to cooking might be directly correlated with the initial density.[6] Such correlations may then influence the quantitative predictions of a Monte Carlo risk calculation.

We consider first how to describe such correlations. Then we illustrate the effect that such correlations have on the risk calculation per se.

---

[6] If, for example, persons who correctly perceived a particular portion to be more vulnerable would be more fastidious in food preparation, thereby increasing removal in the cooking operation.

Fundamental Meaning of Ordinary Correlation. When two distributions[7] are independent, it means that the bivariate density function can be given as the simple product of the univariate density functions, that is,

$$h(x_2,x_2) = f_1(x_1)f_2(x_2) \tag{8-13}$$

where f_1 and f_2 are the univariate density functions (also termed the *marginal density functions*) of the two random variables, and h is the bivariate density function. The latter function gives the joint probability density of observing the pair (x_1,x_2).

If the relationship defined in equation 8-13 does not strictly hold, there exists some correlation between the random variables. For example, in the case of a bivariate normal distribution, where each of the two variables is given by normal marginal density functions, the ordinary Pearson product moment correlation coefficient, ρ, completely specifies the interaction between the two random variables. The product moment correlation coefficient may be specified independent of the parameters of the two normal marginals (i.e., their means and standard deviations).

Unfortunately, although the use of the product moment correlation coefficient is the most natural way to describe the relationship between two normally distributed random variables, it is mathematically inconvenient when one or both of the related variables are distributed other than by a normal distribution. An alternative approach is the use of a rank correlation coefficient (or Spearman coefficient), for which we use the symbol τ.

Concept of Rank Correlation. The rank correlation may be illustrated most readily by a numerical example. Consider the sets of observations given in Table 8-5. The first two columns give the raw observation values. If $x_{i,1}$, and $x_{i,2}$ are the values of the two variables for observation i (i.e., row i in the table), the ordinary correlation coefficient can be computed as

$$\rho = \frac{\sum\limits_{i=1}^{k} (x_{i,1} - \overline{x_{.,1}})(x_{i,2} - \overline{x_{.,2}})}{\sqrt{\sum\limits_{i=1}^{k} (x_{i,1} - \overline{x_{.,1}})^2 \sum\limits_{i=1}^{k} (x_{i,2} - \overline{x_{.,2}})^2}} \tag{8-14}$$

where $\overline{x_{.,1}}$ and $\overline{x_{.,2}}$ are the arithmetic means of the x_1 and x_2 variables, respectively.[8]

---

[7] The nomenclature in this section is based on continuous distributions. An analogous exposition for discrete distributions is also possible.

[8] In the EXCEL environment, the function PEARSON( ), which takes two array arguments, can be used to obtain this correlation coefficient directly.

TABLE 8-5 Hypothetical Data Set (x_1 and x_2) with Ranked Values

x_1	x_2	r_1	r_2
7.17	17.65	8	10
0.73	1.73	1	1
3.48	4.36	3	4
4.37	10.03	6	9
1.09	2.26	2	2
11.79	7.01	10	8
4.43	4.39	7	5
4.10	5.40	5	6
3.60	3.87	4	3
10.12	6.73	9	7

To obtain the rank correlations, the numerical observations are replaced by their ranks with respect to that variable. If two (or more observations) are *tied*, they are each assigned the average of the ranks that that group of observations would have been assigned were they not tied.[9] For example, in the third row, the value of 0.73 (for x_1) is the smallest observation among the set of $x_{i,1}$ values, so it is assigned the rank of 1. The observation 11.79 for x_1 is the largest among the set of $x_{i,1}$ values, so it is assigned the rank of 10 (since there are 10 observations). The observation 4.36 for x_2 is the fourth in order among the $x_{i,2}$ values, so it is assigned the rank of 4; and so on. From the ranks, the Spearman correlation may be obtained simply by using the ranks in place of the raw observations, to obtain

$$\tau = \frac{\sum_{i=1}^{k} (r_{i,1} - \overline{r_{.,1}})(r_{i,2} - \overline{r_{.,2}})}{\sqrt{\left[\sum_{i=1}^{k} (r_{i,1} - \overline{r_{.,1}})^2\right]\left[\sum_{i=1}^{k} (r_{i,2} - \overline{r_{.,2}})^2\right]}} \tag{8-15}$$

This may, however, be dramatically simplified [since for k observations the mean rank will always be equal to $(k + 1)/2$] to the following (43):

$$\tau = 1 - \frac{6 \sum_{i=1}^{k} (r_{i,1} - r_{i,2})^2}{k(k^2 - 1)} \tag{8-16}$$

[9] Unfortunately, although there is a RANK function in EXCEL, it does not handle ties according to this rule, so it is not acceptable for use directly in computing the rank correlation coefficient in the presence of ties.

In general, the larger the absolute value of the rank correlation, the greater the dependency between the paired random variables.

Example 8-4. For the data in Table 8-5, compute the Pearson and the Spearman correlations.

SOLUTION. Application of the PEARSON function in EXCEL yields a value of ρ equal to 0.484. To evaluate the Spearman correlation, we first need to compute $\Sigma_{i=1}^{k}(r_{i,1} - r_{i,2})^2$. From Table 8-5, this is equal to 28.0. Since $k = 10$, using equation 8-16, the value of τ is computed to be 0.830.

Not all paired observations are of necessity correlated. For example, consider the data set (obtained from Table 8-5 by permuting the rows of observations of x_2) in Table 8-6. Computation of the Spearman correlation coefficient for the data in Table 8-6 yields a value of τ of 0.030.

The impact of the correlation on derived functions of random variables can be illustrated with reference to the two hypothetical data sets in Tables 8-5 and 8-6. Both data sets have identical marginal distributions for x_1 and x_2 and differ only with respect to the correlation between these variables. In other words, the column of x_1 in Table 8-6 has the same distribution as the column of x_1 in Table 8-6, and similarly for the columns of x_2. Consider a particular derived function consisting of the ratio $z_i = x_{i,1}/x_{i,2}$. For each of the two data sets, we compute the ratio and plot the empirical cumulative distribution for the correlated (Table 8-5) and relatively uncorrelated (Table 8-6) data. In addition, we can compute a ratio for all combinations of $x_{i,1}$ and $x_{i,2}$ (i.e., the $10 \times 10 = 100$ combinations). These plots are given in Figure 8-6.

The distribution of the derived function computed from the relatively uncorrelated data set (Table 8-6) is in close agreement with the distribution from

TABLE 8-6   Rearranged Hypothetical Data

x_1	x_2
7.17	17.65
0.73	10.03
3.48	6.73
4.37	5.40
1.09	1.73
11.79	2.26
4.43	4.36
4.10	4.39
3.60	3.87
10.12	7.01

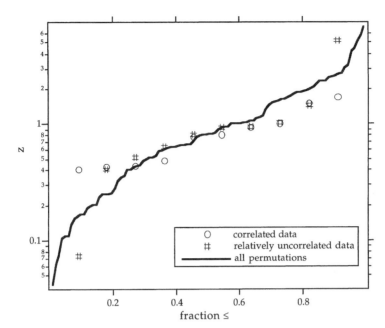

Figure 8-6 Distributions of derived functions from correlated (Table 8-5) and relatively uncorrelated (Table 8-6) data.

all permutations of the data (which is expected, since the latter computation is based on complete independence). However, there is a substantial difference, particularly at the upper and lower extremes, between the distribution of the derived function for correlated data and the other cases.

In general, the presence of any correlation between random variables affects the distribution of functions derived from such random variables. As suggested by the example in Figure 8-6, the magnitude of such effects frequently will become more important at the extremes of the distribution (and in cases where the distribution of the derived quantity is highly skewed, perhaps also in the arithmetic mean). Therefore, the potential presence of correlations between input quantities must be considered when performing a Monte Carlo risk assessment.

The effect of correlation upon the potential bias in estimation of the mean and potential inflation or deflation of the variance of functions of random variables has been studied by Smith et al. (52) in the particular case of lognormally distributed variables. In this case, analytical relationships can be used to assess the potential error in neglecting correlations. The relative impact of distributional shape on the estimation of the errors associated with neglecting correlation appears to be fairly minor (13). Based on this work, the proposal by Burmaster and Anderson (14) (Table 8-3, rule 9) to incorporate consideration of correlations into an analysis if a rank correlation with absolute value in excess of 0.6 is suspected seems like a reasonable rule of thumb.

Modeling Correlations. Now the problem occurs of how to quantitatively describe bivariate distributions in which correlations exist. Currently available software (@Risk, CrystalBall) has the capability to include rank correlations between nonindependent inputs. Unfortunately, the precise algorithms used are not well documented, nor is the mathematical structure for describing these relationships well explicated in the literature.

Essentially, the problem is modification of equation 8-13 in a manner to facilitate introduction of rank correlation as a parameter. The mathematical theory most appropriate to the problem is the theory of copulas. To develop this theory, we first note that we can integrate equation 8-13 from the origin to some point in the two-dimensional plane, which would yield the integrated bivariate cumulative distribution function (H) in terms of the marginal cumulative distribution functions F_1 and F_2:

$$H(x_1,x_2) = F_1(x_1)(F_2(x_2)) \tag{8-17}$$

H gives the probability that a random pair (x_1^0,x_2^0) would be in the region bounded by $\langle -\infty \cdots x_1, -\infty \cdots x_2\rangle$.

A convenient way to generalize the structure of the independent model is by the use of copulas, which are structures to generate bivarate distributions with fixed marginals (24). A copula is a bivariate probability model whose function H is expressed in terms of its marginals $F(x)$ and $G(y)$ (24). A copula is termed *Archimedean* if it can be expressed as $\psi(H) = \psi(x) + \psi(y)$ (34). For example, if $\psi(u) = \ln(u)$, application to equation 8-17 demonstrates that the independence relationship is Archimedean. Frank's copula is the only Archimedean copula that includes the complete range of admissible dependence described by the Frechet bounds (34):

$$H = -\frac{1}{\alpha} \ln\left[1 + \frac{(e^{-\alpha F_1} - 1)(e^{-\alpha F_2} - 1)}{e^{-\alpha} - 1} \right] \tag{8-18}$$

Note that the copula is written directly in terms of the marginal cumulative distribution functions. When α is equal to 0, there is no interaction between the two compounds that was earlier defined as independence, and it can be shown that by solving equation 8-19 for the limit as α approaches 0; the parameter α can assume any real value.

$$\lim_{\alpha \to 0} H = F_1 F_2 \tag{8-19}$$

By differentiating equation 8-18 once each with respect to the underlying variables (x_1 and x_2), the bivariate density function can be found and written in terms of the interaction parameter of Frank's copula (α) and the underlying density and cumultive distribution functions as follows:

$$h(x_1,x_2) = \frac{\alpha(1 - e^{-\alpha})e^{-\alpha(F_1(x_1)+F_2(x_2))}}{[e^{-\alpha} + e^{-\alpha(F_1(x_1)+F_2(x_2))} - e^{-\alpha F_1(x_1)} - e^{-\alpha F_1(x_1)}]^2} f_1(x_1)f_2(x_2)$$

(8-20)

Note that the first term on the right represents the contribution of the copula, while the remainder of equaiton 8-20 is identical to the independent formulation of equaiton 8-13.

The parameter α in the copula model can be directly related to the Spearman rank correlation coefficient. The formal evaluation is performed by numerically evaluating a double integral as written below (53):

$$\tau = 12 \int_0^1 \int_0^1 \frac{\partial^2 H}{\partial F_1 \, \partial F_2} \left(F_1 - \frac{1}{2}\right)\left(F_2 - \frac{1}{2}\right) dF_1 \, dF_2$$

(8-21)

In the case of Frank's copula, the derivative in equation 8-21 is given by the first term on the right in equation 8-20, that is,

$$\frac{\partial^2 H}{\partial F_1 \, \partial F_2} = \frac{\alpha(1 - e^{-\alpha})e^{-\alpha(F_1+F_3)}}{[e^{-\alpha} + e^{-\alpha(F_1+F_2)} - e^{-\alpha F_1} - e^{-\alpha F_1}]^2}$$

(8-22)

For Frank's copula, this relationship is evaluated and may be plotted as in Figure 8-7. As noted in the graph, the relationship is closely approximated by the following equation:

$$\frac{\tau}{\alpha} = 0.026633 + \frac{0.17732(1 - \tau)^{1.0262}}{0.22377 + (1 - \tau)^{1.0262}}$$

(8-23)

This graph is constructed for positive values of τ and α. For negative values, the same relationship can be used, except that the quantity $(1 - \tau)$ should be replaced by $(1 - |\tau|)$ (i.e., when the sign of τ is negative, the sign of α is also negative).

Example 8.5. Two random variables have a Spearman correlation of -0.7. Compute the corresponding value of Frank's α.

SOLUTION. Since the correlation is negative, we must apply equation 8-23 by first taking absolute values. So we first find

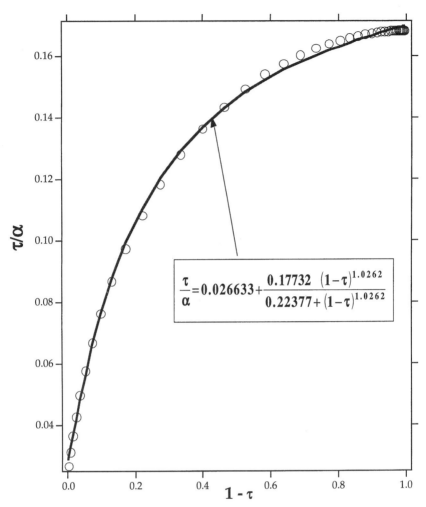

Figure 8-7 Relationship between Spearman correlation and the parameter in Frank's copula (for positive values). Points indicate exact computations of numerical integral. Curve shows the approximate analytical equation.

$$\frac{\tau}{\alpha} = 0.026633 + \frac{0.17732(1 - 0.7)^{1.0262}}{0.22377 + (1 - 0.7)^{1.0262}}$$

$$= 0.1268$$

Since the correlation is negative, the value of α is also negative. Therefore, we have

$$\alpha = \frac{-0.7}{0.1268} = -5.52$$

From the Spearman rank correlation, by using the Frank copula, we can make inferences about the proportion of a bivariate population expected to be present in a given range. This is particularly easy in the case of a rectangular region (on F_1 versus F_2 coordinates). The approach to the problem is illustrated by reference to Figure 8-8.

It is desired to evaluate the total probability of observing a bivariate pair of random variables in rectangle IV, which is bounded as follows:

$$u_1 \leq F_1 \leq u_2 \quad \text{and} \quad v_1 \leq F_2 \leq v_2$$

The cumulative copula function $H(u,v)$ directly provides the probability in a rectangle bounded by

$$0 \leq F_1 \leq u \quad \text{and} \quad 0 \leq F_2 \leq v$$

Application of this property in light of the definition sketch in Figure 8-8 permits an evaluation of region IV by addition and subtraction of combinations of regions. We consider four combinations that can be directly evaluated by application of the cumulative copula function (H):

$$A = I + II + III + IV$$

$$B = I + II$$

$$C = I + III$$

$$D = I$$

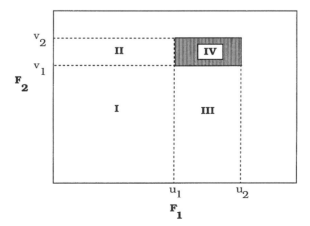

Figure 8-8   Definition sketch for evaluation of bivariate probability in a rectangular region. The probability in region IV is sought.

Using these four regions, whose probabilities in terms of the copula are given in Table 8-7, the probability in region IV can be evaluated as

$$IV = (A + D) - (B + C) \tag{8-24}$$

By substituting the functions in Table 8-7 into equation 8-23, the area of rectangle IV in Figure 8-8 can be written explicitly in terms of a copula as follows:

$$IV = [H(u_2,v_2) + H(u_1,v_1)] - [H(u_1,v_2) + H(u_2,v_2)] \tag{8-25}$$

Example 8-6.   A pair of bivariate variables (e.g., water consumption and microbial density) are correlated with a Spearman coefficient of 0.4. What is the probability that both variables will be observed simultaneously at greater than their individual 90th percentile if the correlation is neglected and the correlation is considered using Frank's copula?

SOLUTION. By virtue of equation 8-25, we evaluate area in a region bounded by 0.9 and 1.0 on each side. Hence the wanted probability (P) is given by

$$P = [H(1,1) + H(0.9,0.9)] - [H(0.9,1) + H(1,0.9)]$$

Note that regardless of the value of α, $H(1,1) = 1$, and $H(0.9,1) = H(1,0.9) = 0.9$ (which can be verified by substitution into equation 8-18). Therefore, we have

$$P = [1 + H(0.9,0.9) - [0.9 + 0.9]$$
$$= H(0.9,0.9) - 0.8$$

First assume independence. From equation 8-17, $H(0.9,0.9) = 0.9 \times 0.9 = 0.81$. Therefore, under independence, the requested probability is $0.81 - 0.8$, or 0.01.

TABLE 8-7   Areas of Regions in Figure 8-8 in Terms of Copulas

Region	Area
A = I + II + III + IV	$H(u_2,v_2)$
B = I + II	$H(u_1,v_2)$
C = I + III	$H(u_2,v_1)$
D = I	$H(u_1,v_1)$

Under the assumption of a Spearman correlation of 0.4, we must first compute the corresponding α value. Applying the relationship of equation 8-22, the corresponding value of α is 2.575. Now, with this value of α, the copula function may be computed (via equation 8-18) as

$$H(0.9,0.9) = 0.8117$$

Therefore, under this extent of correlation, the probability of joint exceedance of the 90th percentile is $0.8117 - 0.8 = 0.0117$, which is only about 12% greater than that under an assumption of independence.

Given the characterization of the correlation among variables distributed according to a bivariate distribution, the associated distribution of a derived quantity can be determined by Monte Carlo methods, provided that a random number generator capable of providing correlated variables is available. This is illustrated by the following example.

Example 8-7. The consumption of drinking water by a certain population is lognormally distributed with a mean of 0.8 L/day and a standard deviation of 0.9 L/day. The water has a density of a particular pathogens that is gamma distributed with mean of 0.5 L^{-1} and a standard deviation of 1 L^{-1}. The water consumption and microbial densities are bivariately distributed with a Spearman correlation of -0.7. Assuming that the number of organisms is Poisson distributed, compute the distribution of number of organisms consumed. (Note: This example requires one of the packages for Monte Carlo analysis.)

SOLUTION. Let V be the volume consumed and N be the density consumed. Then the applicable Poisson average (from which we compute a random number of actual organisms consumed) has a mean of NV. Therefore, the Monte Carlo approach requires the computation of a random pair of $\langle N,V \rangle$, then insertion of the product into a random number generator for the Poisson distribution. For the gamma distribution, the parameters are computed (from the mean and standard deviation) as

$$\alpha = 0.25$$

$$\beta = 2$$

From 5000 trials (using CrystalBall), the distribution of organisms per portion shown in Table 8-8 is obtained. For comparison, the last column in the table gives the distribution of counts if the correlation (between water intake and microbial densities) was set at zero (i.e., the distributions are assumed to be independent). Qualitatively, note that the presence of the (negative) correlation substantially diminishes the likelihood of ingesting larger numbers of organisms compared to an independent assumption.

TABLE 8-8   Results for Example 8-7

Organisms Consumed	Number of Trials (of 5000)	
	Correlated Inputs	Uncorrelated Inputs
0	3907	3668
1	886	929
2	171	259
3	32	80
4	3	30
5	1	13
6		5
7		7
8		1
9		4
10		1
11		2
12		1

The use of a copula framework (or similar approach) for the description of correlations between variables is an attractive and flexible one for use in Monte Carlo computations. The discussion above has focused on the use of Frank's copula, since it is a one-parameter model that can describe the complete range of correlations (from a Spearman coefficient of -1 to a Spearman coefficient of $+1$). Unfortunately, it is not the only copula with this property (24,25), and there are undoubtedly an infinite number of such copulas (many with more than one parameter). For example, Frank (personal communication) has described a two-parameter generalization of which the one-parameter Frank copula is a special case. In addition, Joe (37) has outlined several two-parameter families that include the entire range of correlations as special cases. Additional work needs to be performed to assess the magnitude of effect of assuming particular forms for the bivariate dependence; that is, for a given correlation coefficient and marginal distributions, what are the differences in the bivariate distributions assuming different copulas?

We note in passing that while a number of risk analysis software packages implement bi- and multivariate rank correlations in their random number generators, the algorithms employed are heuristic in nature and not well documented. The primary source of such procedures is the empirical procedure of Iman and Conover (35).

Considering Model Uncertainty or Uncertainty of Distributional Form for Variability

As noted earlier in this chapter (e.g., in Figure 8-2), the inputs to a risk assessment assume distributions due to their possessing elements of variability (intrinsic, irreducible elements that describe actual heterogeneity) or to their possessing elements of uncertainty (elements that result from imperfect knowledge, which can be reduced by further data-gathering efforts). A special type of uncertainty is the uncertainty due to model form or distributional form.

When inputs are uncertain due to model form,[10], this represents a unique type of uncertainty above and beyond parameter uncertainty (see Figure 8-2). There are several approaches to this problem. One approach is to run simulations with each of the particular model alternatives selected as the choice and then to compare the particular approach. A second approach is to assign a *degree of belief* to each of the alternative models such that the sum of the degree of beliefs equals 1. Using this approach, the result is similar to that of an ordinary mixture, or compound distribution, and is amenable to a Monte Carlo approach.

Consider an input variable (x) and an output variable (y) which are related to each other by a equation involving some additional parameters (denoted by a vector $\bar{\beta}$). We have two choices for the deterministic relationship, denoted by the functions f_1 and f_2. In other words,

$$y = f_1(x,\bar{\beta}) \tag{8-26}$$

or

$$y = f_2(x,\bar{\beta}) \tag{8-27}$$

For these two relationships, we have a degree of belief α that the first model (f_1) is applicable, and $1 - \alpha$ that the second model (f_2) is applicable. Then on a particular trial of a Monte Carlo simulation, we select a numerical value for the input x, and if specified by the problem we select a numerical value for the parameter vector. Then a random fraction (between 0 and 1) is obtained from a uniform distribution. If this random fraction is less than or equal to α, equation 8-26 is used to compute the output, y; otherwise equation 8-27 is used.

The preceding paragraph is cast in terms of a deterministic relationship between an input and an output. However, if the model uncertainty is cast in terms of distributional form, equations 8-26 and 8-27 can be viewed as processes for selecting a random variable for y, and the random fraction is used (by comparison with the degree of belief α) to select among alternatives. Note that whether used for a pure deterministic relationship or as a selection rule between distributional alternatives, this approach can readily be generalized to three or more competing alternatives.

---

[10] The term *model form* is used here in a highly general sense to encompass both a choice in a deterministic relationship between an input and an output variable (e.g., between a first- and second-order kinetic equation for microbial die-off), and a choice between a particular distributional form used to describe the nature of variability or uncertainty itself, as for example between an inverse Gaussian versus a lognormal distribution for describing water consumption.

Considering Uncertainty in Dose–Response Parameters

The examples (using Monte Carlo) so far have concentrated on assessing distributions of exposures. However, for a full risk characterization, we need to couple the variability and uncertainty in the exposure assessment with the uncertainty (and possible variability) in the dose–response assessment. Therefore, using the methods of Chapter 7, the distribution of dose–response parameters (which for two or more parameter models may intrinsically be correlated) must be developed as one of the *inputs* to a full risk characterization.

One of the most readily applied methods for obtaining the distribution of dose–response parameters is the bootstrap approach described in Chapter 7. Using this approach, three methods of coupling the uncertainty in dose–response information with the exposure information may be envisioned:

1. For each iteration of a Monte Carlo computation at the risk characterization step, a bootstrap sample of the dose–response experiment is constructed and a set of dose–response parameters obtained for use in *that particular iteration.*

2. A large number of bootstrap samples of the dose–response curve is used to generate a large number of sets of dose–response parameters. For each iteration of a Monte Carlo computation, a single random point (consisting of a set of dose–response parameters) is selected.

3. From a large number of bootstrap samples, a bivariate distribution is fit to the parameter distribution (for a dose–response relationship with two data points; for a single-parameter dose–response relationship, such as the exponential, a univariate distribution would suffice). Then, from this information, at each Monte Carlo iteration, a random "point" would be obtained (from an algorithm to generate a random vector from a bivariate distribution).

Methods 1 and 2 are expected to be equivalent as long as the number of bootstrap replicates used in method 2 is sufficiently large (some computational guidance needs to be developed here). Method 3 depends on the ability to find a good fit to the empirical bootstrap distribution, which may be tedious if at all possible. In the computations in this chapter, method 2 will be employed.[11]

---

[11] Although it is possible to perform the bootstrap computation in a spreadsheet environment, the execution times are generally slower than when the optimization is conducted in a mathematical programming environment such as MATLAB. Hence, from an implementation point of view, at present our preference is to perform the bootstrap computations, which require a nonlinear optimization in MATLAB and then to export the resultant points, representing the bootstrap estimate of the bivariate dose-response parameter distribution to the EXCEL spreadsheet environment for the full risk characterization. Alternatively, the full risk characterization can be conducted in the EXCEL environment (although this lacks some of the more polished postprocessing capability of the more dedicated-purpose risk packages).

The results of a full risk characterization, using single individual risk from a single exposure as an endpoint, the problems posed in the *Salmonella* "chicken" problem (for which the distributional assumptions with respect to exposure were outlined in Table 8-4) will be extended by combination with dose–response information. To do this it is necessary to have dose–response information for *Salmonella* .

The dose–response information for nontyphoid *Salmonella* from oral exposure has been reviewed by Fazil (21). This is reviewed in greater detail in Chapter 9. However, the analysis suggests that the dose–response relationship from a diversity of strains may be coupled into a single beta-Poisson dose–response curve. For a total of 46 data points, the maximum likelihood estimates of the beta-Poisson dose–response parameters are $\alpha = 0.294$ and $N_{50} = 16,211$. The goodness-of-fit statistic (residual deviance) is 51.52, which yields a p value of 0.26, and hence the fit is regarded as acceptable.

From the available dose–response data, the bootstrap sampling distribution was obtained (1000 bootstrap iterations) using the methods outlined in Chapter 7. Figure 8-9 provides a scatter plot of the empirical bootstrap sampling distribution. Note the great tailing toward the lower left of the graph. Also, there is a clear correlation between the two dose–response parameters, which shows the need to consider this as a bivariate input. The Spearman rank correlation coefficient for the bootstrap sample in Figure 8-9 is 0.746, which is clearly large enough to necessitate consideration (by virtue of the Burmaster

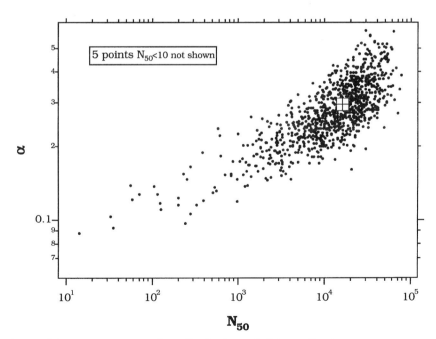

Figure 8-9 Bootstrap sampling distribution of *Salmonella* dose–response curves (1000 bootstrap iterations). Central crosshairs indicate maximum likelihood values.

and Anderson rules). Using as an input for the risk characterization the boot-strap distribution of dose–response parameters in Figure 8-9, a risk characterization for the scenario using the distributions for the exposure in Table 8-4 was conducted.

The summary statistics for the computed risk are given in Table 8-9. The mean of the computed risk (0.0219) is substantially greater than the median (1.86×10^{-3}), indicating a highly skewed distribution. Figure 8-10 provides the cumulative and complementary cumulative distribution curves for the simulation. On the basis of this we might decide to convey the results in a form such as "the estimated 95% confidence region to the computed risk is between 9.30×10^{-5} and 0.24." Note the extreme breadth of the confidence region, which is not atypical for many (chemical as well as microbial) risks. Given this great range in the computed risk, it is desirable to assess the potential sources of heterogeneity in the input parameters. This can be done by examining the rank correlations between the input values (among the individual Monte Carlo trials) and the outputs. These correlations are depicted in Figure 8-11.

Examination of the correlations between inputs and the output (Figure 8-11) shows that the dominant factors responsible for the spread in the output risk distribution are, in order, the removal due to cooking, the initial density, and the dose–response parameters. Comparing the information in Figure 8-11 to that in Figure 8-5 shows that the incorporation of the dose–response uncertainty reduces the magnitude of the correlation involving the exposure varies. Nevertheless, the exposure variability (particularly the variability in removal due to cooking and the variability in initial density) remains dominant with respect to the uncertainty associated with the dose–response information. This finding suggests to the risk manager that for the reduction in risk, it would be of prime importance to concentrate efforts on the removal due to cooking. Perhaps educational intervention or use of adequate temperature conditions would provide a trimming of the distribution for removal due to cooking (preferably with a reduction among those who practice less than desirable time–temperature cooking practices). Note, however, that the correlation with respect to dose–response parameters is still reasonably high, suggesting that the acquisition of new dose–response information might enable a more pre-

TABLE 8-9 Summary Statistics for *Salmonella* Risk Characterization

Statistic	Value
Mean	0.0219
Median	0.00186
Standard deviation	0.0734
Range minimum	8.79×10^{-8}
Range maximum	0.696

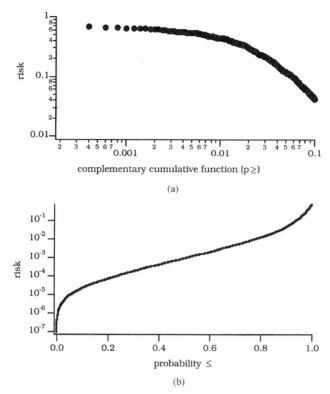

Figure 8-10 Cumulative and complementary cumulative (*a*) distributions for the estimated risk (*b*) in the *Salmonella* case study.

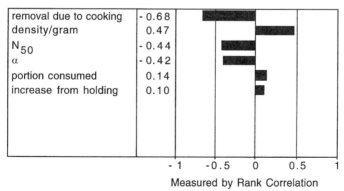

Figure 8-11 Sensitivity chart for *Salmonella* case study.

cise evaluation of risk. It would not appear to be justified to expend additional resources in consideration of regrowth during holding periods or portions consumed.

Presentation of Results: Two-Dimensional Random Variables

The analysis above lumps variability and uncertainty into a pooled cause of heterogeneity in the risk distribution. To some degree this may be undesirable since the two factors arise from different causes: intrinsic factors (for variability) versus factors that are, in principle, reducible given additional information. Hence some workers have proposed the distinct separation of variability and uncertainty in presentation of information (15). This requires a two-dimensional treatment and a two-dimensional presentation of data.

The inputs to a risk estimate are classified into a set of variables (model parameters, etc.) characterized as uncertain (typically, dose–response parameters, for example) and another set of variables characterized as descriptive of variability (intake levels, for example). These are denoted as Θ_U and Θ_V, respectively. Then vectors of random samples of each of these sets of variables are drawn, for example

$$\begin{bmatrix} \Theta_V^1 \\ \Theta_V^2 \\ \vdots \\ \Theta_V^m \end{bmatrix} \qquad \begin{bmatrix} \Theta_U^1 \\ \Theta_U^2 \\ \vdots \\ \Theta_U^n \end{bmatrix}$$

where the superscript indicates the particular random sample (and there are m random draws from variable inputs and n random draws from uncertain inputs). For each combination, the risk, which can be denoted generically as $f(\Theta_U, \Theta_V)$, is computed. This yields a matrix, such as

$$\begin{bmatrix} f(\Theta_U^1, \Theta_V^1) & f(\Theta_U^2, \Theta_V^1) & \cdots & f(\Theta_U^n, \Theta_V^1) \\ f(\Theta_U^1, \Theta_V^2) & & \ddots & \\ \vdots & & & \\ f(\Theta_U^1, \Theta_V^m) & & \cdots & f(\Theta_U^n, \Theta_V^m) \end{bmatrix}$$

Now each column of the matrix above is sorted independently. Therefore, the first row gives the risks that are the lowest (among m variability trials), the second row gives the second lowest, and so on. For each row, the median and various percentiles (in that row) can be computed. This leads to a band, such as in Figure 8-12 (for the *Salmonella* example) in which the distribution with respect to variability is illustrated along the x-axis, and the bands give the distribution with respect to uncertainty. This computation is computationally (particularly memory) intensive, since a large number of iterations in each dimension requires storage of large intermediate matrices. For example, to generate Figure 8-12, the number of iterations in uncertainty (m) was 300 and

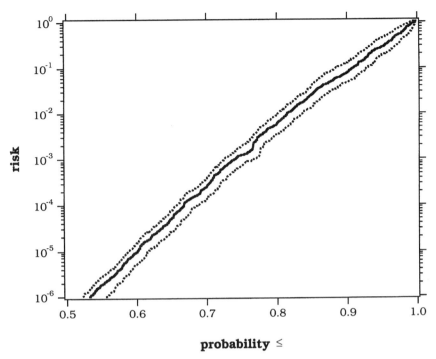

Figure 8-12 Two-dimensional Monte Carlo presentation of *Salmonella* case study. The *x* = axis is the variability dimension. The lines provide ranges based on uncertainty [the upper (97.5) and lower (2.5) percentiles are shown by the dotted lines].

the number of iterations in variability (*n*) was 1000, requiring intermediate storage and manipulation of matrices containing 300,000 elements. This was accomplished in MATLAB. To the authors' knowledge, at present, there is no simple software (such as @Risk or CrystalBall) that permits this two-dimensional analysis.

Examination of Figure 8-12 confirms the sensitivity information obtained by rank correlation (in Figure 8-11 that the components of variability (initial density, reduction from cooking, regrowth, and portion consumed) dominate the components due to uncertainty (dose–response). The range of risk between the median and the upper 97.5 percentile is several orders of magnitude (obtained by looking at the intercept of the central curve at the 0.5 and 0.975 probability). However, at the median variability (*p* = 0.5), looking at the range of risk that encompasses 95% of the uncertainty (the distance between the dashed curves), the latter range is much less than one order of magnitude. Hence the idea that the influence of uncertainty is relatively small for this particular problem, is confirmed.

Figure 8-13 illustrates the type of situation that may occur if the role of uncertainty is relatively insignificant (in the left panel) and if the role of

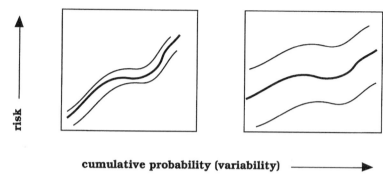

risk

cumulative probability (variability) ⟶

Figure 8-13   Schematic two-dimensional Monte Carlo plots indicating results with low (left) and high (right) importance of uncertainty.

uncertainty is relatively significant (right panel) vis-à-vis variability. In the former case, the slope is relatively large compared to the range encompassed by the uncertainty bands, while in the latter case, the slope is relatively small compared to the uncertainty range.

RISKS TO POPULATIONS: MODELS FOR POPULATION AND COMMUNITY ILLNESSES

To this point, the focus on risk characterization has been on the use of single-person/single-exposure risk (primarily of infection) as an endpoint. In certain circumstances, the use of other endpoints may be desirable. By incorporating information on the morbidity or mortality ratios, or their possible dose dependencies, other endpoints (to single-person/single-exposure risks) might be used. However, in particular cases it may be desirable to examine in detail the peculiarities of dynamics of occurrence of illness in *communities* rather than simply individuals. For example, in situations of outbreaks, which represent an unusually high clustering of illness cases, the cost of response, investigation, and remediation may be much greater on a per case basis than the ordinary course associated with non-outbreak-associated illness (e.g., there is a substantial social cost, as well as real economic cost in the imposition of a "boil water" order for a community water supply). This requires that the detailed dynamics of disease spread in a community be assessed.

There are three basic items of information that must be understood in the description of community-level infectious disease dynamics:

1. The incubation time of the infectious agent, (i.e., the duration between initial ingestion or exposure and the onset of symptoms)
2. The duration of the disease state as well as of the carrier state (the latter being defined as the period in which individuals are capable of inducing secondary cases)

3. The rate at which secondary cases occur from direct or indirect contact with primary cases (or with persons in an asymptomatic carrier state)

The basic aspects of these three processes are reviewed below, together with their synthesis into a community-level model.

Incubation Period

Sartwell (51) made the first systematic analysis of incubation time distributions for common source infectious diseases. From studying the time between the exposure (to an infectious agent) in a well-defined outbreak (one resulting from a relatively short duration exposure at a known time, such as a common meal) and the onset of symptoms, he deduced that the incubation time distribution could frequently be approximated as a lognormal distribution (although he did not check the applicability of alternative distributions, and his fitting procedure was qualitative and graphical in nature). The information for several agents (of particular concern for food and water) from the analysis of Sartwell is abstracted in Table 8-10.

There is a wide variation between the median incubation times that characterize diverse agents, as suggested by this table. Furthermore, as suggested by the dispersion factor, (e.g., in the case of *Salmonella typhimurium*), although the median incubation time is 2.4 days, it would be expected that 5% of the cases would begin only after 4.5 days postexposure, there is substantial heterogeneity among the exposed individuals as to duration of the incubation period.

There has been interest in describing the incubation time distribution from mechanistic considerations. The general framework is that a microorganism, once ingested, will be subject to growth and decay processes within the host. Disease occurs upon the buildup of the internal body burden of organisms above some critical level. Hence the distribution of incubation times is a

TABLE 8-10   Summary of Incubation Time Distribution Information Reported by Sartwell (51)

Infectious Agent	Original Report	Nature of Outbreak	Median Incubation Time (days)	Dispersion Factor[a]
Salmonella typhimurium	47	Foodborne	2.4	1.47
Typhoid fever	49	Foodborne	13.3	1.58
Amebic dysentery	2	Waterborne	21.4	2.11

[a]Defined as the ratio of the median incubation time to the lower 16th percentile of incubation time [i.e., ln(2 × dispersion factor)] equals the standard deviation of logarithms for a lognormal distribution.

reflection of the distribution of times to increase from the initial inoculum to the threshold level for disease (3,44).

Williams (57,58) has formulated mathematical models for analysis of this problem. If k is the death rate (1/time) of organisms in vivo, μ is the birth rate (1/time) in vivo, and N is the number of organisms at any instant of time, the following transition rules can be used to determine the transition from N to either $N - 1$ or $N + 1$ organisms[12]:

- If N is zero, it cannot change (i.e., once organisms are lost from the host, they cannot reappear—unless another exposure occurs). Therefore, the time to an infection here is infinite.
- Otherwise, at an exponentially distributed time interval, with expected value equal to $1/N(k + \mu)$, one of the following transitions occurs:

$$N \rightarrow N + 1 \text{ with probability } \frac{\mu}{k + \mu}$$

$$N \rightarrow N - 1 \text{ with probability } \frac{k}{k + \mu}$$

Note also that the initial number of organisms in a particular host, N_0, is a Poisson-distributed random variable (where the mean dose is specified by the exposure). By running this model for a large number of "hosts," the distribution of times required for the body burden of pathogens to exceed a particular threshold can be ascertained.

The model presented by Williams essentially predicts both the incubation time and the probability of ultimate infection (i.e., that the time to exceed the threshold is finite). The dose–response model implied by the Williams formulation is exponential; that is, the probability of infection given a mean initial dose among a collection of hosts, $\overline{N_0}$, is given by

$$\pi = 1 - \exp\left[-\overline{N_0} \left(1 - \frac{k}{\mu} \right) \right] \tag{8-28}$$

There has not been much work coupling incubation models with dose–response models; however, by positing some between-host variability to the ratio of growth rate to death rate, it may be possible to derive a beta-Poisson model from this formulation.

In any event, using the transition rules specified above, an incubation time distribution can be computed by numerical simulation. An example of this is shown in Figure 8-14, where five trajectories (representing five hosts) are shown. These receive organisms with a Poisson dose distribution and mean

---

[12] We have changed the symbols used here for birth and death rates from the original papers by Williams to make them more consistent with general practices in microbial growth modeling.

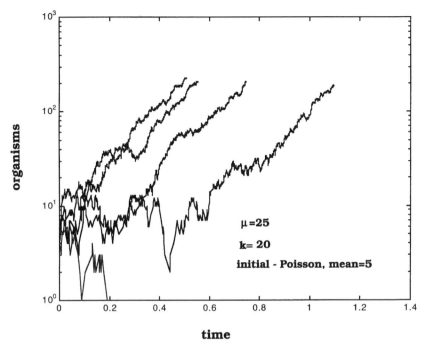

Figure 8-14   Sample trajectories for microbial dynamics in vivo (five trajectories illustrated).

dose of 5 organisms. The growth and death rates are assumed as 25 and 20, respectively. Note that one of the four trajectories reflects organism disappearance from the host, while the other four (at least over the time interval computed) show increasing numbers. If the critical threshold was specified, a horizontal line intersecting the curves would provide points on the incubation time distribution (this particular simulation was stopped when all trajectories either went to zero or intersected a threshold of 20 organisms).

By repeating this with a large number of trajectories, the distribution of incubation times can be obtained numerically. Figure 8-15 shows the results of 100 simulations (36 of the 100 trials resulted in microbial extinction). Note that the distribution of incubation times is highly skewed. Considering the empirical finding of Sartwell (51) with respect to fitting the lognormal distribution to incubation time data, this is not surprising. In fact, the histogram in Figure 8-15 fits a lognormal distribution quite well.

Williams derived asymptotic forms for the incubation time distribution under the condition that the threshold for response (N_c) is sufficiently large. First the incubation time is transformed into a dimensionless form (τ):

$$\tau = (\mu - \lambda)t - \ln\left(N_c \frac{\mu - \lambda}{\mu}\right) \tag{8-29}$$

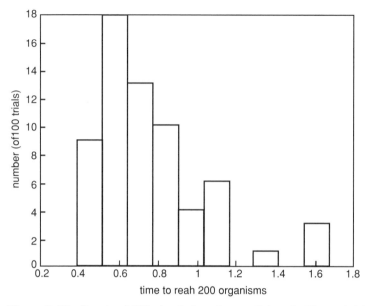

Figure 8-15   Result of 100 simulations for conditions in Figure 8-14.

The probability density function for τ is then given by (note that the range is over the entire real line)

$$f(\tau) = \frac{\sqrt{\delta}}{\exp(\delta) - 1} \left\{ \exp\left[-\frac{1}{2}\tau - \exp(-\tau) \right] \right\} I_1[2\,\exp(-\tau)\sqrt{\delta}]$$

$$\delta = \overline{N_0}\left(1 - \frac{\lambda}{\mu} \right)$$
(8-30)

where $I_1(z)$ is the modified Bessel function of the first kind of order 1 and argument z.

Unfortunately, it does not appear that the work of Williams has received substantial development, nor has the distribution defined by equation 8-30 been studied. However, the shape of this distribution is similar to the lognormal distribution (and also again note Figure 8-15, which has a quite similar lognormal shape), so the general theoretical underpinning is supported by the findings of Sartwell.

The development by Williams led to a general principle that at moderate doses, there should be an inverse logarithmic relationship between dose and mean (or median) incubation time. In other words, at lower doses, the incubation time lengthens (as should the variance of the incubation time) as the log of dose decreases. Although there may be quantitative differences between the simple theory of Williams and observations (3), which may be reconciled by positing time dependency and density dependency to the growth and death parameters (μ and λ), a number of studies have shown qualitative agreement.

For example, Ercolani (19) found a good inverse logarithmic relationship for the infection of bean leaves by the plant pathogen *Pseudomonas syringae*. The growth and death parameters imputed by fitting time to infection data were similar (although not quantitatively identical) to measured growth kinetics from count data.

In studying an outbreak of foodborne hepatitis A, Istre and Hopkins (36), showed that the incubation period was inversely related to the amount of contaminated food that the infected individuals recalled having eaten during exposure at a contaminated meal. While their data were presented on a linear basis, examination of it indicates that a log-linear relationship could also describe the findings.

Hence, in describing community-level spread of infectious disease from a source of primary contamination, the mean incubation time should be regarded as a function of the dose consumed. Thus, in an actual exposed population, in which the dose consumed is itself heterogeneous, the apparent incubation time distribution results from compounding an intrinsic (but dose-dependent) incubation time distribution (such as given by equation 8-28) with the exposure distribution itself. As an empirical modeling approach, therefore, the use of a continuous distribution on the positive line, such as the lognormal, inverse Gaussian, and so on, to describe the apparent outbreak-specific incubation time distribution seems like a reasonable approach.

Incubation time distributions describe the fraction of persons who first become ill at a given time after the initial exposure. However, we first need to define the number of individuals who are on a route to become infected. In a given population at risk, N will be the time-dependent number of persons who are potentially at risk. The parameter $\beta(t)$ is defined as the instantaneous rate of infection at time t. In other words, $N(t)\beta(t)$ is the instantaneous rate of persons entering the "pool" of persons who will ultimately become infected. Immediately, then, we can write a differential equation with respect to N as

$$\frac{dN}{dt} = -\beta N \tag{8-31}$$

The general incubation distribution $F(t)$ is defined as the cumulative fraction of persons who become ill on or before t days after sustaining an exposure, among those who will become ill. This can be differentiated to define the density function of incubation times, $f(t)$. Then the instantaneous rate of illnesses at time t can be written as the convolution

$$Q(t) = \int_0^t \lambda\beta(\tau)N(\tau)f(t - \tau)\, d\tau \tag{8-32}$$

The constant λ is the fraction of infected individuals who become ill, that is, a morbidity ratio (and note that $1 - \lambda$ is the fraction of persons who are infected not becoming ill). This equation is derived by considering that the quantity $\beta(\tau)N(\tau)$ represents the instantaneous rate (number/time) of newly infected persons who enter the infected state at time τ. Of these, a fraction λ will ultimately become ill. These individuals will then enter the diseased state with a frequency distribution $f(t - \tau)$, where t is the time of entry into the diseased state. Hence to determine the total number who enter at time t, it is necessary to integrate over those who entered the pool of infected individuals at all prior times. For the simple case of an instantaneous exposure at time 0, this integral recovers the result that the rate of new illnesses is equal to a constant multiplied by the incubation time itself.[13] In equation 8-32, $N(\tau)$ is the number of persons at time τ who are exposed to the risk of the infectious agent. From these basic concepts, we can formulate the first simple model of community illness (in the following, X is the number of asymptomatic infected individuals, and Y is the number of symptomatic individuals at any one time) as follows:

$$\frac{dN}{dt} = -\beta N$$

$$\frac{dI}{dt} = \beta N - \frac{Q(t)}{\lambda}$$

$$\frac{dX}{dt} = \frac{1 - \lambda}{\lambda} Q(t) \tag{8-33}$$

$$\frac{dY}{dt} = Q(t)$$

This model allows us to estimate the rate at which new cases appear (which is given, by dY/dt) as a function of the instantaneous rate of infection, the proportion of ill persons, and the incubation time distribution. This *does not* allow us to predict the total number of ill or asymptomatically infected individuals in the population at any time, since we have not yet incorporated description of the duration of illness or residence in the asymptomatically infected state.

As a simple example, consider an incident in which exposure occurs over a 1-hour period (e.g., a single meal). The instantaneous rate of infection can then be modeled as a square wave (assuming that the exposed persons ate uniformly over the 1 hour) as follows (assuming a particular microbial level and thus transformed into an instantaneous rate of infection):

---

[13] Since under these conditions, there will be an instantaneous jump in I from the initial state (presumably zero) to its maximum value.

$$0 < t < 1 \text{ h} \qquad \beta = 0.02 \text{ h}^{-1} \qquad \qquad (8\text{-}34)$$

$$t > 1 \text{ h} \qquad \beta = 0$$

For example, if the infectious organism had the characteristics of *Salmonella typhimurium*, the incubation curve might be lognormal with a median of 2.4 (57.6 hours) with a 16th percentile of $57.6/1.47 = 39$ hours based on the analysis by Sartwell (Table 8-10). Using the nomenclature of Chapter 6, the parameters of the lognormal distribution would be (with time in hours) $\zeta = \ln (57.6) = 4.05$ (computed from the median) and $\delta = 0.388$ (computed from the lower percentile).

Using these lognormal parameters, equation 8-33 can be integrated numerically as a function of time. For simplicity, we will take N to be 1000 persons; in other words, 100 persons have consumed the microbially laden meal. Integration of these equations, which represent an integrodifferential equation system, with initial values given (e.g., $N = 1000$ at $t = 0$ and $I = X = Y = 0$ at $t = 0$) must be performed numerically. For the examples of epidemic models in this chapter, we have used the differential equation solvers in MATLAB and modified the program to evaluate the integral in equation 8-32 during each call of the integrator by the trapezoidal rule.

For the assumptions above, this gives the results in Figure 8-16 for the instantaneous rate of cases occurring for the total number of infected persons at any one time and for the cumulative number of cases. Note that the number of total cumulative cases is simply the integral of the product of $\beta(t)$, λ and $N(t)$ with respect to time; and in the example, this is simply equal to $1000 \times 0.02 \times 1 \times 0.5 = 10$.

Clearly, complex shapes for the disease attack rate can be derived if, for example, the infection rate is complex. Consider the preceding scenario with a slight change: Rather than a single 1-hour exposure from 0 to 1 hour, there are two 1-hour exposures, at 0 to 1 and 24 to 25 hours (e.g, a contaminated meal on each of two consecutive days). Figure 8-17 presents the results of this scenario. Comparison with Figure 8-16 shows that the curve of attack rate is displaced somewhat to the right and is also broadened. The latter finding reinforces the point made earlier, namely that the breadth (or variance) of the incubation time distribution observed in an outbreak is influenced by both the intrinsic multiplication of pathogens in vivo and the exposure of the susceptible population to the pathogens. Hence, to determine the intrinsic incubation time distribution from epidemiological data, it is necessary to remove the effect of the exposure distribution. This "inverse" problem is considerably more difficult but is, nonetheless, solvable.

Relationship Between Attack Rate and Dose–Response. The formulation of the model for incubation time contains the central function $\beta(t)$, which we would presume to be related to the exposure levels of organisms (concentration and amount of material ingested) as well as the dose–response

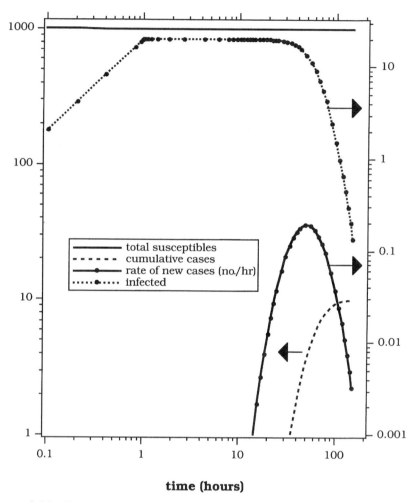

Figure 8-16   Time course of simple outbreak (single 1-hour exposure at $t = 0$–1 hour).

relationship. However, there does not appear to have been a consideration of the quantitative and formal relationship, which we shall now endeavor to present. Unfortunately, as will be noted shortly, there are two distinct ways to consider the problem, which except in the case of exponential dose–response at a constant dose rate led to two functional relationships. Hence further work, which would involve experimental administration of different time courses of pathogens to subjects, is needed to resolve this issue.

The point of departure for the development of the relationship is the use of either equation 8-2 or 8-3. Exposure is represented by the integral of a dose rate d' (organisms per unit time) and time, as follows:

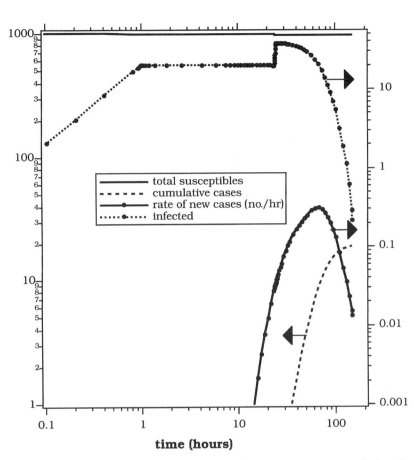

Figure 8-17   Time course of outbreak (two 1-hour exposures at $t = 0$–1 and $t = 24$–25 hours).

$$d = \int_0^t d'\, dt \qquad (8\text{-}35)$$

If the dose rate is constant, the integral simply becomes $d't$; otherwise, it must be evaluated numerically (or if d' is analytic integrable function, analytically). We consider the response from equations 8-2 and 8-3 to two doses, d and $d + \Delta d$. This is then taken to the limit, and from 8-35, $dd = d'\, dt$ is used to get the function β, which is defined as the time derivative of risk that is,

$$\beta = d\pi/dt.$$

For the independent response model (equation 8-2), the predicted response to two doses, d and $d + \Delta d$, is given by

$$\pi(d + \Delta d) = 1 - [1 - \pi(d)][1 - \pi(\Delta d)] \qquad (8\text{-}36)$$

We subtract from this the response to dose d to get $\Delta \pi = (d + \Delta d) - \pi(d)$, which is equal to the following:

$$\Delta \pi = \pi(\Delta d)[1 - \pi(d)] \qquad (8\text{-}37)$$

Now both sides are divided by Δd and the limit taken to yield

$$\frac{d\pi}{dd} = [1 - \pi(d)] \left. \frac{d\pi}{dd} \right|_{d=0} \qquad (8\text{-}38)$$

This is now multiplied by $dd/dt = d'$ to yield the value of β as a function of time:

$$\beta = \frac{d\pi}{dt} = d'[1 - \pi(d)] \left. \frac{d\pi}{dd} \right|_{d=0} \qquad (8\text{-}39)$$

Now, given a particular dose–response function, the value of β versus time can be obtained from equation 8-39 by also computing the cumulative dose from equation 8-35.

For the linear isobole approach, we consider the response to the summation of dose d and Δd, and by applying equation 8-3, we obtain:

$$\frac{d}{\pi^{-1}[\pi(d + \Delta d)]} + \frac{\Delta d}{\pi^{-1}[\pi(d + \Delta d)]} = 1 \qquad (8\text{-}40)$$

Rearranging this, the following identity is obtained:

$$\pi(d + \Delta d) = \pi(d + \Delta d) \qquad (8\text{-}41)$$

and then we proceed directly to the finding of β by

$$\beta = d' \left. \frac{d\pi}{dd} \right|_{d} \qquad (8\text{-}42)$$

Obviously equations 8-42 and 8-39 are different functions. However, as indicated below, for certain dose-rate versus time profiles and dose–response models, they may yield identical results.

Example 8-8. Consider the exponential dose–response model with a constant dose rate d'. Develop the expressions for the rate of infection (β) versus time.

SOLUTION. For the independent model, the derivative of the dose–response function, $\pi = 1 - \exp(-d/k)$, is given by

$$\frac{d\pi}{dd} = \frac{\exp(-d/k)}{k}$$

If we evaluate this at zero dose, the slope becomes $1/k$. Now substituting into equation 8-35, realizing that $d = d't$, we obtain

$$\beta = \frac{d'}{k}\left[\exp\left(-\frac{d't}{k}\right)\right]$$

For the linear isobole model, we can immediately substitute the derivative into equation 8-38 to produce the result:

$$\beta = \frac{d'}{k}\exp\left(-\frac{d't}{k}\right)$$

Note that the two approaches give the same functional form. Also interesting is the finding that the infectivity rate, for constant exposure, under the exponential model is predicted to be intrinsically declining with time.

For other dose–response models and for other dose-rate versus time profiles, the results of the two approaches yield different expressions for infectivity rate versus time. It is quite conceivable that neither of these approaches provide a perfect scenario for the coupling of dose–response models to epidemic curves. However, much further work is needed, including the collection of better data on dose rates during epidemics or in well-designed studies using animal model systems. As an example of the different results that other dose–response models can give, the beta-Poisson model at constant dose rate is noted below.

Example 8-9. Consider a constant-dose-rate situation with the beta-Poisson model. Develop the expressions for the infectivity rate.

SOLUTION. Differentiating the beta-Poisson dose–response model with respect to dose, we obtain

$$\frac{d\pi}{dd} = \left[;\alpha\,\frac{2^{1/\alpha} - 1}{N_{50}}\right]\left[1 + \frac{d(2^{1/\alpha} - 1)}{N_{50}}\right]^{-(\alpha+1)}$$

For the independent model, this derivative is evaluated at $d = 0$, and this is found to be $\alpha(2^{1/\alpha} - 1)/N_{50}$, and therefore the resulting infectivity rate can be computed to be

$$\beta = d' \, \frac{\alpha(2^{1/\alpha} - 1)}{N_{50}} \left[1 + \frac{d't}{N_{50}} (2^{1/\alpha} - 1) \right]^{-\alpha}$$

For the linear isobole model, substitution of the derivative into equation 8-3 produces the result

$$\beta = d' \, \frac{\alpha(2^{1/\alpha} - 1)}{N_{50}} \left[1 + \frac{d't}{N_{50}} (2^{1/\alpha} - 1) \right]^{-(\alpha+1)}$$

It is obvious that the two functions differ with respect to the exponent on the final term. However, like the exponential model, they both describe a diminution of intensity of infection as a function of time in a constant-dose-rate exposure scenario.

Duration of Illness

The integral of equation defines the instantaneous rate of new cases. We may be interested in, or may have data on, the total number of cases at any one time. Hence we need to extend the model to consider the length of time people spend in a disease state. At this point it is useful to start to consider the formal dynamic modeling process itself.

Implicitly, so far, we have considered a particular population to consist of susceptible individuals (N) as a function of time. Let us also define X to be the number of infected individuals at any time and Y to be the number of ill persons at any time. We also consider Z to be the number of persons postinfection or following illness.

Figure 8-18 defines some basic pathways between the compartments. We have already defined $Q(t)$, the rate of which new symptomatic cases occur from the susceptible. Implicitly in the use of the parameter λ, the rate of new asymptomatic infected cases has also been defined. Although we have assumed that the underlying incubation distribution for the conversion to the symptomatic state [$f(t)$] is the same as for the asymptomatic state, this might not be the case and a different distribution function might need to be invoked (any alternative probability density function is a potential candidate). Unfortunately, very few data are available to judge whether or not the two incubation distributions are in fact the same or different, and therefore we show them as being identical—future information may justify the more complex formation.

The symptomatic and asymptomatic individuals convert to the post-infected state. Two rates are indicated here: $R(t)$ and $S(t)$, the number of symptomatic infected individuals per unit time and the number of asymptomatic individuals per unit time who enter the postinfected state, respectively.

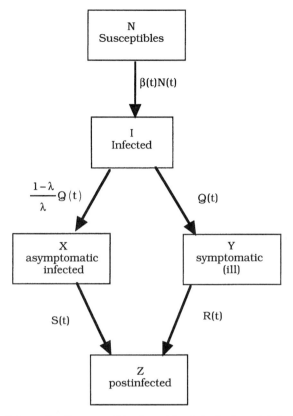

Figure 8-18 Schematic diagram of basic epidemiological model for disease transmission.

It is assumed that there is only a single postinfected state, particular illnesses may require multiple postinfected states.[14]

If functions for $R(t)$ and $S(t)$ are available, it then becomes possible, by setting up a "mass balance" on each of the compartments (N,X,Y,Z), to determine the time dependence of the number of individuals in each compartment. The function $R(t)$ is dependent on the duration of illness.

If we define the cumulative probability distribution function $G(t)$ as the fraction of persons who exit the disease state on or before time t, and if the

---

[14]Consideration of the postinfected state Z becomes important, for example, in the consideration of immunity to subsequent exposures. There may be an additional process in which individuals in state Z transform to state N; in other words, postinfected individuals (by losing some immunity) may become once again susceptible. In this circumstance, one can envision different characteristics with respect to postinfected individuals who were symptomatic and those who were asymptomatic.

corresponding density function is $g(t)$, then by analogy with equation 8-29, the rate of exit from the disease state can be defined by

$$R(t) = \int_0^t Q(\tau)g(t - \tau)\, d\tau \tag{8-43}$$

Information on $g(t)$, at least in a crude sense, is available in standard references (5), typically given as a modal value and a range. In the absence of other information, therefore a triangular distribution might be used to model this distribution.

Information available to support a determination of $S(t)$ is less available. The loss of asymptomatic infections has rarely been measured. In the case of rotavirus infection from children exposed via the day-care environment one such measurement has been reported by Pickering et al. (48) This study measured the excretion of rotavirus by children subsequent to the acquisition of disease. This is not directly a measurement in support of $S(t)$, since it pertains to the loss of infection from initially symptomatic persons. However, it might be reasonable to presuppose that the loss of infectivity in symptomatic individuals is similar to that in asymptomatic individuals. Figure 8-19 summarizes measurements versus time postillness compared to simple exponential fit. For this data set, and exponential loss of infectivity provides good agreement with the data. Hence if we define $H(t)$ as the cumulative fraction of asymptomatic persons who convert to a postinfected state within t days of becoming infected and $h(t)$ is its derivative, we would for this particular case take $h(t)$ to be an exponential function, that is, $h(t) = \exp(-0.24t)/0.24$. Then we can compute $S(t)$ by the following integral:

$$S(t) = \int_0^t \left[\frac{1 - \lambda}{\lambda}\, Q(\tau) \right] h(t - \tau)\, d\tau \tag{8-44}$$

Note that this formulation might be modified if we assume a different incubation time distribution for the asymptomatic state than for the symptomatic state.

With the foregoing definitions of Q, R, and S, we can define formal mass balances corresponding to the model in Figure 8-19. By considering the rates of exit and entry into each of the boxes and equating time derivatives to the net rates of exit and entry, the following five differential equations describe the system dynamics:

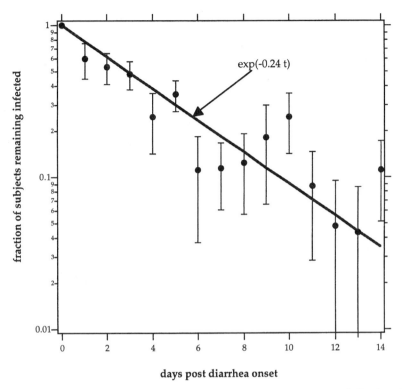

Figure 8-19 Excretion of rotavirus by children subsequent to onset of diarrhea. The straight line is the best-fit exponential. Error bars are standard deviations from binomial sampling. (Data from Ref. 48.)

$$\frac{dN}{dt} = -\beta(t)N$$

$$\frac{dI}{dt} = \beta(t)N - \frac{Q(t)}{\lambda}$$

$$\frac{dX}{dt} = \frac{1-\lambda}{\lambda}Q(t) - S(t) \tag{8-45}$$

$$\frac{dY}{dt} = Q(t) - R(t)$$

$$\frac{dZ}{dt} = R(t) + S(t)$$

This system can be treated as an initial value problem, since in the usual case the values for population size in each of the compartments will be known or assumed at time zero (i.e., $N = N_0$, $X = Y = Z = 0$). However, given the fact that Q, R, and S are integrals (or multiple integrals—i.e., R and S involve integrals of Q, which in turn is another integral), the problem is not a simple differential equation system but rather a system of integrodifferential equations that must be solved numerically.

Figure 8-20*b* is the result of computing the model given by equation 8-41 using the incubation time distributions assumed in earlier examples. A time-variable attack rate (Figure 8-20*a*) over the first 20 hours is assumed. The

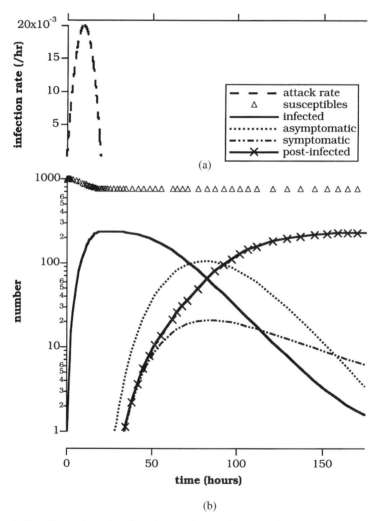

(a)

(b)

Figure 8-20 Dynamics of outbreak considering durations in diseased and asymptomatic states.

distributions for duration in the symptomatic state (density function g) and asymptomatic state (density function h) are both assumed to be lognormal with respective means of 48 and 36 hours, and respective standard deviations of 10 and 72 hours. The value for λ is assumed to be 0.2. Note that the maximum number of symptomatic and asymptomatic individuals is described (about 100 and 20, respectively).

There are a number of advantages to the use of these types of models, although there is a substantial data gap with respect to the parameters (and distributional forms) that are included. First, by depicting the time profile of cases (and asymptomatic individuals), it may allow a more precise estimation of impacts (economic and otherwise) from community-level spread of infectious disease, particularly if the impact is related to both the number of cases and their duration. Second, as illustrated below, a particularly unique aspect of microbial infectious agents (as compared to chemical agents) is their ability to induce secondary cases—additional infections and illnesses in persons who were not exposed to the contaminated water, food, and so on. A modest extension of the framework of equation 8-45 and Figure 8-18 allows formal quantitative consideration of this impact.

Secondary Cases

The use of community-level models of disease transmission allows clear consideration of the possibility of secondary spread. As noted in Table 1-2, there is a clear record of secondary illnesses in outbreaks initiated by ingestion of contaminated food or water or in the day-care setting. Mechanistically, these additional infections and cases may result from person-to-person contact, including transfer of fecal-contained microorganisms, from the transfer of contaminated organisms in a household setting or by sharing of eating or drinking utensils, or from contamination of shared surfaces (e.g., food preparation areas). For example, White et al. (56) documented a foodborne outbreak of Norwalk virus which was prolonged by the maintenance of contaminated food by food service employees who worked in a hotel kitchen during illness and immediately postillness.

There is evidence that the transmission of organisms, at least for some illnesses, may occur preinfection as well as postinfection. Figure 8-19 documented the occurrence of rotavirus in fecal samples from children with rotaviral diarrhea. As part of that same study, the excretion of rotavirus by a cohort of children prior to symptomatic illness was assessed. Figure 8-21 shows the results. As long as 5 days prior to symptomatic illness, more than 10% of the children excreted rotavirus. In fact (not shown), there was one child (of 16) who excreted rotavirus at 13 days prior to developing diarrheal symptoms. This pre-ill excretion of rotavirus represents one route of transmission.

The presymptomatic excretion of microorganisms has not been well documented for many organisms; however, in view of the apparent requirement

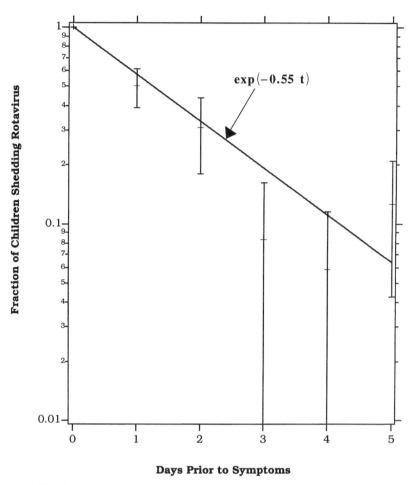

Figure 8-21   Excretion of rotavirus in children prior to symptomatic illness. Error bars are one standard deviation of the observed proportion. (Data from Ref. 48.)

for in vivo multiplication prior to onset of infection and the distribution of such incubation periods (see Figures 8-14 and 8-15), it would not be surprising if many other organisms were excreted prior to symptomatic onset. In addition to transmission from pre- and postsymptomatic individuals, there is the possibility of transmission from asymptomatic individuals. This is documented in the case of salmonellosis (5) and may be true for other illnesses as well.

The occurrence of secondary cases can be accommodated by modification of the prior framework. The treatment of the secondary attack rate was initially expounded by Kermack and McKendrick (38), who described the secondary attack as being proportional to the product of susceptibles and infected individuals. In more complex extensions to this formulation, spatial or temporal heterogeneity to the secondary rates or age dependence (of the susceptible and infected individuals) may be depicted (1). Extending the model of equation 8-45, the secondary infection rate (number of newly infected persons per unit time) is written as $\gamma(t)N(X + Y)$. This model implies that both symptomatic and asymptomatic individuals, for the period of time that they remain in those states, can serve as foci for infection of new individuals. With this additional process, we must also modify the integral that gives the instantaneous rate of newly diseased persons. The resulting model then consists of replacing equations 8-45 and 8-32 by the following:

$$\frac{dN}{dt} = -\beta(t)N - \gamma(t)N(X + Y)$$

$$\frac{dI}{dt} = \beta(t)N + \gamma(t)N(X + Y) - \frac{Q'(t)}{\lambda}$$

$$\frac{dX}{dt} = \frac{1 - \lambda}{\lambda} Q'(t) - S(t) \tag{8-46}$$

$$\frac{dY}{dt} = Q'(t) - R(t)$$

$$\frac{dZ}{dt} = R(t) + S(t)$$

$$Q'(t) = \int_0^t \lambda\{\beta(\tau)N(\tau) + \gamma(t)N(\tau)[X(\tau) + Y(\tau)]\}f(t - \tau)\, d\tau \tag{8-47}$$

The revision of equation 8-47 considers that newly infected persons (I) arise from both primary and secondary cases. Note that if symptomatic and asymptomatic individuals have different propensities to cause secondary cases, this can easily be accommodated by including different specific values of the function γ for each of these categories.

The presence of secondary cases can prolong the overall duration of an outbreak as well as increasing the total disease burden. Unfortunately, for water- and food-borne outbreaks, data necessary to evaluate the secondary attack parameter, γ, are not readily available. However, the impact of the secondary attack process can be illustrated by integrating the model given in equations 8-46 and 8-47.

Using the same assumptions as those underlying Figure 8-20, the model including secondary infection was solved. A secondary attack parameter, γ, of 5×10^{-5} per person-hour was assumed. The resulting instantaneous rate of new illnesses and the total number of infected, ill, and postinfected persons is shown in Figure 8-22. It is clear that the addition of a secondary infection process results in a bimodal distribution of new cases (top panel). It is also

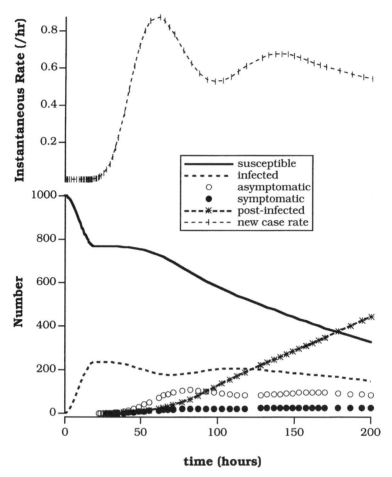

Figure 8-22   Outbreak dynamics with secondary infections.

clear from the total number of ill and asymptomatic individuals at any one time (bottom panel) that a noticeable secondary case burden exists.

If the secondary attack parameter was larger, it would become difficult if not impossible to differentiate a primary from a secondary wave. In other words, the onset of secondary cases would occur at a point before the primary infection process substantially decayed. In this case, examination of new case rates (or total cases) versus time might not visually point out the existence of a secondary process.

Impact of Immunity

The community-level model used previously (either with or without secondary infection) assumes that postinfected persons are no longer susceptible. Although this may be a reasonable assumption, particularly when attack rates are fairly small or in the period immediately subsequent to exposure of a population to a primary source of infection, it breaks down in the long term. For a number of infections (1,4), immunity may be highly transient. In this case, the rate at which previously infected individuals return to the population of susceptibles must be considered.

By extension with our previous approach, it seems reasonable to define another distribution, $i(t)$, which is the density function for residency in the postinfected immune state. This can be used to compute an instantaneous rate at which individuals transform from the postinfected (and immune) state back to the susceptible state by the following convolution:

$$T(t) = \int_0^t [R(\tau) + S(\tau)]i(t - \tau) \, d\tau \qquad (8\text{-}48)$$

$$\frac{dN}{dt} = -\beta(t)N - \gamma(t)N(X + Y) + T(t)$$

$$\frac{dI}{dt} = \beta(t)N + \gamma(t)N(X + Y) - \frac{Q'(t)}{\lambda}$$

$$\frac{dX}{dt} = \frac{1 - \lambda}{\lambda} Q'(t) - S(t) \qquad (8\text{-}49)$$

$$\frac{dY}{dt} = Q'(t) - R(t)$$

$$\frac{dZ}{dt} = R(t) + S(t) - T(t)$$

The repopulation of susceptibles as defined by equations 8-49 may lead to the continued occurrence of new cases via the secondary route, once a source of primary infection has occurred. In other words, we now have a mechanism for the induction of endemic illness from an initial transient source of environmental exposure. Depending on the relative rates and forms of the distributions involved and the parameters of the distributions, this endemic disease burden may be persistent indefinitely or may eventually die out. There is a considerable body of literature in mathematical epidemiology focusing on whether or not (as we have defined it here) secondary transmitted cases may be persistent, or may diminish over time (1,4).

A Final Caveat on Community-Level Models

We conclude with a final comment on community-level models. We have deliberately chosen a deterministic formulation to investigating dynamics. In other words, the number of cases and their changes are described by solutions to coupled (integro-)differential equations. This implicitly treats the number of cases as a continuous variable. This is a reasonable assumption if the total population, as well as the number of cases, is reasonably large. However, for small numbers of cases, it is necessary to reformulate the problem in stochastic terms.

A full stochastic model treats the time evolution of probabilities that the compartments (boxes in Figure 8-18) will assume a set of particular numerical values. For example, $p_{i,j,k,l,m}$ would be the joint probability that the following conditions would all be true simultaneously:

$$N = i \quad I = j \quad X = k \quad Y = 1 \quad Z = m$$

Now each of the individual rates between compartments in the deterministic model can be translated into rates of change of individual probabilities, and a differential equation for each of the $p_{j,k,l,m}$ terms can be written. For example, considering primary infection only, for the differential equation $dp_{i,j,k,l,m}/dt$, we would have terms of form $\beta(i + 1) \Sigma_{j,k,l,m} p_{i+1,j,k,l,m}$, representing the formation of new primary cases. This stochastic model leads to a very large series of coupled differential equations, which to some degree can be simplified by using matrix formulations, however, the solution of complex scenarios using stochastic formulations leads to a more difficult problem than for deterministic formulations. For further discussion of stochastic approaches to modeling community-level disease spread (although focusing on secondary rather than primary spread), Bailey (4) should be consulted.

DETECTABILITY OF OUTBREAKS

As noted in Chapter 1 (see Figure 1-2), the detection of outbreaks occurs when the proximity of cases in time and space exceeds some level that results in notice. This will depend on the number of persons exposed, the dose and duration to which they are exposed, and the intensity of the outcome. For example, a waterborne outbreak of *Cryptosporidium* affecting primarily HIV-positive persons was detected with a relatively low case level, since the symptoms were of quite high severity (high mortality) (27). In the massive Milwaukee outbreak, detection did not really occur until late into the exposure stage, due to the relatively moderate (and relative common) spectrum of symptoms exhibited in normal healthy individuals who became ill (i.e., diarrhea and gastrointestinal distress).

It is therefore clear that the ability to detect an outbreak by monitoring cases will depend on the background or endemic level of that illness and/or those symptoms in the community, as well as factors intrinsic to the microorganism and the exposure itself. In this section we briefly summarize a framework to approach the outbreak detection problem. We are motivated in this problem by an apparent discrepancy between the number of cases recorded in reported food- and waterborne outbreaks, which in the United States are typically under 20,000 per year (see Tables I-1 and I-2) and the relatively

TABLE 8-11 Microbial Daily Intakes Corresponding to Reported U.S. Waterborne Risk

Organism	Dose–Response Parameters	Daily Intake Corresponding to Risk of 9.8×10^{-6}
Rotavirus	Beta-Poisson $N_{50} = 5.60$ $\alpha = 0.265$	1.63×10^{-5}
Salmonella	Beta-Poisson $N_{50} = 16{,}211$ $\alpha = 0.294$	0.0565
Cryptosporidium	Exponential $r = 238$	0.00233

large number of total illnesses estimated to be caused by food and water: 6.5 and 1 million per year in the United States annually (6).

First, we differentiate between what might be termed a *routine outbreak* from an *anomalous outbreak*. The former occur from a coincident clustering of illnesses over a short duration and in geographic proximity during typical conditions (i.e., operation of water treatment plants, production of food under usual conditions).

Doing a very rough computation, if there were 1 million waterborne illnesses a year in the U.S. population (taken as 280 million persons for the purpose of this computation), this results in an annual occurrence rate of 0.0036 per year. Taken over a daily, 365-day basis, this results in a daily occurrence rate of 9.8×10^{-6}. Using dose–response parameters for some infectious organisms, the daily intake of organisms that would correspond to this daily risk can be computed. Results are shown in Table 8-11. The average daily number of organisms ingested to result in an observed risk, of infection, using the information from Bennett, is quite small. Particularly given the great difficulties in quantitatively measuring low levels of organisms in finished water or food, it is highly unlikely that these levels could be measurable.[15] Therefore, the general inability to measure pathogens in food or water as they are being consumed is not necessarily inconsistent with the estimated prevalence of food- and waterborne infectious diseases.

If water- and foodborne illnesses are occurring at a substantial rate (millions of cases per year in the United States), then it may still be that we would only anticipate a small fraction of cases to be associated with detectable

---

[15] For example, in the case of *Cryptosporidium*, it the entire exposure was due to drinking water, and using a 2 L per person per day consumption value, the number in Table 8-11 translates into about 0.001 organisms per liter. Hence, at least 1000 1-L samples would be required to have a substantial probability of detection, even if detection methods were 100% efficient—which they are not.

outbreaks. Aside from the problems of reporting and voluntary surveillance systems noted in the introduction and Chapter 1, an important initial trigger in outbreak identification is the recognition of an unusual pattern of illnesses. For relatively routine symptoms, this may be quite difficult, as will now be illustrated.

We focus attention on a single diagnostic unit. This may be a medical practice, a laboratory, or an individual physician. This unit is the primary recognizer of an unusual aggregation of illnesses. There are assumed to be N_t persons who would present themselves to this unit in the event of severe illnesses. Note that many individuals may not seek such medical attention for mild or moderate illnesses, and hence an unusual cluster of mild illnesses might not be recognized by such a unit. The overall community in which this community operates has an infection prevalence rate of p_0 (cases per person per day). The unit will tend to group cases over a number of days in recognition of unusual events, call this t. Then the expected "usual" number of cases over this time period anticipated by this unit would be $N_t p_0 t$. If there is no unusual clustering, we would expect the case frequency to be Poisson distributed. Hence for different community sizes we can compute the anticipated case frequency that would be anticipated by the equation

$$P(N) = \exp(-N_t p_0 t) \tag{8-50}$$

For various diagnostic unit total populations, equation 8-50 can be used to compute the maximum number of cases (N) such that the probability of observing up to that number is less than or equal to 0.95; this is the upper confidence limit on the number of cases. For a base rate of 0.000294 (taken as 30 days times the daily rate from Bennett), these limits are shown in Table 8-12. Note that the width of the confidence region (ratio of the upper confidence limit to the mean) decreases with population size, as anticipated from the Poisson distribution.[16]

Using this concept we now examine the likelihood that an *exceedance* would be detected if the actual illness rate increased to a particular multiple of the baseline rate in the various populations. This depends on the rule (or algorithm) that is used by the individual units to identify unusual occurrences. There is need of further study. However, we shall illustrate two possibilities: (1) an unusual event is noticed if the cases exceed the 95% upper confidence limit based on the baseline value (the last column in Table 8-11), or (2) an unusual event is noticed if the cases exceed twice the baseline value. Figure 8-23 summarizes the probability of detection of a situation in which the attack rate rises to a multiple of the endemic rate as a function of community size (for the same assumptions as those underlying Table 8-12). There are some

---

[16] Since the variance is equal to the mean for a Poisson distribution, the ratio of standard deviation to the mean equals $1/\sqrt{\text{mean}}$ and hence decreases with increasing unit size.

**TABLE 8-12   Confidence Limits to Number of Observed Illnesses as a
Function of Communiuty Size**[a]

Population	Average Number of Expected Cases	Upper 95% Point
1,500	0.441	1
2,250	0.6615	1
3,500	1.029	2
5,000	1.47	3
7,500	2.205	4
10,000	2.94	5
15,000	4.41	7
25,000	7.35	11
40,000	11.76	17
60,000	17.64	24
90,000	26.46	34
150,000	44.1	54
225,000	66.15	79
400,000	117.6	135

[a] (Constant per capita risk of 0.000294 per person).

important salient observations from this figure that can be summarized as
follows:

- In general, in small diagnostic units, the double-baseline approach detects
  a greater proportion of exceedance events that exceed the 95% limit approach (however, it would also produce a greater proportion of false alarms).
- The double-baseline approach performs quite poorly except when the actual
  attack rate is more than double the baseline. For example, in the largest
  community size, less than 20% of the instances of a 1.8 multiple of baseline
  would be detected using this identification algorithm.
- Although the 95% confidence region approach does perform better for larger
  communities, even in a community of 10,000, if the rate is 1.8 of the baseline, the identification probability is under 60%.

Although there are clearly a great number of assumptions underlying Figure 8-23, many of which may not be rigorously true, what this clearly shows
is that for normal clusters (which result from random fluctuations of underlying performance) the detection of somewhat elevated risks is still quite
imperfect. Hence it should not be surprising that mild exceedances of baseline
rates are not well detected. Therefore, if most infectious disease from water
or food was as a result of these mild rather than abnormal events, there would
not be a high detection likelihood. This remains a substantial reason for as-
serting the lack of inconsistency between the reported outbreak attack rates,
and the estimated overall disease burden.

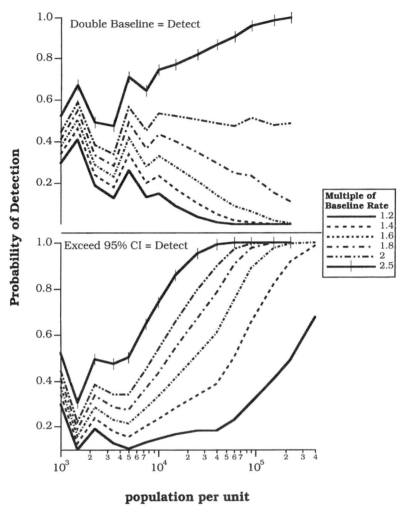

Figure 8-23 Probability of detection of clusters given baseline values from Table 8-12 using two detection rules.

ECONOMIC IMPACT OF INFECTIOUS DISEASE

Once the extent of outcomes, in terms of total infections (or cases) or in terms of community impact (via epidemic models), has been assessed, the risk may need to be quantified in terms of economic impact to a community. One method (although not the only method) of performing risk management is by the use of a risk–benefit calculus. If a number of different interventions (say, levels of water treatment) are evaluated for their relative economic costs, and if the microbial (and other) risks resulting from these alternatives are also evaluated for their economic costs, a balancing approach in commensurable units may be made. It should be noted that there are a number of technical as well as ethical issues in determining the economic value of disease impacts.

Some of the technical issues include how to place an economic value on levels of illness and on lost time other than in employment (e.g., leisure time) (32). In addition, Harrington et al. (32) note that valuation of anxiety, or permanent changes in behavior (or attitude) as a result of illness needs to be better considered. In massive outbreaks, there may be a permanent loss of value to a city, or to a company or employer (the latter as in the case of a foodborne outbreak). Finally, the impact cost estimates that have been reported typically are on a direct cost-incurred basis rather than on a "willingness to pay" to avoid impact basis (with the latter frequently being greater than the former) (32).

Most of the available literature focuses on estimating impacts associated with outbreak situations. There is an unstated presumption that the direct costs to individuals (lost wages, medical costs) from endemic illnesses are similar to those incurred in outbreaks of the same infectious agent. However, in the endemic situation, it is reasonable to expect that a number of avoided costs would be less or nonexistent. For example, in endemic waterborne illness, where the individuals affected did not become aware that their source of illness was from the water, one would not expect behavior-modifying activities such as switching to alternative (e.g., bottled) water supplies or use of home water purifiers to occur. The issue of economic valuation of waterborne disease is therefore one of the most open technical areas. We shall therefore review a number of prior studies, with the aim of providing prototypical values that might be used as a point of departure in subsequent efforts.

A 1981 waterborne outbreak of gastroenteritis in Eagle–Vail, Colorado affected 48.2% of a community of 3540; with an estimated total case burden of 1706 (33). From a telephone survey, the estimated costs incurred by the people affected was determined. The aggregate and per case (of illness) costs incurred are summarized in Table 8-13. The costs incurred by individuals per case of illness were $97.20 (1981 dollars). Apparently no severe illnesses (and no fatalities) were incurred in this event. The long-term costs for plant upgrading were substantial.

Harrington et al. (32) assessed the economic impact of a 1983 outbreak of giardiasis in Luzerne County, Pennsylvania. Of the community of 25,000, there were 6000 total cases and 370 clinically confirmed cases of illness. Surveys were sent to these confirmed cases to estimate incurred costs. The authors, as part of the survey, addressed direct costs incurred by the ill persons as well as costs (such as lost wages) incurred by caregivers. A separate survey of the community was conducted to estimate costs of behavior alterations, particularly switching costs of water supply (during the outbreak). Costs of long-term behavior alterations were not assessed. Alternative supplies were obtained by hauled water or boiling water, with virtually no switching to bottled water. The costs of aversive action is obtained by applying an hourly

TABLE 8-13   Costs of Eagle–Vail Waterborne Gastroenteritis Outbreak (1981 Dollars)

	Total Community	Per Case
Direct costs		
Physician visits	$ 2,423	$ 1.42
Medications	$ 3,119	$ 1.83
Time lost from work	$ 138,281	$ 81.06
Bottled water	$ 22,007	$ 12.90
Total individual costs	$ 165,830	$ 97.20
Emergency repair to treatment plant	$ 92,400	$ 54.16
Total direct costs	$ 258,230	$151.37
Indirect costs		
Long-term repairs	$ 931,200	$545.84
Cost of investigation	$ 8,700	$ 5.10
Total indirect costs	$ 939,900	$550.94
Total costs	$1,198,130	$702.30

Source: Ref. 33.

wage rate to the time required. A summary of the results of these authors is given in Table 8-14, on a total community basis and per case (total) basis. Note that the per case direct costs (medical plus loss of work) are somewhat more substantial than in the Eagle–Vail outbreak (Table 8-13), due to the greater severity of giardiasis (relative to gastroenteritis). What is also very noteworthy about Table 8-14 is the substantially greater cost from averting illness during an outbreak (obtaining water from alternative supplies) than the direct medical and lost work and productivity/leisure costs. As stated before, the costs of aversive behavior would probably not be incurred in a nonoutbreak situation[17]; hence there is a much higher (economic) value to avoiding illnesses in an outbreak situation than in an endemic (nonoutbreak) situation.

The direct medical and productivity costs associated with several foodborne illnesses were estimated by Mauskopf and French (42). These do not include costs of aversive actions; the authors were particularly interested in endemic situations. In addition, an alternative methodology using quality-adjusted life years was employed. For mild, moderate, and severe salmonellosis the methods yielded cost estimates of $197 and $222, $622 and $890, and $86,895 and $743,000 per case, respectively. The much larger impacts from severe salmonellosis result from the relative large (13%) mortality ratio assumed. However, note that the cost impact of mild salmonellosis is similar to that for gastroenteritis (Table 8-13), while the impact of moderate salmonellosis is similar to that for giardiasis (Table 8-14) (excluding aversion costs).

---

[17] Since only in the presence of an outbreak would a "boil water" notice or some other form of public notification be triggered that resulted in aversive behavior.

TABLE 8-14 Economic Costs of Luzerne County, Pennsylvania Giardiasis Outbreak (1984 Dollars)[a]

Loss Category	Total Community Costs			Costs per Case		
	A	B	C	A	B	C
Medical costs	$ 1,070,000	$ 1,050,000	$ 1,030,000	$ 178	$ 175	$ 172
Loss of work	$ 2,150,000	$ 1,630,000	$ 1,250,000	$ 358	$ 272	$ 208
Loss of productivity and leisure time	$ 3,780,000	$ 2,910,000	$ 2,310,000	$ 630	$ 485	$ 385
Loss due to averting actions (best est.)	$38,510,000	$12,940,000	$12,120,000	$6,418	$2,157	$2,020
	$45,510,000	$18,530,000	$16,710,000	$7,585	$3,088	$2,785

Source: Ref. 32.

[a]The three scenarios use a different assumed after-tax wage rate for unemployed, homemakers, and retirees of $6.39 (A), $6.08 (B), and $2.65 (C) per hour.

389

An attempt to assess overall economic costs of foodborne illness was made by the Council for Agricultural Sciences and Technology (17). For four pathogens of concern, their estimates are given in Table 8-15. These are generally higher than the prior tables, since they incorporate directly the value of lives lost due to mortality. The much higher costs associated with *Campylobacter jejuni, E. coli* O157:H7, or *Listeria monocytogenes* are due to the higher severity and (particularly for the latter two organisms) mortality encountered than for *Salmonella.*

TABLE 8-15   Total Medical and Productivity Costs^a Associated with Selected Foodborne Pathogens

Organism	Annual U.S. Case Burden (Midpoint Estimate)	Annual U.S. cost Impact (Midpoint Estimate)	Cost per Case
Salmonella	1,920,000	$1,388,000,000	$ 723
Campylobacter jejuni	2,100,000	$ 961,500,000	458
E. coli O157:H7	14,058	388,000,000	27,600
Listeria monocytogenes	1,550	221,000,000	142,581

Source: Modified from Ref. 17.

^{a} Excluding avoidance and aversion costs but including mortlaity (1992 dollars).

PROBLEMS

8-1. A raw (untreated) drinking water supply contains 10 *Cryptosporidium* oocysts per liter. The process of treatment typically removes 2.5 logs of organisms (i.e., $1/10^{2.5}$ of the initial concentration remains after treatment). Drinking water consumed directly by a person is typically 1 L/day. If the dose–response relationship is exponential, with an *r* value of 0.00419 (see Example 7-3), compute the point estimate for the risk from a single day's ingestion of drinking water.

8-2. If the daily risk of infection by some agent on each of four successive days is 0.1, 0.05, 0.003, and 0.07, what is the overall risk of incurring an infection anytime during the four-day period? Compare this to the simple arithmetic sum of the risks on the four days.

8-3. The following are 10 sets of parameter combinations from the bootstrap of the *Salmonella* dose–response fit. Compute the Spearman rank correlation for the two parameters.

α	N_{50}	α	N_{50}
0.374	47,237	0.200	10,354
0.205	5,888	0.153	1,139
0.215	4,989	0.370	33,831
0.231	6,747	0.240	9,975
0.255	8,240	0.216	5,394

8-4. Consider the following hypothetical data set (x and y):

Observation, i	x_i	y_i
1	1	4
2	2	1
3	2.7	0.09
4	3.3	0.09
5	4	1
6	5	4

[note that this is really an artificial data set obtained by the relationship $y = (x - 3)^2$.]

(a) Compute the Pearson and Spearman correlation coefficients for these data.

(b) For each of the following functions, $z(x,y)$, compute the value of the function evaluated at the means of the inputs, $z(\bar{x}, \bar{y})$, and the mean value of the function over the six observations, $\frac{1}{6} z(x_i, y_i)$:

   (i) $z = xy$
   (ii) $z = x/y$
   (iii) $z = x - y$
   (iv) $z = \exp(x - y)$

8-5. Repeat the' computation in Example 8-7 (probability in a rectangular region) for computation of the joint probability of bivariate variables exceeding their individual 0.99 confidence values.

8-6. Consider the results of Example 8-8.

(a) Compute the mean of the counts from the simulations for both the independent and negatively correlated simulations. If you use this to fit a Poisson distribution (recall Chapter 6, in which the maximum likelihood fit to a Poisson distribution with data of this nature is given by the arithmetic mean), determine if the distribution of counts is adequately fit by a Poisson distribution.

(b) Try fitting these data sets to some of the alternative discrete distributions discussed in Chapter 6. If you can get an acceptable fit, estimate (for both scenarios) the probability that a single exposure would result in ingestion of 20 or more organisms.

8-7. Assume that a particular exposure scenario has an exponential decaying dose rate [i.e., $d' = d'_0 \exp(-\gamma t)$ (where d'_0 and γ are constants)]. For both the independent and linear isobole approaches, derive the expression for infectivity rate versus time:

(a) Using the exponential dose–response model.

(b) Using the beta-Poisson dose–response model.

8-8. (This problem requires access to one of the Monte Carlo programs.) The analysis of the problem posed in Table 8-4 resulted in a conclusion that the most important input (in terms of influencing the variability of the output) was the effect of removal due to cooking. Replace the assumption for this input in Table 8-4 by another beta distribution with the same mean but half of the standard deviation (i.e., a lower amount of variability in removal due to cooking). Using this alternative distribution, compare the results in terms of the mean risk, and the upper 95% confidence limit to the risk, with the results from the distributions in Table 8.4.

REFERENCES

1. Anderson, R. M., and R. M. May. 1991. Infectious diseases in humans: dynamics and control. Oxford University Press, Oxford.

2. Anonymous. 1936. Epidemic amebic dysentary: the Chicago outbreak of 1933. Bulletin 166. National Institute of Health, Washington, DC.

3. Armitage, P., G. G. Meynell, and T. Williams. 1965. Birth–death and other models for microbial infection. Nature 207:570–572.

4. Bailey, N. T. J. 1975. The mathematical theory of infectious diseases and its applications, 2nd ed. Oxford University Press, New York.

5. Benenson, A. S. 1990. Control of communicable disease in man, 15th ed. American Public Health Association, Washington, DC.

6. Bennett, J. V., S. D. Holmberg, M. F. Rogers, and S. L. Solomon. 1987. Infectious and parasitic diseases. Am. J. Prev. Med. 3(5 Suppl.): 102–114.

7. Berenbaum, M. C. 1981. Criteria for analyzing interactions between biologically active agents. Adv. Cancer Res. 35:269–335.

8. Berenbaum, M. C. 1985. The expected effect of a combination of agents: the general solution. J. Theor. Biol. 114:413.

9. Berenbaum, M. C. 1989. What is synergy? Pharmacol. Rev. 41:93–141.

10. Bhaduri, S., C. O. Turner-Jones, R. L. Buchanan, and J. G. Phillips. 1994. Response surface model of the effect of pH, sodium chloride and sodium nitrite on growth on *Yersinia enterocolitica* at low temperatures. Int. J. Food Microbiol. 23: 233–245.

11. Bliss, C. I. 1939. The toxicity of poisons applied jointly. Ann. Appl. Biol. 26: 585–615.

12. Buchanan, R. L., L. K. Bagi, R. V. Goins, and J. G. Phillips. 1993. Response surface models for the growth kinetics of *Escherichia coli* O157:H7. Food Microbiol. 10(4):303–315.

13. Bukowski, J., L. Korn, and D. Wartenberg. 1995. Correlated inputs in quantitative risk assessment: the effects of distributional shape. Risk Anal. 15(2):215–219.

14. Burmaster, D. E., and P. D. Anderson. 1994. Principles of good practice for the use of Monte Carlo techniques in human health and ecological risk assessment. Risk Anal. 14(4):477–481.

15. Burmaster, D. E., and A. M. Wilson. 1996. An introduction to second-order random variables in human health risk assessments. Hum. Ecol. Risk Assessment 2(4):892–919.

16. Cohrssen, J. J., and V. T. Covello. 1989. Risk analysis: a guide to principles and methods for analyzing health and environmental risks. National Technical Information Service, U.S. Department of Commerce, Springfield, VA.

17. Council for Agricultural Science and Technology. 1994. Foodborne pathogens: risks and consequences. Task Force Report 122. CAST, Ames, IA.

18. Duncan, H. E., and S. C. Edberg. 1995. Host–microbe interaction in the gastrointestinal tract. Crit. Rev. Microbiol. 21(2):85–100.

19. Ercolani, G. L. 1985. The relation between dosage, bacterial growth and time for disease response during infection of bean leaves by *Pseudomonas syringae* pv. *phaseolicola*. J. Appl. Bacteriol. 58:63–75.

20. Evans, J., G. Gray, R. Sielken, A. Smith, C. Valdez-Flores, and J. Graham. 1994. Use of probabilistic expert judgment in uncertainty analysis of carcinogenic potency. Regul. Toxicol. Pharmacol. 20:15–36.

21. Fazil, A. M. 1996. M.S. thesis. Drexel University.

22. Ferson, S., and L. R. Ginzburg. 1996. Different methods are needed to propagate ignorance and variability. Reliabil. Eng. Syst. Safety. 54:133–144.

23. Finkel, A. M. 1990. Confronting uncertainty in risk management. Resources for the Future, Center for Risk Management, Washington, DC.

24. Genest, C., and R. MacKay. 1986. The joy of copulas. Am. Stat. 40:280–283.

25. Genest, C., and L.-P. Rivest. 1993. Statistical inference procedures for bivariate archimedean copulas. J. Am. Stat. Assoc. 88(423):1034–1043.

26. Gerba, C. P., J. B. Rose, and C. N. Haas. 1996. Sensitive populations: who is at the greatest risk? Int. J. Food Microbiol. 30(1–2):113–123.

27. Goldstein, S. T., D. D. Juranek, O. Ravenholt, A. W. Hightower, D. G. Martin, J. L. Mesnick, S. D. Griffiths, A. J. Bryant, R. R. Reich, and B. L. Herwaldt. 1996. Cryptosporidiosis: an outbreak associated with drinking water despite state of the art water treatment. Ann. Intern. Med. 124(5):459–468.

28. Haas, C. N., K. Cidambi, S. Kersten, and K. Wright. 1996. Quantitative description of mixture toxicity: effect of level of response on interactions. Environ. Toxicol. Chem. 15(8):1429–1437.

29. Haas, C. N., S. P. Kersten, K. Wright, M. J. Frank, and K. Cidambi. 1997. Generalization of independent response model for toxic mixtures. Chemosphere 34(4): 699–710.

30. Haas, C. N., and B. A. Stirling. 1994. A new quantitative approach for the analysis of binary toxic mixtures. Environ. Toxicol. Chem. 13:149–156.

31. Halstead, S. B. 1982. Immune enhancement of viral infection. Prog. Allergy 31:301–364.

32. Harrington, W., A. J. Krupnick, and W. O. Spofford, Jr. 1989. The economic losses of a waterborne disease outbreak. J. Urban Econ. 25:116–137.

33. Hopkins, R. S., R. J. Karlin, G. B. Gaspard, and R. Smades. 1986. Gastroenteritis: case study of a Colorado outbreak. J. Am. Water Works Assoc. 78(1):40–44.

34. Hutchinson, T. P., and C. D. Lai. 1990. Continuous bivariate distributions, emphasizing applications. Rumsby Scientific Publishing, Adelaide, Australia.

35. Iman, R. L., and W. J. Conover. 1982. A distribution-free approach to inducing rank correlation among input variables. Commun. Stat. Simul. Comput. 11(3):311–334.

36. Istre, G. R., and R. S. Hopkins. 1985. An outbreak of foodborne hepatitis A showing a relationship between dose and incubation period. Am J. Public Health 75(3):280–281.

37. Joe, H. 1993. Parametric families of multivariate distributions with given marginals. J. Multivariate Anal. 46:262–282.

38. Kermack, W. O., and A. G. McKendrick. 1927. A contribution to the mathematical theory of epidemics. Proc. R. Soc. A 115:700–721.

39. Ku, H. 1966. Notes on the use of propagation of error formulas. J. Res. Nat. Bur. Standards C Eng. Instrum. 70C(4):263–273.

40. Loewe, S., and H. Muishchnek. 1926. Uber Kombinationswirkungen. Arch. Exp. Pathol. Pharmakol. 114:313–326.

41. Marking, L. L. 1985. Toxicity of chemical mixtures, pp. 164–176. *In* G. M. Rand and S. R. Petrocelli, eds., Fundamentals of aquatic toxicology: methods and applications. Hemisphere Publishing, Washington, DC.

42. Mauskopf, J. A., and M. T. French. 1991. Estimating the value of avoiding morbidity and mortality from foodborne illnesses. Risk Anal. 11(4):619–631.

43. Mendenhall, W., and T. Sincich. 1988. Statistics for engineering and computer sciences, 2nd ed. Dellen Publishing Company, San Francisco.

44. Meynell, G. G., and B. A. D. Stocker. 1957. Some hypotheses on the aetiology of fatal infections in partially resistant hosts and their application to mice challenged with *Salmonella paratyphi-B* or *Salmonella typhimurium* by intraperitoneal injection. J. Gen. Microbiol. 16:38–58.

45. Moore, D. R. J. 1996. Using Monte Carlo analysis to quantify uncertainty in ecological risk assessment: are we gilding the lily or bronzing the dandelion? Hum. Ecol. Risk Assessment 2(4):628–633.

46. Morgan, M. G., and M. Henrion. 1990. Uncertainty: a guide to dealing with uncertainty in quantitative risk and policy analysis. Cambridge University Press, Cambridge.

47. Mosher, W. E., S. M. Wheeler, H. L. Chant, and A. V. Hardy. 1941. Studies of the acute diarrheal diseases. V. An outbreak due to *Salmonella typhimurium* Public Health Rep. 56:2415.

48. Pickering, L. K., A. V. Bartlett, R. R. Reves, and A. Morrow. 1988. Asymptomatic excretion of rotavirus before and after rotavirus diarrhea in children in day care centers. J. Pediatr. 112(361–365).

49. Ramsey, J. H., C. H. Benning, and P. F. Orr. 1926. An epidemic of typhoid fever following a Church Supper. Am. J. Public Health 16:1101.

50. Roseberry, A. M., and D. E. Burmaster. 1992. Lognormal distributions for water intake by children and adults. Risk Anal. 12(1):99–104.

51. Sartwell, P. E. 1950. The distribution of incubation periods of infectious disease. Am. J. Hyg. 51:310–318.

52. Smith, A. E., P. B. Ryan, and J. S. Evans. 1992. The effect of neglecting correlations when propagating uncertaingy and estimating the population distribution of risk. Risk Anal. 12(4)467–474.

53. Stuart, A., and J. K. Ord. 1987. Kendall's advanced theory of statistics, 5th ed., Vol. 1: Distribution theory. Oxford University Press, New York.

54. U. S. Environmental Protection Agency. 1986. Guidelines for the health risk assessment of chemical mixtures. Fed. Reg. 51(185):34014–34025.

55. Vose, D. 1996. Quantitative risk analysis: a guide to Monte Carlo simulation modeling. Wiley, New York.

56. White, K. E., M. T. Osterbolm, J. A. Mariotti, J. A. Korlath, D. H. Lawrence, T. L. Ristinen, and H. B. Greenberg. 1986. A foodborne outbreak of Norwalk virus gastroenteritis. Am. J. Epidemiol. 124(1):120–126.

57. Williams, T. 1965. The basic birth–death model for microbial infections. J. R. Stat. Soc. B 27:338–360.

58. Williams, T. 1965. The distribution of response times in a birth–death process. Biometrika 52(3–4):581–585.

59. Wilson, R., E. Crouch, and L. Zeise. 1985. Uncertainty in risk assessment. *In* D. G. Hoel, R. A. Merrill, and E. P. Perera, eds., Risk quantitation and regularoy policy. Banbury Report 19. Cold Spring Harbor Laboratories, Cold Spring Harbor, NY.

CHAPTER 9

COMPENDIUM OF DATA

CRITICALLY ANALYZED DOSE–RESPONSE CURVES

The objective of this chapter is a comprehensive review of available information on dose–response relationships. These are based primarily on the work of the authors and their students and to a lesser degree on other authorities. Portions of material in this chapter have been cited earlier. However, in this chapter we present a systematic treatment of particular microorganisms of interest.

Rotavirus

Among the viruses for which dose–response studies are available, human rotavirus appears to be the most infective. It has been used as a prototypical virus in developing U.S. drinking water standards for treatment of surface water (62). The development of potency estimates for this virus originally is outlined below (32). Human rotaviruses have been recognized as pathogens since 1973. They are common infectious agents in all age groups, although predominantly in children and the elderly. Contaminated water is believed to be an important source of infection (35).

Data. Human volunteer studies of rotavirus infectivity were conducted by Ward et al. (82). The experimental results are summarized in Table 9-1.

Results. Preliminary analysis indicated that these data could be explained best by the beta-Poisson model, with the difference from the exponential model being statistically significant. This was demonstrated in Example 7-1.

TABLE 9-1   Infectivity of Human Rotavirus in Volunteers

Dose	Infected Subjects	Negative Subjects	Total Subjects
90,000	3	0	3
9,000	5	2	7
900	7	1	8
90	8	1	9
9	8	3	11
0.9	1	6	7
0.09	0	7	7
0.009	0	7	7

Source: Ref. 82.

The best-fitting beta-Poisson model had point estimates of $\alpha = 0.265$ and $N_{50} = 5.597$. The residual deviance from the likelihood fitting was 6.82, which passes a goodness-of-fit test. Graphical goodness of fit was presented in Figure 7-7. The likelihood-based confidence regions for the parameters are shown in Figure 7-12, and the bootstrap confidence "cloud" is shown in Figure 7-19.

Model Validation. Using the information above, a point risk estimate of exposure to virus may be made. This can be compared to the observed risk of disease found in a prospective intervention study conducted in the Montreal area (58). In that study, an annual risk of illness of 0.24 case/person per year was found. The area in which the epidemiological study was conducted corresponded to that in which prior virus measurements were taken (59) and are believed to characterize the distribution to which persons in the epidemiological study were exposed (P. Payment, personal communication). This is equivalent to a mean daily risk of 0.00082.

Payment et al. (59) found an average of 0.0006 virus/L in finished drinking water in the Montreal area. The overall occurrence distribution was found to be lognormal. For a point risk assessment, to approximate a most exposed individual (MEI), twice this value was used (i.e., 0.0012 L^{-1}). With a tap water consumption of 1.7 L/person per day estimated from the geometric mean found by Roseberry and Burmaster (69), the point risk estimate of the daily probability of disease was found to be 0.000717. This is equivalent to an annual risk of disease of 0.23 (using an independent and identical daily risk model). This compares favorably with the daily and annual risks in the Payment study (58) of 0.00082 and 0.24. Hence it is not implausible that a virus with the infectivity of human rotavirus could have been responsible for the apparent gastroenteritis associated with water.

Salmonella[1]

Infections from nontyphoid *Salmonella* have been increasing dramatically since the 1950s (80). More than 2000 *Salmonella* serotypes are known to exist, with the number of nontyphoid salmonellosis cases in the United States per year estimated to be between 2 million (5) and 5 million (4). These estimates translate to 1:50 to 1:125 odds of contracting salmonellosis per year in the United States. Microbial infections via food and water do occur, and they are believed to occur at a rate much higher than that reflected in statistics collected by governing bodies. The occurrence of pathogenic *Salmonella* in poultry is of increasing concern, and estimated annual costs of the health impacts from poultry-borne salmonellosis in the United States are $64 million to $362 million (10). In the United Kingdom, occurrence rates of *Salmonella* in retail chicken has been found to be 18 to 25% (60).

Data and Methods. Dose–response data were obtained from human feeding studies conducted by McCullough and Eisele (48,49), who investigated the pathogenicity of *Salmonella* species isolated from eggs and egg products. The response in terms of infectivity is summarized in Table 9-2. The maximum likelihood method was used to determine the dose–response model that best described the raw data, both individually and upon pooling the data.

Results. The initial attempt at fitting the pooled data to the beta-Poisson model was unsuccessful, $p = 0.028$ ($p > 0.05$, for statistically significant fit). The log-Probit model was also tested; however, this model was also unable to fit the pooled data. The presence of suspected outliers that may be influencing the pooled data was investigated.

To test for outliers, standardized residuals were calculated for the pooled data set. The chi-squared residuals (55) were calculated using the equation

$$\frac{P_{obs} - P_{pred}}{\sqrt{T\pi(1 - \pi)}} \qquad (9\text{-}1)$$

where P_{obs} is the observed positive, P_{pred} the predicted positive, T the total number of subjects at that dose, and π the predicted proportion.

The standardized residuals calculated were then plotted against the dose. The standardized residuals should be randomly distributed about zero, with most of the points lying between -2 and $+2$. The presence of outliers, however, should also be judged in consideration with the experimental dose–response trend: For instance, if at a lower dose a much larger proportion is

---

[1] This section was written substantially by Aamir Fazil as portion of his M.S. thesis at Drexel University and was presented at the 1996 Annual AWWA Conference.

TABLE 9-2 Infectivity of *Salmonella* (Nontyphoid Strains) in Human Volunteers

Infected with:	Dose	Positive Subjects	Negative Subjects	Total Subjects
S. newport	1.52×10^5	3	3	6
	3.85×10^5	6	2	8
	1.35×10^6	6	0	6
S. derby	1.39×10^5	3	3	6
	7.05×10^5	4	2	6
	1.66×10^6	4	2	6
	6.4×10^6	3	3	6
	1.5×10^7	4	2	6
S. bareilly	1.25×10^5	5	1	6
	6.95×10^5	6	0	6
S. anatum strain I	1.7×10^6	5	1	6
	1.2×10^4	2	3	5
	2.4×10^4	3	3	6
	6.6×10^4	4	2	6
	9.3×10^4	1	5	6
	1.41×10^5	3	3	6
	2.56×10^5	5	1	6
	5.87×10^5	4	2	6
	8.6×10^5	6	0	6
S. anatum strain II	8.9×10^4	5	1	6
	4.48×10^5	4	2	6
	1.04×10^6	6	0	6
	3.9×10^6	4	2	6
	1×10^7	6	0	6
	2.39×10^7	5	1	6
	4.45×10^7	6	0	6
	6.73×10^7	8	0	8
S. anatum strain III	1.59×10^5	2	4	6
	1.26×10^6	6	0	6
	4.68×10^6	6	0	6
S. maleagridis strain I	1.2×10^4	3	3	6
	2.4×10^4	4	2	6
	5.2×10^4	3	3	6
	9.6×10^4	3	3	6
	1.55×10^5	5	1	6
	3×10^5	6	0	6
	7.2×10^5	4	1	5
	1.15×10^6	6	0	6
	5.5×10^6	5	1	6
	2.4×10^7	5	0	5
	5×10^7	6	0	6
S. maleagridis strain II	1×10^6	6	0	6
	5.5×10^6	6	0	6
	1×10^7	5	1	6
	2×10^7	6	0	6
	4.1×10^7	6	0	6
	1.5×10^6	5	1	6
S. malegriis strain III	7.68×10^6	6	0	6
	1×10^7	5	1	6
	1.58×10^5	1	5	6

Source: Data from Refs. 48 to 50.

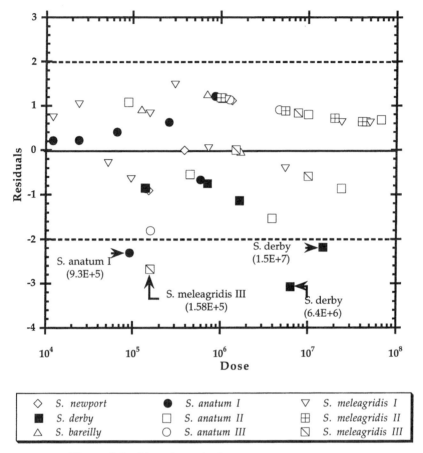

Figure 9-1 Plot of standardized residuals versus dose.

positive than at higher doses for the same study, this low dose response could be a strong candidate for an outlier.

The plot of standardized residuals versus dose is shown in Figure 9-1. There were four outliers identified out of a pool of 50 data points. The outliers are marked on the plot with their respective doses in parentheses.

A standardized residual for *S. derby* corresponding to a dose at 1.5×10^7 appeared to be an outlier. However, upon examination of the raw data, it did not appear to represent nonmonotonic behavior in this particular data set. This point was thus included in the pooled data set. The three other outliers were removed from the data sets and the model was rerun.[2]

---

[2] Removal was done based on their lack of consistency with adjacent doses in the same experiment.

The beta-Poisson model was able to provide a statistically significant fit to the data with the other three outliers removed (Table 9-3). The significance level improved from $p_{fit} = 0.028$ to $p_{fit} = 0.309$. The median infective dose for *Salmonella* in general, represented by N_{50}, was determined to be 2.36×10^4 organisms.

The pooled model was tested to determine if it was statistically indistinguishable from the models for the individual strains. The results of the test are shown in Table 9-4. The p_{add} value (representing a test of improvement in fit of the beta-Poisson model relative to the exponential model) is greater than 0.05; thus the pooled model (with fewer adjustable parameters) is not statistically distinguishable from the models used to describe the individual species. This means that one model can be used accurately with 95% confidence to determine the infectivity of these five *Salmonella* species and six strains.

Figure 9-2 shows the dose–response curve with the lower and upper confidence intervals for the pooled model. This figure is plotted within the experimental dose range, with a linear y-axis. It should be kept in mind that since there were only approximately six subjects tested at every dose, the error bars associated with these points would be relatively large, and 95% of them would overlay the confidence limits of the model.

Model Validation. Prior to using the dose–response model for *Salmonella* to estimate risks associated with the organism, it is necessary to validate the estimates generated by the model. This step allows a comparison to be made between the estimates of the model and real-world situations. If the estimates of the model compare favorably, it lends credibility to the model for estimating risks. Alternatively, an unfavorable comparison may demand further analysis and refinements to the model so as to improve the estimates generated.

To test the predictive abilities of the risk estimates, they can be compared to epidemiological findings in disease outbreaks. The most effective way to validate the results are utilize attack rates and duration of exposure in an outbreak to estimate the most likely dosage in the incriminated medium. This dosage can then be compared with levels detected in the epidemiological investigation. One validation scenario is presented: a waterborne salmonellosis outbreak (9,64).

TABLE 9-3   *Salmonella* Pooled Beta-Poisson Model Parameters[a]

	α	N_{50}	$2 \ln(L)$	p_{fit}
Salmonella pooled model	0.3126	2.36×10^4	49.1877	0.3092

[a]Outliers deleted.

TABLE 9-4 Pooled Versus Individual Species Model Comparison

	Σ 2 ln(L) Individual	2 ln(L) Pooled	p_{fit}	Δ2 ln(L)	p_{add}
S. anatum I, II, III					
S. meleagridis I, II, III	28.015	49.1877	0.3091 at 45 degrees of freedom	21.1722	0.1719 at 16 degrees of freedom
S. derby, S. bareilly, S. newport					

Waterborne Outbreak, Riverside, California. In late May to early June 1965, an epidemic of salmonellosis in which over 16,000 people had symptomatic disease occurred in Riverside, California (9). Epidemiological investigation incriminated the municipal water supply, the causative agent being identified as *Salmonella typhimurium* phage type 2 (64).

Information gathered from local hospitals and a house-to-house survey indicated that the contamination continued either at a fairly constant level or at intermittent intervals for a period of 12 days. Water samples collected from various sites, including city reservoirs and household taps, isolated *S. typhimurium* with a most probable number of 17 salmonellae per liter (9). In addition, *S. typhimurium* was isolated by the Department of Agriculture work-

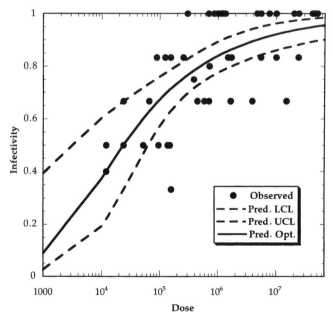

Figure 9-2 Dose–response curve for *Salmonella* in the range observed.

ing independently, suggesting an order of magnitude of 1000 organisms per liter (9).

The average attack rate for persons residing in areas served by the Riverside City water supply was determined to be 13.1%. The highest attack rates were reported in the northeastern part of Riverside. The University of California reported the highest attack rate at 29.5%. The southern parts of the city had the lowest attack rates, with the lowest reported as 5.9% (64).

To predict the dose ranges that would produce attack rates similar to those reported in this outbreak, the daily attack rates were assumed to be independent and identically distributed. Hence the average attack rate of 0.131 over a 12-day period gives a daily attack rate of 0.0116, the highest attack rate of 0.295 gives a daily attack rate of 0.0287, and the lowest attack rate of 0.059 gives a daily attack rate of 0.0051.

To complete the comparison between the dosages predicted by the model and those observed, it is necessary to estimate the amount of water consumed. According to the Riverside County Health Department's report on the outbreak, few residents of Riverside would have drunk as much as 1 L of water a day. The average amount of water consumed per day by the residents of Riverside was estimated to be 0.75 L. The number of organisms per liter predicted by the model were calculated, and the results are summarized in Table 9-5. The ranges predicted by the model are thus well within the ranges detected in samples collected during the outbreak.

Implications

Point Risk Estimates. Once the dose–response model has been validated, it can be applied to the estimation of risks associated with a variety of scenarios. The limiting issue in this aspect would be the availability of additional information, such as organism distribution, occurrence, survival, growth, reduction, and other information necessary for generating risk estimates with a minimum of uncertainty. The model can also be applied to assess compliance, for instance, with regulations and standards or to assist in an effective monitoring program.

TABLE 9-5 Predicted Dose Ranges for 0.75 L/day Water Consumption

12-Day Attack Rate	Daily Attack Rate	Predicted Concentration Range[a] (No. organisms/L)		
		LCL	Optimum	UCL
0.295	0.0287	15.76	375.62	1459.41
0.131	0.0116	6.03	146.69	464.83
0.059	0.0051	2.56	62.86	247.17

[a]Concentrations observed: 17 to 1000 organisms/liter.

There are estimated to be a total of 2 million cases of salmonellosis annually in the United States (5). With a population of 250 million, this represents an annual risk of 8×10^{-3}, or a chance of 1 in 125 of contracting salmonellosis. The average individual daily risk is approximately 2.17×10^{-5}; this corresponds in the dose–response model to a dose range of 0.8 to 0.08 organism. For a 7-day period, the daily risk is approximately 1.1×10^{-3}, giving a dose range of 0.4 to 41 organisms. The dose range increases to 3 to 292 organisms if the daily risk is calculated for a 1-day exposure period.

The U.S. waterborne salmonellosis caseload has been estimated to be 60,000 per year (5), which represents a risk of approximately 1 in 4000 of contracting salmonellosis via the waterborne route. This caseload translates to an annual attack rate of 2.4×10^{-4}. The average individual daily risk, assuming independent exposure and a uniform daily risk, is approximately 6.58×10^{-7}, which corresponds to a dose range of 2×10^{-4} to 0.02 organism. The concentration range corresponding to these doses is 0.01 to 1 organism per 100 L assuming water consumption of 2 L/person per day (62). If the exposure period is reduced (as is more likely to be the case during a contamination event) to a 7-day period, the average individual daily risk is approximately 3.4×10^{-5}, giving a dose range of 0.01 to 1.3 organisms and a concentration of 0.6 to 62 organisms per 100 L.

The surface water treatment rule (46) suggests a safety goal of 10^{-4} (1: 10,000) annual risk of infection. To meet this annual risk level, the corresponding concentrations in finished water can be calculated. The uniform daily risk producing an annual risk of 10^{-4} is calculated to be 2.74×10^{-7}. If the daily water intake is assumed to be 2 L/person per day, the concentration of salmonellae in water should be maintained at less than 0.13 organism per 100 L (95% C.I., 0.005 per 100 L to 0.49 per 100 L). From a monitoring standpoint, this corresponds to the detection of 1 organism in approximately 792 L of water.

Assessing Risks Using Monte Carlo Analysis. The infectivity dose–response assessments were used to develop an indication of potential risks from *Salmonella* in finished drinking water. To investigate the relative contribution of variability and uncertainty, Monte Carlo computations were conducted keeping all potential sources of uncertainty and variability as distributions, as well as a partial Monte Carlo assessment.

The partial Monte Carlo analysis used the best-fit point estimates of infectivity (α, N_{50}) and probability distributions for the other inputs. The full Monte Carlo analysis incorporates the uncertainty distribution rather than the point estimates of the parameters (α, N_{50}) in the dose–response model as an input into the analysis. The Monte Carlo risk calculations and the bootstrap estimation of the uncertainty distribution of the dose–response parameters were performed using CrystalBall.

The uncertainty distribution of the parameters for the beta-Poisson model were determined using a bootstrapping approach. Figure 9-3 shows a scatter-

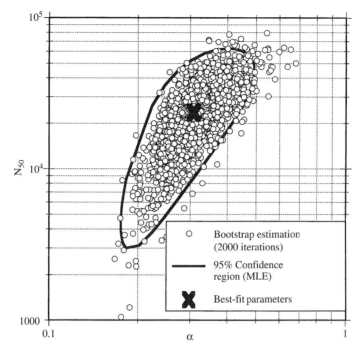

Figure 9-3 Confidence regions and bootstrap estimation.

plot of a bootstrap sample of 2000 points which was used as the input for the Monte Carlo analysis.

To perform a Monte Carlo risk analysis, several additional input distributions must be defined. This can be done by consulting the literature, or when data are lacking, approximations to the distribution can be made. The goal of this Monte Carlo risk analysis was to analyze the risk of infection from *Salmonella* associated with drinking treated water drawn from source waters of varying quality.

To use the beta-Poisson model to assess the risk of infection, two ultimate inputs are required. The first input is the set of parameters for the model. These values were simulated, as illustrated previously, to generate a set of 1000 pairs of parameter values. The second input is the dose ingested. The dose ingested is composed of several variables that have to be simulated. Figure 9-4 shows a flowchart for the Monte Carlo simulation process.

To model source waters of varying quality, the concentrations of coliforms in the source water were used as a starting point. Mui (56) measured the frequency distribution of coliforms in Lake Michigan and tested the negative binomial, Poisson, and lognormal distribution for their ability to describe the data. The lognormal distribution with parameters $\mu = 2$ and $\sigma = 1.41$ [units: in(number/100 mL)] described the data best. This input assumption could

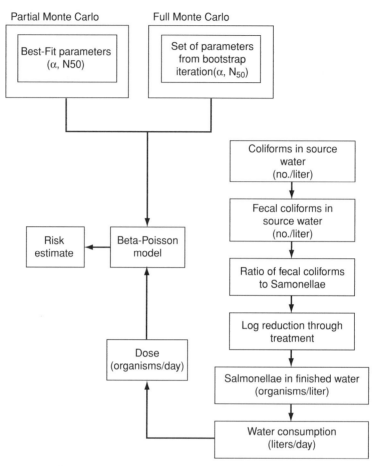

Figure 9-4 Monte Carlo simulation flowchart.

easily be changed to simulate a specific source water after the distribution of coliforms at that location has been determined.

To simulate the frequency distribution of fecal coliforms in the source water, a 5:1 ratio of coliforms to fecal coliforms was used (6). This ratio was used to directly modify the parameters of the lognormal coliform distribution. As a result, a lognormal distribution for fecal coliforms in source water with parameters $\mu = 0.3905$ and $\sigma = 1.41$ [units: ln(number/100 mL)] (geometric mean = 1.47 per 100 mL) was employed directly. To simulate a poor-quality water source, the concentration of fecal coliforms in the source water was assumed to be 2000 per 100 mL; thus a lognormal distribution with parameters $\mu = 7.6$ and $\sigma = 1.41$ [units: ln(number 100 mL)] simulated the high-pollution water source. In some states this is the maximum allowable level of fecal coliform bacteria allowed in a source water used for a public water

(57). However, there is no federal standard on a maximum allowable level for fecal coliforms in source water prior to treatment. Four intermediate-quality water sources were also simulated, and the parameters used for all six water sources are shown in Table 9-6.

The ratio of fecal coliforms to salmonellae can vary from location to location. Ratios of 1 : 100 and 1 : 540 *Salmonella* to fecal coliforms have been reported (6). To describe this relationship and account for the uncertainty associated with the ratio, a triangular distribution was selected with a modal value of 1 : 540 and maximum and minimum values of 1 : 5400 and 1 : 54 respectively.

A triangular distribution was also used to simulate the log reduction through treatment. The minimum and modal values for this distribution were selected to be a 4-log reduction and a maximum log reduction value of 5. This assumption can be refined based on the specific treatment plant being simulated. Furthermore, individual treatment processes can also be simulated in stages.

The final step is to simulate the water consumption rate. The lognormal distribution reported by Roseberry and Burmaster (69) was used for this input. A simulation of 5000 iterations was run for all six water quality scenarios. The risk estimates associated with each water source are tabulated in Table 9-7. The range of the risks estimated in the simulation are relatively large. This range can be reduced dramatically if more precise information becomes available. However, a certain portion of the range arises from irreducible population variability (e.g., in water consumption). The distributions of the inputs were based on broad assumptions. For instance, if the distribution of fecal coliforms at a specific source water is determined and the ratio of the fecal coliform to *Salmonella* for the site is measured, the range of risks would decrease dramatically.

It is nonetheless worthwhile to note that the upper range of risks estimated for all the source waters are in excess of the 1 : 10,000 risk of infection

TABLE 9-6   Lognormal Parameters for Fecal Coliform Frequency Distributions

Source Water Quality	Lognormal Distribution Parameters		
	μ	σ	Geometric Mean (No./100 mL)
Poor	7.6	1.41	2000
Intermediate 1	6.91	1.41	1000
Intermediate 2	6.21	1.41	500
Intermediate 3	4.61	1.41	100
Intermediate 4	3.91	1.41	50
Good	0.391	1.41	1.47

TABLE 9-7 Full Monte Carlo Simulation Results for Annual Risk

Source Water Quality	Fecal Coliforms in Source Water Geometric Mean (No./100 mL)	Annual Risk Range		
		Range Minimum (Approximate Odds 1:?)	Range Mean (Approximate Odds 1:?)	Range Maximum (Approximate Odds 1:?)
Poor	2000	1.12×10^{-7} (8,900,000)	3.09×10^{-4} (3,200)	1.27×10^{-1} (8)
Intermediate 1	1000	3.86×10^{-8} (25,900,000)	1.62×10^{-4} (6,100)	5.64×10^{-2} (10)
Intermediate 2	500	1.02×10^{-8} (98,000,000)	8.14×10^{-5} (12,000)	2.4×10^{-2} (40)
Intermediate 3	100	3.94×10^{-9} (250,000,000)	1.33×10^{-5} (75,000)	5.13×10^{-3} (100)
Intermediate 4	50	1.31×10^{-9} (700,000,000)	7.82×10^{-6} (120,000)	1.62×10^{-3} (600)
Good	1.47	6.07×10^{-11} (16,000,000,000)	2.86^{-7} (3,400,00)	1.94×10^{-4} (5,100)

suggested by the surface water treatment rule. In addition, the mean values of the annual risk for the poor quality (fecal coliform geometric mean 2000 per l00 mL) and first intermediate (fecal coliform geometric mean 1000 per l00 mL) source waters are also in excess of the 1 : 10,000 risk. These two simulations also produce results that compare favorably with estimates (5) of a 1:4000 risk of infection from *Salmonella* via the waterborne route.

It is important to note that the risk to the consumer from water drawn from these sources, especially the two lower-quality waters might be unacceptably great. The odds for infection, at least once in any year, for the simulation using a fecal coliform concentration of 2000 per l00 mL, was determined to be approximately 1 : 3200. In addition, since there are no national regulations on source water quality (with respect to coliforms or to pathogenic microorganisms), the risk to the consumer in some cases may be even greater. It is obvious that the quality of the source water plays an important role in the final risk imposed on the consumer. This illustrates the importance of sewage treatment requirements and regulations, and the implementation of other watershed protection measures, such as the control of agricultural runoff.

The risks estimated by the Monte Carlo analysis need further refinements in the uncertainty associated with the inputs. With such wide confidence bands, it is difficult to convey reliable risk estimates for use in risk management. The two inputs (parameters and dose) of the dose–response model can be evaluated to assess the source of the greatest uncertainty. The partial Monte Carlo analysis, using the best fit estimates for the parameters will generate a range of risks based on the variation in the exposure estimates. This range can then be compared to the range generated using the full Monte Carlo analysis. From a comparison of the partial and full Monte Carlo analysis shown in Figure 9-5, it can be seen that the uncertainty associated with the dose–response parameters in risk estimates is minimal in comparison with the uncertainty associated with estimating the exposure itself.

Summary. The dose–response relationship for nontyphoid *Salmonella* can be described by the beta-Poisson model. This model was able to describe sufficiently the dose–response relationship of five *Salmonella* species and six strains. The median infective dose for *Salmonella* in general was estimated to be 2.36×10^4.

Point risk estimates performed using the model determined the average exposure for infection from *Salmonella* to range from 3 to 292 organisms, based on a 1-day exposure period and an estimated total U.S. caseload of 2 million cases per annum (5). Estimates for the infecting exposure in water, based on an estimated 60,000 cases per annum (5), water consumption of 2 L/person per day, and a 7-day exposure period, ranged from 0.6 to 62 organisms per 100 L. It was also determined that based on the 1:10,000 risk of infection suggested as a safety goal in the surface water treatment rule, the detection of less than one organism in 791 L of water would represent an acceptable level of risk.

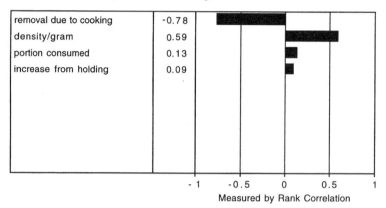

Sensitivity   Chart

Target Forecast: organisms consumed

Figure 9-5   Comparison of partial and full Monte Carlo analysis for annual risk of infection estimates.

A full and a partial Monte Carlo analysis were performed to determine the probability distribution of risk. The risk associated with drinking treated water drawn from source waters with a range of qualities was estimated. The results showed that the 1 : 10,000 risk of infection was exceeded at the upper risk range for all source waters simulated; in addition, the two lesser-quality source waters (fecal coliform concentration 2000 per 100 mL and 1000 per 100 mL) had a mean risk of infection in excess of the suggested risk level. However, the range of risks estimated in the full Monte Carlo analysis were very large. To assess the source of the greatest uncertainty (dose–response parameters and exposure), the full and partial Monte Carlo risk estimates were compared. It was determined that the uncertainty associated with the dose–response parameters in the risk estimates were minimal compared with the uncertainty and variability associated with estimating the exposure itself.

The dose–response model developed can be used to assess the risks associated with *Salmonella* transmission through the water- and foodborne route. The risk assessment model can be utilized by governing bodies as a tool for setting regulations and standards.

In the food industry, risks can be assessed as a preliminary step in hazard analysis and critical control point (HACCP) programs in manufacturing facilities. In addition, a quantitative risk assessment can be used to determine points in the food production chain in which improvements or modifications to the process can reduce the risk to the consumer.

Water utilities can employ the methods outlined to set treatment levels based on predetermined risk levels, for decision making, to set goals for monitoring programs, and to determine compliance with regulatory requirements such as the Safe Drinking Water Act and the Water Pollution Control Act.

It should also be noted that outbreaks of salmonellosis frequently involve *S. typhimurium* or *S. enteriditis,* especially in food. The body of dose–response information does not include these strains (Table 9-2), so some uncertainty about applicability of strain-to-strain extrapolation remains. However, confidence in the applicability of the general nontyphoid *Salmonella* model is enhanced by the comparison with the Riverside CA outbreak. It is also comforting that the majority of nontyphoid strains can be adequately depicted by a single dose–response model. Whether this will continue to be the case, particularly as more data become amassed using a greater number of subjects, is not foreseeable.

Shigella[3]

Dating as far back as 1933 there have been documented cases of deaths caused by shigellosis in the United States (24). Dupont and Hornick asserted that ingestion of only 10 to 100 *Shigella* cells can lead to infection. Since this dosage is much lower than the estimated doses of 10^5, 10^8, and 10^8 estimated for probable infection by *Salmonella, E. coli,* and *V. cholerae,* respectively (19), *Shigella* can easily be transmitted by person-to-person contact, food, and water to create adequate exposure for infection. In the United States there are an estimated 300,000 cases and 600 deaths per year attributed to infection by *Shigella,* of which approximately 40% of the cases are due to ingestion of contaminated food or water (5).

Data Background. The data from dose–response studies involving two species and three strains of pathogenic *Shigella* used in this paper are shown in Table 9-8. All the studies involve the use of healthy male adults. The doses and number infected in each study are indicated in Table 9-8. In all there are 13 possible data points that could be used to fit a model. The *Shigella* species denoted 2A## and 2A# are from the same species and strain, but the differences in notation are used for ease of separation of data results and evaluation in this paper between the two studies performed at two different dates.

Data Analysis. From the data sets in Table 9-8, the best-fit values of the parameters were found using the maximum likelihood estimation (MLE) procedure (29,32,62). Each column of Table 9-9 shows the dose–response parameter estimate values for each of the data sets for the two models. Some columns show the dose–response parameter estimate values for specific strains where a specific point was removed for an acceptable fit. Also, Table 9-9 shows the dose–response parameter estimate values when the data sets

---

[3] This section is modified from Crockett et al. (16).

TABLE 9-8 Experimental Data Sets for *Shigella*

Infected with[a]: Strain	Dose	Positive Subjects[b]	Negative Subjects	Total Number of Subjects	Reference
S. dysenteriae					
Pan M-131	10	1	9	10	43
Pan M-131	200	2	2	4	43
Pan M-131	2,000	7	3	10	43
Pan M-131	10,000	5	1	6	43
S. flexneri 2A##					
2457T	10,000	1	1	4	20
2457T	100,000	3	1	4	20
2457T	1×10^6	7	1	8	20
2457T	1×10^7	13	6	19	20
2457T	1×10^8	7	1	8	20
S. flexneri 2A#					
2457T	180	6	30	36	21
2457T	5,000	33	16	49	21
2457T	10,000	66	21	87	21
2457T	100,000	15	9	24	21

[a] *S. flexneri* 2A## represents data from Dupont and Hornick's 1969 test; *S. flexneri* 2A# represents data from Dupont and Hornick's 1972 test. This notation is used for distinction between data sets.
[b] Positive subjects were defined as having signs of illness or infection, such as fever, diarrhea, positive stool isolations, or antibody positive.

TABLE 9-9 Dose–Response Parameters for *Shigella*

	Species/Strain					
	S. dys 131	*S. flexneri 2A##*	*S. flexneri 2A#*	*S. flexneri 2A#*	Pooled	Pooled
Dose removed	None	None	1E + 7	None	None	1E + 7 (2A##)
Exponential model						
k	2.210	9,080,000	9,520,000	14,908	14,414	14,415
$2 \ln(L)^a$	13.21	73.6	73.55	168.59	551.64	299.03
p_0^b	0.0042	7.2×10^{-16}	7.35×10^{-16}	2.5×10^{-36}	2.2×10^{-110}	1.4×10^{-57}
Beta-Poisson model						
α	0.277	0.144	0.258	0.265	0.162	0.2099
N_{50}	238	35,400	29,400	1,482	1,127	1,120
$2 \ln(L)^a$	0.032	3.44	1.26	8.73	24.0	16.85
p_0^b	0.9844	0.3288	0.5337	0.0127	0.0127	0.0778

[a]The $2 \ln(L)$ value is compared to the chi-squared statistic for number of degrees of freedom equal to the number of data points minus the number of parameters used. The $2 \ln(L)$ value is twice the log-likelihood value determined using the MLE method for each specified data set and model.

[b]If $p_0 > 0.05$, the fit is accepted; if $p_0 < 0.05$, the fit is rejected.

are pooled to determine whether the *Shigella* species could be characterized by a single dose–response function.

Initial examination of the data (Figure 9-6) suggested that the *S. flexneri* 2A## data contained an "outlier" at the 10^7 dose level. Several additional analyses supported the hypothesis that this dose level might represent an aberrant observation. First, the contribution of this dose level to the lack-of-fit likelihood ratio statistic was the greatest of all the dose levels. Second, when a monotonic test for trend was performed (42), inclusion of all data points resulted in failure to reject the null hypothesis of no trend ($p = 0.057$), while deletion of this dose level resulted in a statistically significant ($p = 0.014$) trend. Based on this analysis, further quantification of the dose–response information was performed both including all data points and excluding the 10^7 dose from the *S. flexneri* 2A## data.

When the data sets were pooled together, the combined data sets did not provide a statistically significant fit to the models($p > 0.05$). However, removal of the 10^7 dose from the *S. flexneri* 2A## data set resulted in a statistically significant fit to the dose–response models ($p > 0.05$). In later statistical comparisons between the pooled and separated models, only the results from the separated data sets with the outlier dose removed were used.

In most cases the beta-Poisson model fit the data, while the exponential model did not fit at all, even when certain doses where removed. Table 9-10 shows the significance in improvement from the exponential to the beta-

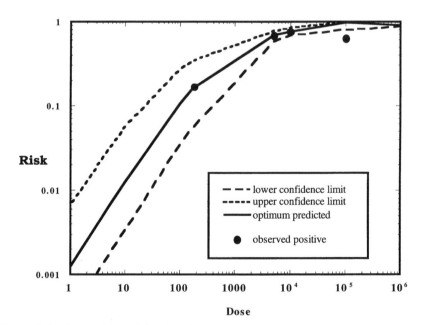

Figure 9-6 Comparison of fitted and observed dose–response data for *Shigella flexneri* 2A##.

TABLE 9-10 Improvement in Fit with Beta-Poisson Model Versus Exponential Model

			Strain			
	Pan M-131	2A##	2A##	2A#	Pooled	Pooled
Dose removed	None	None	1×10^7	None	None	1E + 7 (2A##)
Exp 2 ln(L)	13.21	73.6	73.6	168.6	551.64	299.03
Beta 2 ln(L)	0.032	3.44	1.26	8.73	24	16.85
$\Delta 2$ ln(L)	13.178	70.16	73.55	159.86	527.64	282.18
p_0	0.003	5.5×10^{-17}	9.8×10^{-18}	1.2×10^{-36}	9.2×10^{-117}	2.5×10^{-63}
Improvement	Yes	Yes	Yes	Yes	Yes	Yes

Poisson model. Table 9-11 also performs a test for improvement in fit by pooling the *Shigella* data versus the data fitted separately for each *Shigella* species. The value of p_0 was greater than 0.05, with a value of 0.1452. Therefore, the null hypothesis that the data sets are indistinguishable cannot be rejected.

The 95% upper and lower likelihood-based confidence limits for the pooled model for *Shigella* and the individual *Shigella* species predicted infectivity curves are shown in Figure 9-7. From this graph one can see that the pooled *Shigella* model definitely overlaps the other *Shigella* species-predicted infectivity curves except for the 2A## strain. However, the 2A## strain 95% upper confidence limit was found to equal 1, creating an overlap to some extent between the 2A## and pooled models. These overlaps confirmed the adequacy of the pooled *Shigella* model to determine a representative predicted infectivity. If the highest 95% upper confidence limit (in our case, Pan M-131) and lowest 95% lower confidence limit (in our case, 2A##) of the data sets are plotted as shown in Figure 9-7, and the upper confidence limit of 2A## is excluded, the pooled *Shigella* model infectivity curves are contained within these boundaries, again confirming the adequacy of the pooled *Shigella* model to determine a representative predicted infectivity.

If the 95% confidence regions are determined and plotted for the acceptable models for each data set and the pooled data. sets as shown in Figure 9-8, some interesting observations can be made. The *Shigella* pooled model confidence region overlaps all of the other data set confidence regions and encompasses the best-fit parameter of the data set for *Shigella 2A#*. The considerable overlap in the confidence regions shows that the fitting technique used is fairly robust to the removal of the outliers from the pooled data set. The pooled *Shigella* confidence region is fairly small compared to the other

TABLE 9-11   Test for Statistical Indifference: Pooled and Separate Data Sets[a]

Strain	Beta 2 ln(L)	Number of Parameters
Pan M-131	0.0315	2
2A##[b]	1.2556	2
2A#	8.731	2
Pooled	16.8464	2
$\Delta 2 \ln(L)$	6.8283	
p_0^c	0.1452	4 degrees of freedom (6 separate, 2 pooled)

[a] Chi-squared value at 4 degrees of freedom = 9.488 ($p = 0.05$).
[b] Outlier removed from 2a## ata set. [c] data sets are statistically the same only if $p_0 > 0.05$. If $p_0 < 0.05$, data sets are statiscially different from one another.

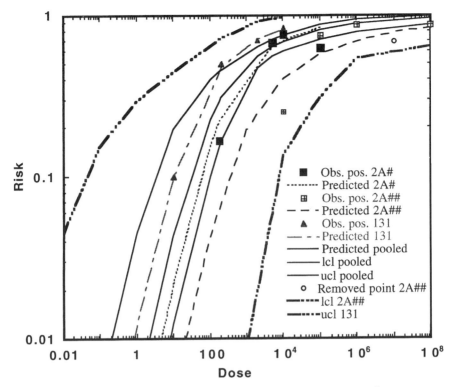

Figure 9-7 Overlap of 95% confidence limits and predicted curves of separate species with all species pooled.

Shigella species confidence regions plotted in Figure 9-8 and when ranges of N_{50} are compared in Table 9-9. This is expected because an increased number of observations will decrease the estimated variance. The upper confidence limit in the range of α does not go past 1.5, and the lower confidence limit goes substantially below an α value of 1, indicating that there is a deviation in the data from the exponential model. The uncertainty in the estimation of α for the pooled model is about a factor of 3 and N_{50} is about a factor of 8. Also, in Figure 9-8 one can see that the confidence region for 2A## is open to the left, meaning that a median infective dose could be much less than 10 organisms indicated by Dupont and Hornick (19).

Discussion. Based on an estimated 300,000 annual cases in the United States for *Shigella* and a range of 30 to 40% due to food- and waterborne infections and assuming that the entire U.S. population of 250 million people is at risk uniformly throughout an entire year, this would create an average individual daily risk ranging from 9.9×10^{-7} to 1.3×10^{-6}. Using the dose–response relationship from the pooled *Shigella* model, the average daily exposure

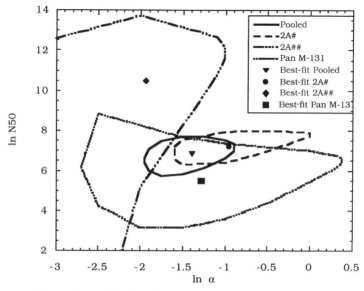

Figure 9-8   95% Confidence regions for *Shigella* species.

would be 2×10^{-4} and 3×10^{-4} organisms for the minimum and maximum, with a range of 1.8×10^{-5} to 1×10^{-3} organism. Since some reported cases of outbreaks have occurred at annual local events such as picnics, swimming, and concerts, or at a vacation spot or a cruise (7,8,41,44,47,61,70,76,77), it is plausible to assume that exposure may occur only sporadically, for example, for 7 days out of the year, with the risk on the other days being zero. The daily risk then determined over this 7-day period is between 5.1×10^{-5} to 6.9×10^{-5} for the 30 to 40% range of estimated cases. Using the pooled *Shigella* model, this would correspond to an average exposure of 0.01 to 0.014 organism, with a range of 0.001 to 0.05 organism. These results for the pooled *Shigella* model are consistent with prior statements that it only takes ingestion of 1 *Shigella* cell to cause infection (19).

Since *Shigella sonnei* and *Shigella flexneri* are considered as the two main species of *Shigella* responsible for outbreaks (21), the assumption that *S. flexneri* or a combination of *S. flexneri* and *S. dysentariae* is as or more infective may not be true. However, the data presented here show an overlap in the *S. dysentariae* and *S. flexneri* confidence regions and limits as well as the acceptable fit of the pooled data set. This suggests that more than one species of *Shigella* can be pooled to give an accurate prediction for *Shigella* in general.

In additional work not shown here, the dose–response relationships were demonstrated to be in concordance with two outbreaks of *Shigella* (16).

Pathogenic *E. coli*

Since the mid 1960s, it has been recognized that in addition to being a normal component of the human fecal flora, some strains of *Escherichia coli* may be human pathogens (72). In the United States, there are estimated to be 200,000 cases of infection and 400 deaths per year attributable to pathogenic *E. coli* (5). It is also estimated that half the cases result from exposure to contaminated food or water. In a recent dramatic example, the 1993 foodborne outbreak caused by the O157:H7 strain in the northwestern United States affected thousands of people and resulted in at least four deaths (39).

Note that the O157:H7 strain of *E. coli* may be more pathogenic than many other (nonenterohemmorhagic) strains. Therefore, this section would not (perhaps) give a realistic portrayal of the risk from this strain. In this section, an analysis of the available data on response of humans to ingestion of oral doses of pathogenic *E. coli* is made. This is used to develop a dose–response relationship that would be suitable for assessment and management of risks in food, water, and via other vehicles of exposure.

Data Sets. There are four studies that have been conducted on human dose–response to pathogenic *E. coli*. These encompass seven strains of organisms. All involve oral administration of the organism to healthy adults. The dosing levels and strains are summarized in Table 9-12 (these data were also shown previously in Table 7-9). There are a total of 19 different combinations of strains and dose levels administered. In all cases, the response noted was illness (e.g., diarrhea, fever, gastrointestinal distress). In all cases, microbial levels were determined by viable plating techniques. The organisms used in these studies may be classified into one of three groups (22): enteropathogenic, enteroinvasive, and enterotoxigenic, and this classification is also noted in the table.

Results and Discussion. For the data set, the best-fit values of the parameters of both the exponential and beta-Poisson dose–response models were determined using the method of maximum likelihood (32,55). Comparison of the two models showed a statistically significant improvement in fit using the beta-Poisson model, which therefore was adopted for depicting these organisms.

The first row of Table 9-13 shows the values of the dose–response parameters obtained using all data in Table 9-12. The goodness-of-fit statistic [likelihood ratio, 2 ln (L)] is statistically acceptable.

It was also desirable to test whether or not all the strains in the database could be characterized by a single dose–response function. To do this, the maximum likelihood procedure was repeated a number of times, deleting the data of one strain at a time. The second and subsequent rows in Table 9-13 show the effect of deleting individual strain data sets on the fit. The difference

TABLE 9-12 Experimental Data Sets for *E. coli*

Strain	Dose (No. organisms)	Affected Subjects	Unaffected Subjects	Total Subjects	Classification (22)	Reference
1624 (O124)	1.00×10^4	0	5	5	EI	18
	1.00×10^6	1	8	9	EI	18
	1.00×10^8	3	2	5	EI	18
4608 (O143)	1.00×10^4	0	5	5	EI	18
	1.00×10^6	0	5	5	EI	18
	1.00×10^8	5	3	8	EI	18
B2C (O6:H16)	1.00×10^8	2	3	5	ET	18
	1.00×10^{10}	3	2	5	ET	18
B7A (O148:H28)	1.00×10^8	1	4	5	ET	18
	1.00×10^{10}	4	1	5	ET	18
H10407 (O78:H11)	2.70×10^8	9	7	16	ET	27
O111	7.00×10^6	7	4	11	EP	26
	5.30×10^8	8	4	12	EP	26
	6.50×10^9	11	0	11	EP	26
	9.00×10^9	12	0	12	EP	26
O55	1.40×10^8	6	2	8	EP	37
	1.70×10^9	5	2	7	EP	37
	5.00×10^9	6	2	8	EP	37
	1.60×10^{10}	7	1	8	EP	37

[a]EI, enteroinvasive; ET, enterotoxigenic; EP, enteropathogenic.

TABLE 9-13 Effect of Deleting Strains from the Analysis[a]

	α	N_{50}	$2 \ln(L)$	Points	p_a 19.4%	p_0 (%)
All Data	0.1952	3.01×10^7	21.7542	19		
Strain omitted						
1624	0.1886	2.7×10^7	21.6445	16	8.6	99.06
O55	0.1993	3.11×10^7	20.8350	15	7.6	92.18
4608	0.1806	2.41×10^7	20.0487	16	12.9	63.57
H10407	0.2043	2.57×10^7	20.7825	18	18.7	32.42
B2C	0.2127	2.56×10^7	19.0341	17	21.2	25.66
B7A	0.2028	2.45×10^7	18.1238	17	25.6	16.28
O111	0.1778	8.6×10^7	6.5558	15	92.4	0.43

[a] $2 \ln(L)$, Log-likelihood statistic for goodness of fit; p_a, goodness of fit of the model to the data (model is rejected if $<5\%$); p_0, outlier probability (the omitted strain is rejected if $<5\%$).

in the log-likelihood statistic (from that obtained considering all data sets—that is, in the first row of Table 9-13) was compared against a chi-squared distribution with degrees of freedom equal to the number of data points from a particular strain. The upper tail probability, p_0, in Table 9-13, was regarded as the probability that an improvement in fit as great could have occurred if there was in fact no difference between the excluded strain and the remaining strains. In other words, p_0 is the probability that deletion of the same number of data points would result in as great an improvement in fit if in fact there was no underlying difference between the removed and retained points.

In a similar manner, although not shown, a test was made of the difference between enterotoxigenic strains, enteroinvasive strains, and enteropathogenic strains. This test revealed no significant difference among these three groups of organisms ($p > 0.05$, by likelihood ratio test; results not shown). On this basis it was concluded that the O111 data set was statistically different from the other data sets. Therefore, the remainder of this analysis was conducted both with the entire data and the subset excluding the O111 data. There does not appear to be a significant difference in experimental design in the study involving O111, and hence the statistical difference could suggest the existence of a strain difference in intrinsic pathogenicity. Since there was no statistically significant difference per se between the groups of organisms based on mechanism of pathogenicity, the nature of the discrepancy involving O111 would appear to be a strain-specific (or experiment-specific) factor.

Figure 9-9 compares the experimental observations to the fitted dose–response curve using the entire data set ($\alpha = 0.1952$, $N_{50} = 3.01 \times 10^7$). Also shown on this figure are the likelihood-based 95% confidence limits using a likelihood ratio approach. Note that three of the four data points for O111 were outside the confidence region, visually demonstrating the point made by the analysis of Table 9-13. A secondary effect of including the O111 data points was the pulling upward of the dose–response curve (predicting greater risk) and the forcing of other data points out of the confidence region (B2C and H10407).

Figure 9-10 presents the analogous comparison for the case where the O111 data are excluded from the analysis ($\alpha = 0.1778$, $N_{50} = 8.6 \times 10^7$). There was better "coverage" of the observed data by the dose–response relationship, which results from the exclusion of the apparently different information provided by the O111 experiments.

Figure 9-11 compares the 95% confidence regions and the maximum likelihood estimates of the parameter values for the full data set and for the subset excluding O111. As would be anticipated, exclusion of the apparent outliers widened the confidence region. However, interestingly, there was considerable overlap in the regions, which shows that the fitting technique appears to be relatively robust to the apparent outlying data. The upper confidence limit to the parameter α was substantially below 1, which indicates a major deviation from the exponential model and reflects the present of heterogeneity in the host–pathogen interaction process. The uncertainty in the estimation of the

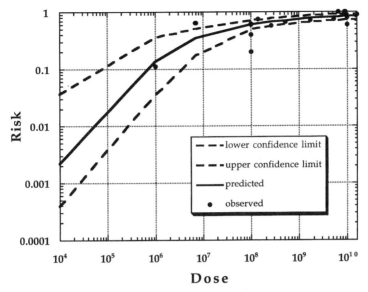

Figure 9-9 Comparison of fitted and observed dose–response data: all points included. Legend indicates strain examined.

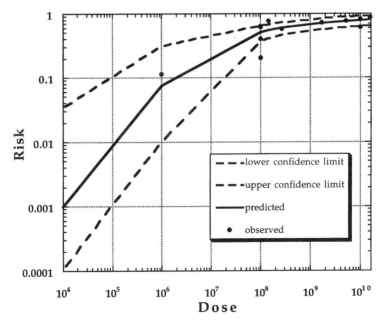

Figure 9-10 Comparison of fitted and observed dose–response data: O111 data excluded. Legend indicates strain examined.

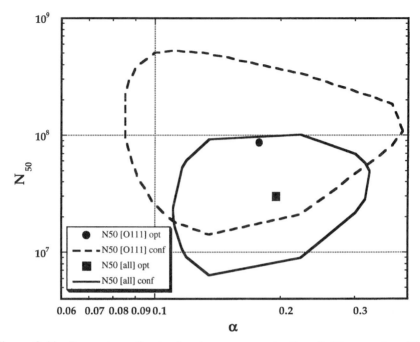

Figure 9-11 Parameter estimates for dose response to *E. coli*. Lines enclose 95% confidence regions. Points mark maximum likelihood estimates.

median effective dose was less than an order of magnitude in either direction, while the uncertainty in the estimation of α was about a factor of 2 in either direction.

As noted above, there are 100,000 estimated *E. coli* cases per annum attributed to food and water exposure in the United States (5). This may be translated into an annual attack rate of $p_y = 4 \times 10^{-4}$ based on a total U.S. population of 250 million. If this represents the result of exposure to uniform daily risk, and if the exposures are independent, the daily risk may be computed from the assumption of independent identical risks.

Assuming that the entire U.S. population is at risk uniformly for the entire year, this leads to an average daily attack rate of 1.1×10^{-6}. If the dose–response relationship (based on excluding the 0111 data) is extrapolated downwards, this implies an average daily exposure to 11 organisms (range 0.3 to 105). Very likely, the risk to pathogenic coliform organisms is not uniform in time. If, instead, it is assumed that the risk occurs during only 7 days each year (the risk on other days being zero), the daily risk over this 7-day period is 5.7×10^{-5} (this is computed by substituting i = 7 in equation 8-2, which corresponds to an average exposure during this period of 527 organisms (range 14 to 5500). As the estimated number of days for exposure decreases and the corresponding daily risk level during the vulnerable period increases, the number of organisms would also increase. Although there does not appear to be a systematic estimate of the number of enteropathogenic coliforms that individuals are exposed to daily from food and water (a total daily intake number), the levels implied by this use of the dose–response relationship appear to be plausible.

Several interesting observations emerge from analysis of these data. First, in this work, the median effective dose of enteropathogenic *E. coli* is estimated at 2.5×10^6 organisms. Our estimates for the median effective doses of *Vibrio cholerae* and *Shigella* are approximately 250 (for organisms taken simultaneously with sodium bicarbonate) and 200 to 40,000 (depending on strain), respectively. Hence the coliforms are of considerably less infectivity (as judged by the median effective dose) than the more classical pathogenic agents.

Second, it is interesting that with the exception of a single strain (O111), all the other strains examined (both invasive and toxigenic) could be described by a single dose–response relationship. The O111 data could not be described by either an exponential or a beta-Poisson dose–response relationship ($p < 0.05$ for goodness-of-fit statistic), although (not shown) a fit of these data to a lognormal dose–response relationship could be obtained. This suggests that this early data set may have contained excessive variability.

Whether a relatively consistent dose–response relationship may characterize other strains of *E. coli* is not known, since human dose–response information is lacking. In particular, enterohemorrhagic strains such as O157:H7, because they act by different mechanisms (22), may have somewhat different dose–response relationships. From the few published epidemiological investigations, however, the pathogenicity of the hemorrhagic coliforms appears greater than that of nonhemorrhagic strains, for which dose–response information is available (38,83). Therefore, the dose–response relationships developed in this section may represent only upper limits to safe dose determination for hemorrhagic strains. Clearly, further work is necessary on this point.

It should be noted that unlike most other data sets examined, these data utilized disease as an endpoint rather than infection. The variability in morbidity ratio may introduce a greater degree of "noise" in dose–response information. It has been shown that adults may be infected with enteropathogenic *E. coli* without exhibiting symptoms (65). Quantitative characterization of the dose–response relationship for morbidity ratio with pathogenic *E. coli* awaits additional information. In this respect, animal models may be useful.

Summary. The risk associated with exposure to ingested pathogenic *E. coli* can be described by a beta-Poisson dose–response relationship. This relationship fits data from all but one strain of organisms for which data exist. The beta-Poisson parameters show significant evidence of heterogeneity, and from this it is estimated that the current U.S. caseload of pathogenic *E. coli* illness would occur based on a mean daily exposure to 10 (confidence limits 0.8 to 100) organisms per day.

E. coli O157:H7

The particular strain of *E. coli* known as O157:H7 has been of increasing concern, particularly since a multistate foodborne outbreak in the United States in 1994 from hamburger (3). First identified as a foodborne pathogen in 1982, approximately three foodborne outbreaks per year have been identified in the United States since then (25). A major waterborne outbreak in Cabool, Missouri, occurred in 1989 and resulted in over 240 illnesses and over 32 deaths (79). There have also been swimming-associated outbreaks of this organism (38).

The severity of infection with this organism results from its possession of "shiga-like toxins" having commonality with the genus *Shigella* (25,28). The prevalence of this organism in foodstuffs, and the apparent low median infectious dose, may be due to the acid tolerance that is manifest in this serotype. It has been isolated in prepared foods with pH levels as low as 3.6, at which levels it can apparently persist (23,25). Although no known human dose–response studies of *E. coli* O157:H7 have been conducted, it has been suggested that on the basis of similarity of pathogenic mechanism, the dose–response relationship for *Shigella* species (Table 9-9) may be a suitable surrogate (12).

Cryptosporidium

Human dose–response data on pathogenic *Cryptosporidium parvum* has been presented in Table 7-7. Example 7-3 illustrated data fitting to the exponential dose–response model, with a graphical comparison of goodness of fit in Figure 7-9.

Water. *Cryptosporidium parvum* is a recently emerging pathogen of concern in drinking water. The largest outbreak reported to date also serves to provide confirmation of the reasonabless of the experimental dose–response curves.

Confirmation: Milwaukee Outbreak.[4] The largest known waterborne outbreak of disease occurred in March–April 1993, resulting from an apparent breach of treatment in one Milwaukee, Wisconsin, water treatment plant, resulting in a widely disseminated exposure to *Cryptosporidium* in finished water. This event is believed responsible for over 400,000 cases of illness (45). Using the information presented in the outbreak investigation, in conjunction with the

---

[4] The comparison with the Milwaukee outbreak was previously reported in Haas and Rose (31).

dose–response curve, we examine whether the occurrence of the Milwaukee outbreak was consistent with the infectivity as noted in controlled laboratory investigations.

Based on the investigation of the Milwaukee outbreak, the following information can be gleaned (45):

- Based on the distribution of onset cases, the most likely duration of contamination (t) appears to have been about 21 days, with a possible range of 15 to 30 days. A triangular distribution is used to model this uncertainty.
- The attack rate (r) (based on the entire metropolitan area) determined from an epidemiological survey was 0.21. Based on the sample size employed, the attack rate distribution was described as normal with a standard deviation of 0.01.

To complete the analysis, information is needed on the water ingestion rates of the exposed population. In the absence of site-specific information, the distribution of Roseberry and Burmaster (69) is used, in which the daily water ingestion rate (q) (in mL) is lognormally distributed with a mean of 1948 mL and a standard deviation of 827 mL. One further detail must be assumed to complete the calculation. As with the assumptions used earlier, it is assumed that each day of exposure constituted an individual and identical risk.

From these assumptions, we wish to determine what the average oocyst concentration would be during the exposed period (assuming level exposure) consistent with the attack rate, duration of exposure, and dose–response information. Given $r = 0.21$, we determine that an average daily risk (based on 21 days of exposure) would be

$$p = 1 - (1 - r)^{1/t}$$
$$= 1 - (1 - 0.21)^{1/21} \qquad (9\text{-}2)$$

or $p = 0.0112$. Now substituting this daily risk into the exponential dose–response model and using the best-fit value of the exponential dose–response parameter $k = 238$, the daily dose can be estimated to be

$$d = k \ln(1 - p)$$
$$= -238 \ln(1 - 0.0112) = 2.7$$

From the mean daily water consumption of 1.948 mL, the estimated mean concentration during the outbreak is determined to be 2.7/1.948, or 1.4 oocysts/L.

During the course of the outbreak, a number of samples were taken for protozoan analysis. In eight samples of finished and distribution system water, four positive samples were obtained with a geometric mean among the positive samples of 0.025 L^{-1}; however, these samples were taken during the latter stages of the outbreak (66). Samples of ice manufactured during the outbreak were obtained and analyzed as well. Based on two sets of duplicate samples analyzed by membrane filter concentration, a geometric mean of 0.079 L^{-1} was obtained (66); however, it is believed that as much as a 90% loss could have occurred in oocyst concentration from freezing and thawing (67). Applying this correction, the geometric mean oocyst concentration could have been 0.79 L^{-1}. With this correction, it would appear that the level of *Cryptosporidium* intrusion into the distribution system necessary to cause the observed outbreak in Milwaukee—assuming the validity of the dose–response relationship—is consistent with the observed levels found by measurement.

Confirmation: Bradford U.K. Outbreak. In late 1992, an outbreak of cryptosporidiosis was reported in Bradford, in the north of England. A detailed epidemiological investigation was conducted and reported by Atherton et al. (2). The salient features of this incident are as follows:

- In a supply region of 50,000 persons, 125 excess cases of cryptosporidiosis (laboratory confirmed) were noted (resulting in an attack ratio of 0.0025.
- The period of exposure was estimated to be 7 days.
- Shortly after the onset of illnesses, oocyst levels were measured in treated water and in distributed water, and the average concentration was 0.0187 L^{-1}.

We use the approach of the preceding section to analyze this incident. First, using equation 9-2, the daily attack rate is estimated to be 3.58×10^{-4}. Now we can use the dose–response relationship for *Cryptosporidium parvum* to compute the requisite daily dose that would yield this attack rate. The resulting imputed daily oocyst consumption rate is 0.083 oocyst/day. Hence the resulting risk would be consistent with the reported oocyst level, and a daily water consumption of about 1 L/day. It is believed that in the United Kingdom, the direct water consumption rates are lower than in the United States.[5] Hence the observed attack rates are consistent with the observed oocyst levels.[6]

---

[5] Due to greater consumption of tea and other boiled fluids than in the United States.

[6] Two other issues that we have not formally considered in these "validation" examples are the efficiency of the oocyst measurement methodologies and the difference between infection (as the primary response in the DuPont dosing studies (17). It has been well established that the efficiency of measuring oocysts of *C. parvum* in drinking water is quite low, frequently under 50% (13). Furthermore, as noted earlier, the morbidity ratio in the DuPont study was about 50% among volunteers (i.e., only half of the infected subjects became ill). These two sources of error would be compensating, and hence they have not been considered explicitly. There is an additional source of potential error as well, since it has been asserted that a considerable fraction of oocysts in drinking water are nonviable (40), although the inability to measure viability directly in water samples makes this observation somewhat less than conclusive.

Food. We note in passing that the presence of oocysts of *Cryptosporidium parvum* in food may also be a matter of public health concern. These organisms have been found in samples of herbs and vegetables that are ordinarily consumed raw (54). Furthermore, there have been outbreaks associated with the consumption of apple cider containing oocysts (presumably from fruit that had contacted fecally contaminated soil) (52).

Vibrio cholerae

Vibrio cholerae, the etiologic agent of cholera, is a major worldwide pathogen. Although there have been studies of the virulence of *V. cholerae,* there does not appear to have been a study to date of the quantitative dose–response relationships that may characterize human risk associated with exposure to this organism. Here we present an analysis of previous data (33) on the infectivity and morbidity associated with oral ingestion of this organism using methods described in previous chapters.

Underlying Data Set. Hornick et al. (33) exposed healthy human volunteers to various doses of *V. cholerae* Inaba and Ogawa strains with and without simultaneous ingestion of sodium bicarbonate. The latter agent served to buffer intestinal tract pH and protect against acid inactivation of the microorganisms. We will discuss the Inaba 569B data with ingestion of sodium bicarbonate, since it was found that the administration of bicarbonate resulted in substantial increase in infectivity and pathogenicity and also since too few data points (dilutions) for the Ogawa experiments were taken to permit thorough quantitative analysis.

In this study the number of respondents was measured using three distinct endpoints:

1. Presence of *V. cholerae* in stools, or presence of positive antibody response
2. Occurrence of diarrhea, with microorganisms present, but of severity not requiring rehydration
3. Severe watery diarrhea requiring intravenous rehydration

For the purpose of the present analysis, all individuals showing any of the foregoing responses are regarded as positive cases. People showing that none of the foregoing responses are regarded as negative cases. Positive cases showing only positive stool or antibody reaction, without diarrhea, are termed asymptomatic, while all other cases are termed symptomatic.

Table 9-14 summarizes the results presented by Hornick et al. (33) using the definitions above. Based on this information, a dose–response curve for infectivity, as well as a relationship for morbidity ratio, could be obtained.

TABLE 9-14 Experimental Data of Hornick et al. (33): Response to
***V. cholerae* Inaba 569B with Simultaneous Ingestion of Bicarbonate**

			Positive Subjects	
Dose	Total Subjects	Negative Subjects	Symptomatic Cases	Asymptomatic Cases
10	2	2	0	0
1,000	4	1	0	3
10,000	13	2	9	2
100,000	8	1	6	1
1,000,000	23	2	20	1
100,000,000	2	0	2	0

Infectivity Dose–Response Relationship. From the data, the total number of cases as a function of dose was used to develop a dose–response relationship; that is, if T_i is the total number of subjects administered a dose d_i and $p_{i,inf}$ is the number of total cases (symptomatic plus asymptomatic), the observed proportion of infectives is obtained by this ratio.

Preliminary investigations found that the beta-Poisson model provided superior fits to the exponential model. The best-fit values of the beta-Poisson dose–response parameters can be determined using the method of maximum likelihood. From the data, these are found to be $\alpha = 0.25$ and $N_{50} = 243$. By a likelihood-ratio goodness-of-fit test, the fit is deemed acceptable, and 95% likelihood-based confidence regions can be computed for the dose–response relationship (Figure 9-12) and for the dose–response parameters themselves (Figure 9-13).

Several aspects of this dose–response relationship deserve comment. First, the data show substantial heterogeneity (i.e., the value of α is fairly small, and the beta-Poisson model provides a substantial improvement of fit over the exponential. This is seen by the fact that the 95% confidence region is closed on the right. Second, the median infective dose is rather small, and in particular the confidence region for N_{50} is quite broad, and extends well below 1 organism. Therefore, one cannot rule out the possibility that very small numbers of organisms may result in substantial levels of infection. In the experimental procedure used by Hornick, the microbial dose was ascertained by use of a spectrophotometric standard curve produced by calibration against colony counts on brain heart infusion agar. It is well established that a number of bacteria can form viable nonculturable forms or may be imperfectly enumerated on standard culture media (11,36,71,73–75). Particularly relevant to the present report is the finding that after simple preparation and washing, *V. cholerae* 569B can contain 26 to 35% injured organisms, increasing up to 91 to 92% following chlorine exposure (75). Hence it is possible that the low median infective dose found in this study may have been due to the presence of viable nonculturable organisms in the inoculum.

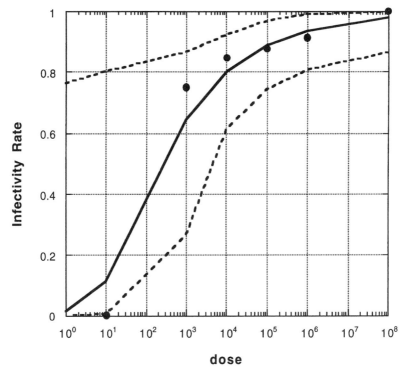

Figure 9-12 Comparison of observed and model predicted total response rates, with upper and lower confidence limits shown.

Morbidity Ratio. The Hornick data set also permits exploration of the morbidity ratio. We define this as the conditional probability of being symptomatic (regardless of the degree of severity) among those who are show any positive response to the organism. There is a clear dose dependency to the morbidity ratio (nonparametric test for monotone trend, $p < 0.001$). Therefore, it was of interest to determine whether this dose dependency of the morbidity ratio could be described quantitatively.

The exponential and beta-Poisson models were used as candidate descriptors of the morbidity dose–response distribution. By analogy to analysis of dose–response curves for infectivity, the conditional morbidity probability was described as a function of dose and the optimum parameters obtained by maximum likelihood estimation. Application of this method showed that the exponential model was unsatisfactory, while the beta-Poisson model provided an acceptable fit. The optimum parameter values were $\alpha = 0.49$ and $N_{50} = 3365$. Figure 9-14 presents a comparison of the observed morbidity ratio to the fitted model, with 95% confidence limits, and Figure 9-15 presents the likelihood-based parameter confidence regions.

Somewhat surprisingly, the morbidity ratio confidence regions are narrower than the infection confidence regions (compare Figures 9-13 and 9-15). The

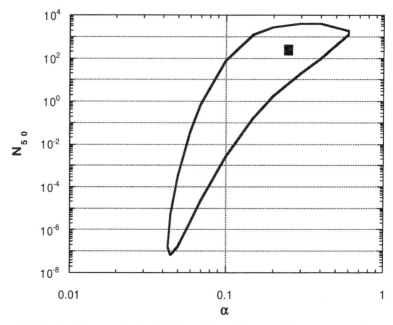

Figure 9-13 Confidence limits (95%) to infectivity model parameter estimates for infectivity dose–response of *Vibrio*. The maximum likelihood estimate is indicated by the square.

dose–response parameters for morbidity are also greater than the corresponding values for infectivity.

Discussion. This report appears to be the first in which a quantitative dose–response model was fit to human data on *V. cholerae*. The fitting of these data with a beta-Poisson model indicates that there may not be a "threshold" for either infection or frank illness. In particular, the relatively low levels of organisms required to initiate an infection are quite interesting and may play a role in maintaining this organism in an endemic state in many portions of the world. This analysis may also be the first in which the morbidity ratio of a microorganism has been quantitatively related to the level of exposure.

The dose–response models presented here can be used to develop a risk assessment for exposure to *V. cholerae*. Using the maximum likelihood parameters, the expected fraction of asymptomatic and symptomatic cases can be computed. From the analysis above, the probability of infection ($\pi_{\text{infection}}$) and the conditional probability of morbidity given infection ($\pi_{\text{M|I}}$) can be computed from the two dose–response relationships. Based on this, the fraction of symptomatic and asymptomatic cases can be obtained as follows:

$$\pi_{\text{symptomatic}} = \pi_{\text{infection}} \pi_{\text{M/I}}$$

$$\pi_{\text{asymptomatic}} = \pi_{\text{infection}}(1 - \pi_{\text{M/I}})$$

$$(9\text{-}3)$$

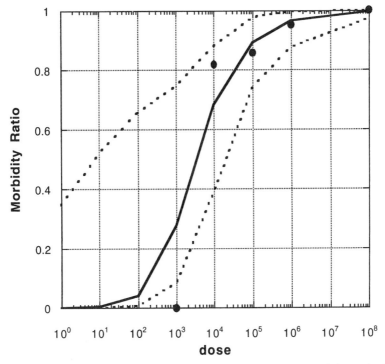

Figure 9-14 Comparison of observed and model predicted morbidity ratio for *V. cholerae*, with confidence limits shown.

Based on this approach, it is concluded that to maintain the risk of total infections at a low level (e.g., < 0.0001 per exposure), a very low level of microorganisms is allowable in the medium via which exposure occurs. This has some ramifications with respect to the design and implementation of sampling and surveillance programs to assure minimal risk. To maintain the level of symptomatic cases low, such as < 10^{-9} per exposure, as might be justifiable in a vehicle (water, food) to which a population is frequently exposed, it is clear that a similarly low level of microbial loading must exist. As a corollary to this, the failure to detect *V. cholerae* in a system, when the minimal detection limit is fairly high, is not necessarily sufficient to ensure protection against undesirably high risk.

Other Organisms

There are a number of other organisms for which human dose–response information is available and for which dose–response parameters have been computed. In Table 9-15 these parameters are summarized (along with those

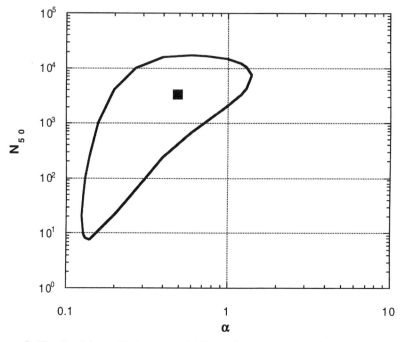

Figure 9-15   Confidence limits to morbidity ratio parameters. The maximum likelihood estimate is indicated by the square.

already discussed). As noted, the fits are either to the exponential or beta-Poisson dose–response models. There are no organisms for which human infectivity data are presently available sufficient to fit dose–response curves which have been definitively found either not to fit these two models or to fit some other (i.e., one with nonlinearity at a low dose) model in a statistically superior fashion.

Many of the virus data fit the exponential model, due to the paucity of doses used or the small number of subjects (thereby giving insufficient power to detect the difference between an exponential fit an alternative models). The protozoans, however (particularly *Giardia* and *Cryptosporidium*) fit an exponential model despite having a large number of subjects and doses. On the other hand, the enteric bacteria (*Salmonella, Shigella, Vibrio, Campylobacter,* and *E. coli*) quite strongly require the beta-Poisson model to fit the data adequately. It is noteworthy that for the latter group of organisms, the parameter α of the best fit remains quite small (0.108 to 0.31), reflecting a great heterogeneity in the distribution of host–pathogen infection probability.

TABLE 9-15   Best Fit Dose–Response Parameters (Human)

Organism	Exponential, Beta-Poisson Parameters			References
	k	N_{50}	α	
Poliovirus I (minor)	109.87			53
Rotavirus		6.17	0.2531	32,82
Hepatitis A virus[a]	1.8229			81
Adenovirus 4	2.397			15
Echovirus 12	78.3			1
Coxsackie[b]	69.1			14,78
Salmonella[c]		23,600	0.3126	
Salmonella typhosa		3.60×10^6	0.1086	34
Shigella[d]		1,120	0.2100	34
				34
Escherichia coli[c]		8.60×10^7	0.1778	
Campylobacter jejuni		896	0.145	51
				51
Vibrio cholera		243	0.25	
Endamoeba coli		341	0.1008	63
Cryptosporidium parvum	238			17,30
Giardia lamblia	50,23			68

[a] Dose in grams of feces (of excreting infected individuals).
[b] *Flexnerii* and *dysenteriae* pooled.
[c] Multiple (nontyphoid) pathogenic strains (*S. pullorum* excluded).
[d] B4 and A21 strains pooled.
[e] Nonenterohemmorhagic strains (except O111).

REFERENCES

1. Akin, E. W. 1981. Presented at the U.S. EPA Symposium on Microbial Health Considerations of Soil Disposal of Domestic Wastewaters.

2. Atherton, F., C. P. S. Newman, and D. P. Casemore. 1995. An outbreak of waterborne cryptosporidiosis associated with a public water supply in the UK. Epidemiol. Infect. 115(1):123–131.

3. Bell, B. P., M. Goldoft, P. M. Griffin, M. A. Davis, D. C. Gordon, P. 1. Tarr, C. A. Bartleson, J. H. Lewis, T. J. Barrett, J. G. Wells, R. Baron, and J. Kobayashi. 1994. A multistate outbreak of *Escherichia coli* O157:H7: associated bloody diarrhea and hemolytic uremic syndrome from hamburgers. J. Am. Med. Assoc. 272(17):1349–1353.

4. Benenson, A. S. 1990. Control of communicable disease in man, 15th ed. American Public Health Association, Washington, DC.

5. Bennett, J. V., S. D. Holmberg, M. F. Rogers, and S. L. Solomon. 1987. Infectious and parasitic diseases. Am. J. Prev. Med. 3(5 Suppl.):102–114.

6. Berg, G. 1978. Indicators of viruses in water and food. Ann Arbor Science Publishers, Ann Arbor, MI.

7. Black, R. E., G. F. Craun, and P. A. Blake. 1978. Epidemiology of common-source outbreaks of shigellosis in the United States, 1961–1975. Am. J. Epidemiol. 108(1):47–52.

8. Blostein, J. 1991. Shigellosis from swimming in a park pond in Michigan. Public Health Rep. 106(3):317–321.

9. Boring, J. R., W. T. Martin, and L. M. Elliot. 1971. Isolation of *Salmonella ty-phimurium* from municipal water, Riverside, California, 1965. Am. J. Epidemiol. 93(1):49–54.

10. Bryan, F. L., and M. P. Doyle. 1995. Health risks and consequences of *Salmonella* and *Campylobacter jejuni* in raw poultry. J. Food Prot. 58(3):326–344.

11. Camper, A. K., and G. A. McFeters. 1979. Chlorine injury and the enumeration of waterborne coliform bacteria. Appl. Environ. Microbiol. 37(3):633–641.

12. Cassin, M. H., A. M. Lammerding, E. C. D. Todd, W. Ross, and R. S. McColl. Quantitative risk assessment for *Escherichia coli* O157:H7 in ground beef ham-burgers. Unpublished.

13. Clancy, J. L., W. Gollnitz, and Z. Tabib. 1994. Commercial labs: how accurate are they? J. Am. Water Works Assoc. 86(5):89–97.

14. Couch, R. B., T. Cate, P. Gerone, W. Fleet, D. Lang, W. Griffith, and V. Knight. 1965. Production of illness with a small-particle aerosol of Coxsackie A21. J. Clin. Invest. 44(4):535–542.

15. Couch, R. B., T. R. Cate, P. J. Gerone, W. F. Fleet, D. J. Lang, W. R. Griffith, and V. Knight. 1966. Production of illness with a small-particle aerosol of ade-novirus type 4. Bacteriol. Rev. 30:517–528.

16. Crockett, C., C. N. Haas, A. Fazil, J. B. Rose, and C. P. Gerba. 1996. Prevalence of Shigellosis in the U.S.: consistency with dose–response information. Int. J. Food Microbiol. 30(1–2):87–100.

17. Dupont, H., C. Chappell, C. Sterling, P. Okhuysen, J. Rose, and W. Jakubowski. 1995. Infectivity of *Cryptosporidium parvum* in healthy volunteers. N. Engl. J. Med. 332(13):855.

18. Dupont, H. L., S. B. Formai, R. B. Hornick, M. J. Snyder, J. P. Libonati, D. G. Sheahan, E. H. Labrec, and J. P. Kalas. 1971. Pathogenesis of *Escherichia coli* diarrhea. N. Engl. J. Med. 285(1):1–9.

19. Dupont, H. L., and R. B. Hornick. 1973. Clinical approach to infectious diarrheas. Medicine 52(4):265–270.

20. Dupont, H. L., R. B. Hornick, A. T. Dawkins, M. J. Snyder, and S. B. Formal. 1969. The response of man to virulent *Shigella flexneri* 2a. J. Infect. Dis. 119:296–299.

21. Dupont, H. L., R. B. Hornick, M. J. Snyder, J. P. Libonati, S. B. Formal, and E. J. Gangarosa. 1972. Immunity in shigellosis II: protection induced by oral live vaccine or primary infection. J. Infect. Dis. 125(1):12–16.

22. Dupont, H. L., and J. J. Mathewson. 1991. *Escherichia coli* diarrhea, pp. 239–254. *In* A. S. Evans and P. S. Brachman, eds., Bacterial infections of humans: epidemiology and control, 2nd ed. Plenum Medical Book Co., New York.

23. Erickson, J. P., J. W. Stamer, and L. A. Van Alstine. 1995. An assessment of *Escherichia coli* O157:H7 contamination risks in commercial mayonnaise from pasteurized eggs and environmental sources, and behavior in low-pH dressings. J. Food Prot. 58(10):1059–2064.

24. Ewald, P. W. 1991. Waterborne transmission and the evolution of virulence among gastrointestinal bacteria. Epidemiol. Infect. 106:83–119.

25. Feng, P. 1995. *Escherichia coli* serotype O157:H7: novel vehicles of infection and emergence of phenotypic variants. Emerg. Infect. Dis. 1(2):47–52.

26. Ferguson, W. W., and R. C. June. 1952. Experiments on feeding adult volunteers with *Escherichia coli* 111 B$_4$: a coliform organism associated with infant diarrhea. Am. J. Hyg. 55:155–169.

27. Graham, D. Y., M. K. Estes, and L. O. Gentry. 1983. Double-blind comparison of bismuth subsalicylate and placebo in the prevention and treatment of enterotoxigenic *Escherichia coli* induced diarrhea in volunteers. Gastroenterology 85: 1017–1022.

28. Griffin, P. M., and R. V. Tauxe. 1991. The epidemiology of infections caused by *Escherichia coli* O157:H7, other enterohemorrhagic *E. coli* and the associated hemolytic uremic syndrome. Epidemiol. Rev. 13:60–98.

29. Haas, C. N. 1983. Estimation of risk due to low doses of microorganisms: a comparison of alternative methodologies. Am. J. Epidemiol. 118(4):573–582.

30. Haas, C. N., C. Crockett, J. B. Rose, C. Gerba, and A. Fazil. 1996. Infectivity of *Cryptosporidium parvum* oocysts. J. Am. Water Works Assoc. 88(9):131–136.

31. Haas, C. N., and J. B. Rose. 1994. Presented at the Annual Conference of the American Water Works Association, New York.

32. Haas, C. N., J. B. Rose, C. Gerba, and S. Regli. 1993. Risk assessment of virus in drinking water. Risk Anal. 13(5):545–552.

33. Hornick, R. B., S. I. Music, R. Wenzel, R. Cash, J. P. Libonati, and T. E. Woodward. 1971. The Broad Street pump revisited: response of volunteers to ingested cholera vibrios. Bull. N.Y. Acad. Med. 47(10):1181–1191.

34. Hornick, R. B., T. E. Woodward, F. R. McCrumb, A. T. Dawkin, M. J. Snyder, J. T. Bulkeley, F. D. L. Macorra, and F. A. Corozza. 1966. Study of induced typhoid fever in man. I. Evaluation of vaccine effectiveness. Trans. Assoc. Am. Physicians 79:361–367.

35. Hrdy, D. B. 1987. Epidemiology of rotaviral infection in adults. Rev. Infect. Dis. 9(3):461–469.

36. Huq, A., R. R. Colwell, R. Rahman, A. Ali, and M. A. R. Chowdhury. 1990. Detection of *Vibrio cholerae* O1 in the aquatic environment by fluorescent-monoclonal antibody and culture methods. Appl. Environ. Microbiol. 56(8):2370–2373.

37. June, R. C., W. W. Ferguson, and M. T. Worfel. 1953. Experiments in feeding adult volunteers with *Escherichia coli* 55 B$_5$: a coliform organism associated with infant diarrhea. Am. J. Hyg. 57:222–236.

38. Keene, W., J. McAnulty, F. Hoesly, L. Williams, K. Hedberg, G. Oxman, T. Barrett, M. Pfaller, and D. Fleming. 1994. A swimming associated outbreak of hemorrhagic colitis caused by *Escherichia coli* O157:H7 and *Shigella sonnei*. N. Engl. J. Med. 331(9):579–584.

39. Knight, P. 1993. Hemorrhagic *E. coli:* the danger increases. ASM News 59(5): 247–250.

40. Lechevallier, M. W., W. D. Norton, and R. G. Lee. 1991. *Giardia* and *Cryptosporidium* spp. in filtered drinking water supplies. Appl. Environ. Microbiol. 57: 2617–2621.

41. Lee, L. A., S. M. Ostroff, H. B. McGee, D. R. Johnson, F. P. Downes, D. N. Cameron, N. H. Bean, and P. M. Griffin. 1991. An outbreak of shigellosis at an outdoor music festival. Am. J. Epidemiol. 133:608–615.

42. Lee, Y. J. 1983. Trend in proportions, test for, pp. 328–334. *In* S. Kotz and N. Johnson, eds., Encyclopedia of statistical sciences, Vol. 9. Wiley, New York.

43. Levnie, M. M., H. L. Dupont, S. B. Formal, R. B. Hornick, A. Takeuchi, E. J. Gangarosa, M. J. Snyder, and J. P. Libonati. 1973. Pathogenesis of *Shigella dysenteriae* 1 (shiga) dysentery. J. Infect. Dis. 127(3):261–269.

44. Lew, J. F., D. L. Swerdlow, M. E. Dance, P. M. Griffin, C. A. Bopp, M. J. Gillenwater, T. M. Mercatente, and R. I. Glass. 1991. An outbreak of shigellosis aboard a cruise ship caused by a multiple-antibiotic-resistant strain of *Shigella flexneri* Am. J. Epidemiol. 134:413–420.

45. Mac Kenzie, W. R., N. J. Hoxie, M. E. Proctor, M. S. Gradus, K. A. Blair, D. E. Peterson, J. J. Kazmierczak, K. R. Fox, D. G. Addias, J. B. Rose, and J. P. Davis. 1994. Massive waterborne outbreak of *Cryptosporidium* infection associated with a filtered public water supply, Milwaukee, Wisconsin, March and April 1993. N. Engl. J. Med. 331(3):161–167.

46. Macler, B. A., and S. Regli. 1993. Use of Microbial risk assessment in setting United States drinking water standards. Int. J. Food Microbiol. 18(4):245–256.

47. Makintubee, S., J. Mallonee, and G. R. Istre. 1987. Shigellosis outbreak associated with swimming. Am. J. Public Health 77(2):166–168.

48. McCullough, N. B., and C. W. Eisele. 1951. Experimental human salmonellosis. I. Pathogenicity of strains of *Salmonella meleagridis* and *Salmonella anatum* obtained from spray dried whole egg. J. Infect. Dis. 88:278–289.

49. McCullough, N. B., and C. W. Eisele. 1951. Experimental human salmonellosis. III. Pathogenicity of strains of *Salmonella newport, Salmonella derby* and *Salmonella bareilly* obtained from spray dried whole egg. J. Infect. Dis. 89:209–213.

50. McCullough, N. B., and C. W. Eisele. 1951. Experimental human salmonellosis. IV. Pathogenicity of strains of *Salmonella pullorum* obtained from spray dried whole egg. J. Infect. Dis. 89:259–265.

51. Medema, G. J., P. F. M. Teunis, A. H. Havelaar, and C. N. Haas. 1996. Assessment of the dose–response relationship of *Campylobacter jejuni.* Int. J. Food Microbiol. 30(1–2):101–112.

52. Millard, P., K. Gensheimer, D. G. Addiss, D. M. Sosin, G. A. Beckett, A. Houck-Jankoski, and A. Hudson. 1994. An outbreak of cryptosporidiosis from fresh-pressed apple cider. J. Am. Med. Assoc. 272(20):1592–1596.

53. Minor, T. E., C. I. Allen, A. A. Tsiatis, et al. 1981. Human infective dose determination for oral poliovirus type I vaccine in infants. J. Clin. Microbiol. 13:388.

54. Monge, R., and M. Chinchilla. 1996. Presence of *Cryptosporidium* Oocysts in fresh vegetables. J. Food Prot. 59(2):202–203.

55. Morgan, B. J. T. 1992. Analysis of quantal response data. Chapman & Hall, London.

56. Mui, B. G. 1986. M.S. thesis. Illinois Institute of Technology.

57. Ohio River Valley Water Sanitation Commission. 1993. Pollution Control standards for discharges to the Ohio River, 1993 revision, notice of requirements. Appendix 6. Ohio River Valley Water Sanitation Commission.

58. Payment, P., L. Richardson, J. Siemiatycki, and R. Dewar. 1991. A randomized trial to evaluate the risk of gastrointestinal disease due to consumption of drinking water meeting current microbiological standards. Am. J. Public Health 81:703.

59. Payment, P., M. Trudel, and R. Plante. 1985. Elimination of viruses and indicator bacteria at each step of treatment during preparation of drinking water at seven water treatment plants. Appl. Environ. Microbiol. 49:1418.

60. Plummer, R. A. S., S. J. Blissett, and C. E. R. Dodd. 1995. Salmonella contamination of retail chicken products sold in the UK. J. Food Prot. 58(8):843–846.

61. Reeve, G. D. L., J. Martin, R. E. Pappas, Thompson, and K. D. Greene. 1989. An outbreak of shigellosis associated with the consumption of raw oysters. N. Engl. J. Med. 321(4):224–227.

62. Regli, S., J. B. Rose, C. N. Haas, and C. P. Gerba. 1991. Modeling risk for pathogens in drinking water. J. Am. Water Works Assoc. 83(11):76–84.

63. Rendtorff, R. C. 1954. The experimental transmission of human intestinal protozoan parasites. I. *Endamoeba coli* cysts given in capsules. Am. J. Hyg. 59:196–208.

64. Riverside County Health Department, California State Department, Center for Disease Control, and National Center for Urban and Industrial Health. 1971. A waterborne epidemic of salmonellosis in Riverside, California (1965): epidemiologic aspects. Am. J. Epidemiol. 93(1):33–48.

65. Robins-Browne, R. M. 1987. Traditional enteropathogenic *Escherichia coli* of infantile diarrhea. Rev. Infect. Dis. 9:28–53.

66. Rose, J. B. 1993. Results of the samples collected from Milwaukee associated with a waterborne outbreak of *Cryptosporidium*.

67. Rose, J. B., and C. N. Haas. 1994. Presented at the Annual Conference of the American Water Works Association, New York.

68. Rose, J. B., C. N. Haas, and S. Regli. 1991. Risk Assessment and the control of waterborne giardiasis. Am. J. Public Health 81:709–713.

69. Roseberry, A. M., and D. E. Burmaster. 1992. Log-normal distributions for water intake by children and adults. Risk Anal. 12(1):99–104.

70. Rosenberg, M. L., K. K. Hazlet, J. Schaefer, J. G. Wells, and R. C. Pruneda. 1976. Shigellosis from swimming. J. Am. Med. Assoc. 236(16):1849–1852.

71. Roszak, D. B., and R. R. Colwell. 1987. Survival strategies of bacteria in the natural environment. Microbiol. Rev. 51(3):365–379.

72. Sack, R. B. 1975. Human diarrheal disease caused by enterotoxigenic *Escherichia coli*. Annu. Rev. Microbiol. 29:333–353.

73. Singh, A., et al. 1990. Rapid detection of chlorine-induced bacterial injury by the direct viable count method using image analysis. Appl. Environ. Microbiol. 56:389–394.

74. Singh, A., and G. A. McFeters. 1990. Injury of enteropathogenic bacteria in drinking water, pp. 368–379. *In* G. A. McFeters, ed., Drinking water microbiology. Springer-Verlag, New York.

75. Singh, A., P. Yu, and G. A. McFeters. 1990. Rapid detection of chlorine-induced bacterial injury by the direct viable count method using image analysis. Appl. Environ. Microbiol. 56(2):389–394.

76. Sorvillo, F. J., S. H. Waterman, J. K. Vogt, and B. England. 1988. Shigellosis associated with recreational water contact in Los Angeles County. Am. J. Trop. Med. Hyg. 38:613–617.

77. Spitka, J. S., F. Dabis, N. Hargrett-Bean, J. Salcedo, S. Veillard, and P. A. Blake. 1987. Shigellosis at a Caribbean resort: hamburger and North American origin as risk factors. Am. J. Epidemiol. 126(6):1173–1180.

78. Suptel, E. A. 1963. Pathogenesis of experimental Coxsackie virus infection. Arch. Virol. 7:61–66.

79. Swerdlow, D. L., B. A. Woodruss, and R. C. Brady. 1989. A waterborne outbreak in Missouri of *Escherichia coli* O157:H7 associated with bloody diarrhea and death. Ann. Intern. Med. 117:812–819.

80. Tauxe, R. 1991. *Salmonella:* a postmodern pathogen. J. Food Prot. 54:563.

81. Ward, R., S. Krugman, J. Giles, M. Jacobs, and O. Bodansky. 1958. Infectious hepatitis: studies of its natural history and prevention. N. Engl. J. Med. 258(9): 402–416.

82. Ward, R. L., D. L. Bernstein, E. C. Young, J. R. Sherwood, D. R. Knowlton, and G. M. Schiff. 1986. Human rotavirus studies in volunteers: determination of infectious dose and serological response to infection. J. Infect. Dis. 154(5):871.

83. Willshaw, G. A., J. Thirlwell, and M. Hickey. 1994. Vero cytotoxin-producing *Escherichia coli* O157 in beefburgers linked to an outbreak of diarrhoea, haemorrhagic colitis and haemolytic uraemic syndrome in Britain. Lett. Appl. Microbiol. 19(5):304–307.

INDEX